Archives of Virology

Supplementum 7

O.-R. Kaaden, W. Eichhorn,
C.-P. Czerny (eds.)

Unconventional Agents
and Unclassified Viruses

Recent Advances in Biology
and Epidemiology

Springer-Verlag Wien New York

Prof. Dr. O.-R. Kaaden
Dr. W. Eichhorn
Dr. C.-P. Czerny
WHO Collaborating Centre for Collection and Evaluation of Data on Comparative
Virology, Institute for Medical Microbiology, Infectious and Epidemic Diseases,
Veterinary Faculty, University of Munich, Munich, Federal Republic of Germany

Typesetting: Best-set Typesetter Ltd., Hong Kong

Printed on acid-free and chlorine-free bleached paper

With 79 partly coloured Figures

ISSN 0939-1983
ISBN-13:978-3-211-82480-1   e-ISBN-13:978-3-7091-9300-6
DOI: 10.1007/978-3-7091-9300-6

Foreword

The First Munich Symposium on Microbiology was held from October 14–17, 1975 during the inauguration of the WHO Collaborating Centre for Collection and Evaluation of Data on Comparative Virology here in Munich. Since then seven other symposia were organized and the topics covered such different fields like

- Experimental Animals and in vitro Systems in Medical Microbiology (1976)
- Natural History of Newly Emerging and Re-emerging Viral Zoonoses (1978)
- Mechanisms of Viral Pathogenesis and Virulence (1979)
- Leukemias, Lymphomas and Papillomas (1980)
- New Horizons in Diagnostic Virology (1982).

It was an honour and my privilege to welcome the participants on behalf of the organizing committee at the 8th Munich Symposium on Microbiology entitled "*Unconventional Agents and Unclassified Viruses: Recent Advances in Biology and Epidemiology*" held from October 21 to 22, 1992, here in Munich.

The Munich Symposia once established by Professor Anton Mayr were later organized by the late Peter Bachmann. His sudden death in 1985 terminated the meetings for a couple of years. This conference is the revival of the once very well received tradition. Together with my co-organizers and all staff members of the Institute of Medical Microbiology, Infectious and Epidemic Diseases at the University of Munich we acknowledge and appreciate that the invited speakers and chairmen accepted our invitation even though it was conveyed only seven months before.

In particular, we gratefully appreciate the financial support by the Federal Ministry of Health which made this symposium possible and we do hope that this support will be continued for the next planned conferences as part of the scientific activity of our WHO Reference Centre for Comparative Virology. The titles of the papers presented during the conference dealt with newly emerging diseases like transmissible spongiform encephalopathies in man and animals, non-A-non-B hepatitis viruses or positive-strand RNA virus infections amongst others thus emphasizing the comparative character of virology for medical and veterinary medicine and molecular biology.

The meetings were always hosted by the generous help of the Carl Friedrich von Siemens Stiftung (Foundation) in its lovely surrounding and have been known for long for its private and intimate atmosphere. Therefore, we want to extent our gratitude to the Carl Friedrich von SIEMENS Stiftung. Thanks to the co-operation with Dr. H.-D. Klenk and the Springer-Verlag, the contributions of the symposium can be made accessible to the scientific community world-wide as a Supplementum to Archives of Virology.

The organizers are also thankful to Marita Bongarts and Andrea Aigner for their efficient secretarial skills.

O.-R. Kaaden

Contents

Brief Reports

Arch Virol (1993) [Suppl] 7: 1–14

Molecular characterization of hepatitis C and E viruses

D.W. Bradley, M.J. Beach, and **M.A. Purdy**

Hepatitis Branch, Division of Viral and Rickettsial Diseases, National Center for
Infectious Diseases, Centers for Disease Control, Atlanta, GA, U.S.A

Summary. The molecular features of each of the major viruses of non-A,
non-B hepatitis, namely hepatitis C virus (HCV) and hepatitis E virus
(HEV) are briefly described. The organization of the genome of each of
these viruses is discussed and compared to those of other related or
distantly related viruses that contain single-stranded, positive-sense RNA
genomes. HCV has been tentatively classified as a separate genus within
the *Flaviviridae*, whereas HEV has been loosely associated with caliciviruses and subsequently assigned to the *Caliciviridae*, although it does
possess unique genetic features not found in other caliciviruses.

Introduction

Parenterally-transmitted non-A, non-B hepatitis (PT-NANB), and
enterically-transmitted non-A, non-B hepatitis (ET-NANBH) are now
known to be causally associated with infection by hepatitis C virus
(HCV) and hepatitis E virus (HEV), respectively. Fundamental studies
leading to the isolation, characterization, and molecular cloning of these
two extraordinary viruses have been reviewed recently [5]. The extremely
high rate of progression of hepatitis C to chronic liver disease (>70%)
and the worldwide extent of this disease makes it a significant cause of
morbidity and mortality in man [1]. Hepatitis E, once thought to be a
disease confined to underdeveloped countries, is also recognized as a
disease of widespread geographic distribution [6]. According to World
Health Organization statistics, HEV appears to be most common cause
of acute viral hepatitis observed among young to middle-aged adults
living in "third-world" countries.

Hepatitis C virus (HCV)

The recent cloning and sequencing [10, 22] of the primary agent responsible for PT-NANB, now designated hepatitis C virus (HCV), has

resulted in a wealth of new information about the virus and the disease. HCV has a single-stranded, 9.4 kb, positive-sense RNA genome which contains one large open reading frame (ORF) capable of encoding a polyprotein of 3010/11 amino acids [11, 22]. These facts, plus sequence comparisons and similar hydrophobicity plots, indicate that the HCV genome is most closely related to those of the pesti- and flavi-viruses [3, 11, 29]. The pesti- and flavi-viruses also contain a similarly organized genome encoding a large polyprotein that is subsequently cleaved into mature viral proteins by a combination of host and viral-encoded proteases.

Genome characterization

Figure 1 illustrates the putative genomic organization of HCV. The 341 nucleotide 5'-untranslated region (5'-UTR) resembles that found in the pestiviruses in several aspects. First, the lengths are almost identical (flaviviruses have a significantly shorter 5'-UTR). Secondly, both contain multiple short ORF's upstream of the authentic initiation codon that may play a role in translational regulation [17]. Thirdly, the 5'-UTR's of HCV and pestiviruses contain multiple conserved blocks of sequence which may be important in stabilizing the predicted secondary structure through nucleotide pairing [39].

The 5' end of the large ORF encodes the putative structural proteins of the virus (Fig. 1). In vitro translation of this region indicates that it is processed into a minimum of 4 proteins [18, 20]. The amino terminal end is cleaved to produce an unglycosylated, basic, 19–22 kDa protein (p22) which is the putative core protein. The core protein is followed by 2 putative envelope glycoproteins of 33–35 kDa and 70–72 kDa, respectively, (18–21 kDa and 38 kDa protein backbone) designated E1 and E2. However, the possibility that E2 may be a nonstructural component analogous to the glycoprotein NS1 of the flaviviruses has not been excluded. The amino terminal end of E2 contains a 28 amino acid segment, referred to as the hypervariable region (HV), that exhibits up to 58% variation among geographically distinct isolates [41]. This may indicate that HCV is capable of rapid mutation in order to evade immune detection by the host and may complicate efforts towards the development of a vaccine. Variation throughout the HCV genome, although less than in E2, has already led to proposals for separation of HCV into three groups (one U.S., 2 Japanese) prompting speculation that HCV may be a group of related viruses rather than a single entity. Following E2 is the extremely hydrophobic NS2 region. This nonstruc-

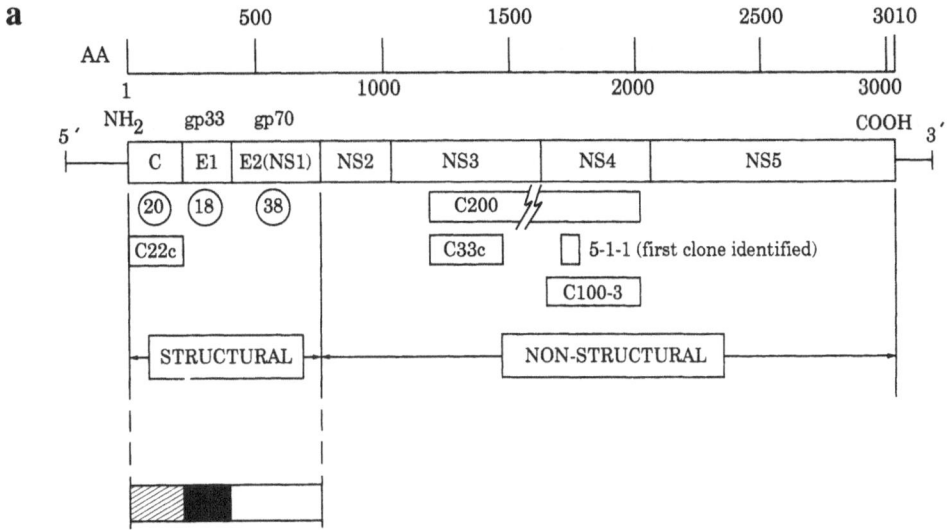

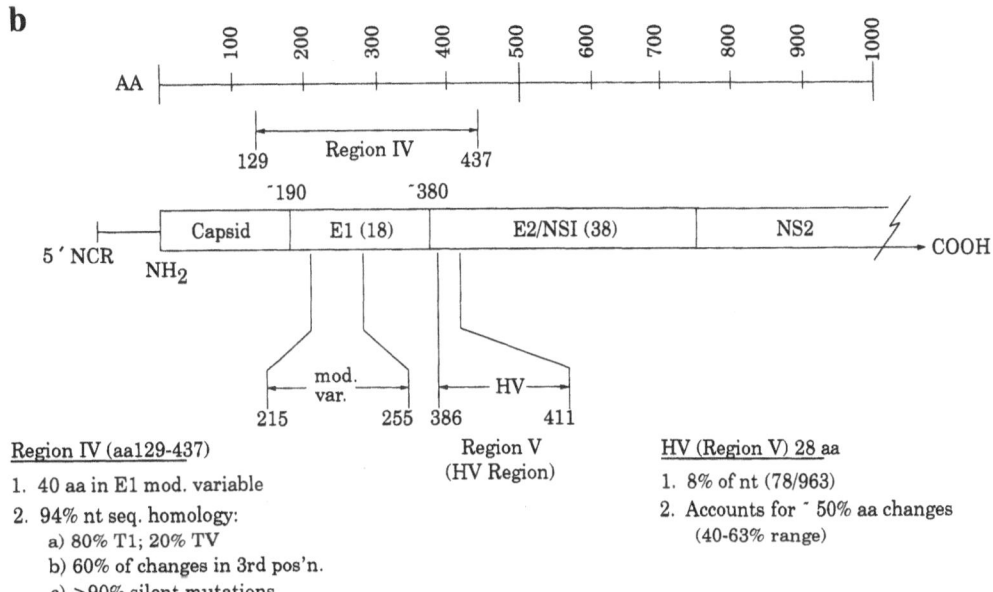

Fig. 1. a General organization of the HCV genome. Location of expressed proteins used in diagnostic assays. **b** Variable and hypervariable regions of genome. Variable regions in genome of hepatitis C virus (HCV)

tural protein has been identified in flavivirus-infected cells although no function has yet been ascertained [9].

The putative NS3 region (Fig. 1) encodes a protein of approximately 60 kDa that contains consensus sequences for two distinct enzymatic functions. The amino terminal end contains consensus sequences found in serine proteases, including the catalytic triad (H, D, S) and the consensus sequence GxSGxP surrounding the active-site serine residue [3, 11, 15]. The spacing and order of the residues is similar to that found in the flavi- and pesti-viruses for which NS3 has been demonstrated to be a viral protease involved in polyprotein processing [30]. Adjacent to the protease segment of NS3 is a domain containing consensus sequences indicative of superfamily 2 helicases [16] that are found in the same order and location as those in pesti- and flavi-viruses (25–30% sequence similarity) [3, 11, 29]. Although helicase (NTP-dependent unwinding of double-stranded RNA replicative intermediates) function has not been demonstrated for these viruses, NS3 has been shown to exhibit NTPase activity [42]. In addition, the analogous 280 amino acid region of tobacco etch virus, a plant potyvirus which also displays 30% sequence similarity with HCV [3], has been demonstrated to possess both ATPase and RNA helicase activity [27]. The NS4 region, as with NS2, is extremely hydrophobic and, as yet, has no known function.

The NS5 region (Fig. 1) encodes a protein of approximately 116 kDa that contains the consensus GDD and other sequences indicative of RNA-dependent RNA polymerases [2, 3, 11, 29]. These sequences are all found in the same relative location and order as the flavi- and pesti-viruses. As with the helicase sequence, NS5 also exhibits a significant degree of similarity with the RNA-dependent RNA polymerases of plant carmoviruses suggesting that each may be functionally derived from ancestral helicase or polymerase genes [3, 29].

Based on similar biophysical characteristics, genome organization, hydrophobicity plots, and consensus sequences, it is clear that HCV is most closely related to the pesti- and flavi-viruses. Consistent with this is the most recent recommendation of the ICTV to place HCV, pesti-, and flavi-viruses in the family *Flaviviridae* as separate genera.

Detection of HCV Ag and RNA

Krawczynski et al. have recently developed an HCV-specific fluorescent antibody assay for detection of HCV antigen in liver tissue [25]. The procedure has also been adapted into a competitive binding format for assaying patient sera for anti-HCV. The competition assay allows detection of antibodies to "native" HCV antigens expressed in infected liver tissue.

Detection of HCV RNA in serum [40] and liver tissue [13] has been accomplished by combining RNA extraction, cDNA synthesis, and amplification utilizing the polymerase chain reaction. These studies have indicated that viral RNA can be detected within one week following infection and may persist for years after the initial infection [4, 12, 36].

Clearly, past indicators of convalescence, such as normal liver enzyme levels, must be reevaluated in light of the fact that many infected animals and patients thought to be convalescent still have detectable levels of circulating HCV RNA. Application of a PCR assay for HCV-RNA in serum and use of commercial assays for the detection of circulating antibodies to structural and non-structural proteins of HCV in chimpanzees that have been cross-challenged or re-challenged (homologous inoculum), have provided some evidence for a weak immunity to HCV [31]. Many of the animals that lacked detectable antibodies or HCV-RNA by PCR (either during or after their bout of acute disease) were shown to seroconvert after re-challenge or become positive for HCV-RNA, respectively. Between one-fourth and one-third of the animals either developed recurrent disease or electron microscopic (EM) evidence of reinfection after challenge. These findings are consistent with the growing consensus that infection with HCV most often leads to persistent infection and/or weak immunity to the virus. Other studies have also indicated that HCV circulates as a population of closely-related genomes [28]. The "quasispecies" nature of HCV was supported by data that showed one-half of the circulating HCV RNA molecules in a single patient (sequence analysis of 5′-UTR and NS3/NS4 regions of genome) were identical, but that the other half consisted of a spectrum of "mutants" that differed from each other by one to four nucleotides, resulting in either silent mutations or in-frame stop codons. These findings suggest that defective HCV particles (genomes) may provide a supplementary mechanism (in addition to the above described HV region within the putative E2 portion of the HCV genome) for the development of persistent infection. It is well known that defective RNA genomes can decrease the efficiency of virus replication, modulate the clinical course of disease, and facilitate persistent virus infection [19]. The role(s) of the HV region and "quasispecies" variability of HCV-RNA in the development of persistent disease and/or infection remain to be fully elucidated.

Hepatitis E virus (HEV)

Hepatitis E virus (HEV), the major etiologic agent of enterically transmitted non-A, non-B hepatitis (ET-NANBH), is a 32 nm diameter, non-enveloped virus with physicochemical properties similar to many well-characterized caliciviruses or calici-like viruses [5, 7]. HEV has a sedimentation coefficient of 183S in sucrose and is labile in the presence of high salt concentrations [7, 8]. Cynomolgus macaques (cynos) have been shown to be most susceptible to infection with human-origin HEV and have been used extensively for the generation of virus-positive source materials and antisera reactive with HEV particles and antigen(s). Molecular cloning of HEV was approached with the assumption that it had an RNA genome. Molecular clones of HEV were identified by two separate procedures using lambda gt_{10} and gt_{11} as cloning and expression vectors, respectively. cDNA was prepared from both uninfected and HEV (Burma isolate) infected cyno bile and inserted into lambda gt_{10} for "plus-minus" screening using cDNA probes (from both cDNA libaries) labelled by random priming with ^{32}P NTPs. Differential hybridization studies, using radiolabelled probes, revealed the presence of several candidate clones derived from infected cyno bile, including ET1.1, the first virus-specific clone to be identified [33]. ET1.1 was shown to, (1) hybridize specifically to infected source cDNA, (2) hybridize specifically to infected liver RNA (approx. 7.5 Kb polyadenylated molecule), (3) encode a portion of an RNA-dependent RNA polymerase consensus motif, (4) be exogenous to both infected and uninfected primate DNA, and (5) have a sequence unique from any contained in GenBank. Clone ET1.1 was also shown to specifically hybridize to cDNAs prepared from HEV-infected stools collected from outbreak-related cases in Borneo, Mexico, Pakistan, Somalia, and the USSR. The latter cDNAs were prepared by a novel cDNA amplification procedure (SISPA) [34] that permitted the expansion of the limited amount of cDNA originally prepared from the source stools.

The second approach to HEV clone identification relied on the use of immunoscreening of cDNA libraries (prepared from a Mexican outbreak case stool) prepared in lambda gt_{11}, a highly efficient expression vector. Virus-specific proteins, expressed as beta-galactosidase fusion proteins, were identified using a high-titer convalescent serum from another well-documented case of hepatitis E from a separate outbreak in Mexico. Two of these virus-specific clones, 406.3-2 and 406.4-2, were used to identify additional virus-specific messages of 2.0 and 3.7 Kb in polyA-RNA extracted from HEV infected cyno liver, indicating the possible presence of sub-genomic RNAs involved in virus replication. Paneling of expressed proteins against a variety of paired acute and convalescent phase human sera from geographically distinct regions of the world, as well as pre-inoculation, acute-phase, and convalescent-phase sera from infected cynos, demonstrated the specificity of these expressed proteins for anti-HEV antibodies.

Genome organization

Analysis of the entire nucleotide sequence of the HEV-Burma and HEV-Mexico isolates has revealed the presence of several consensus sequences, including one associated with an NTP-binding, helicase-like function 5′ to the putative RNA-dependent RNA polymerase [37] (see more extensive discussion below). These nonstructural elements are contained within a single open reading frame (ORF) of approximately 5 Kb, 28 nucleotides from the 5′ end of the genome. Clone 406.3-2 (Mexico isolate) was contained within a second ORF that extended another 2 Kb to a point 68b upstream from the polyA tail. ORF2 is thought to encode for the major structural protein(s) of HEV. The second immunoreactive clone, 406.4-2 (Mexico isolate), was identified within a third ORF (plus 2 frame) that contained only 369 nucleotides. It overlapped ORF1 by one nucleotide and ORF2 by 328 nucleotides. HEV appears to be substantially different from picornaviruses, including HAV, in that it has an RNA genome that encodes for structural and nonstructural proteins through the use of discontinuous, partially overlapping ORFs (Fig. 2).

As noted above, the genomes of both Burma and Mexico HEV isolates have now been cloned and completely sequenced; they have been shown to possess the basic genomic features described above. We believe that a fundamental understanding of the key elements of the genetic organization and expression strategy of HEV (still tentative, pending sequencing of authentic viral proteins) have been elucidated and shown to set it apart from other previously defined hepatotrophic viruses of man. Recent studies of the genome of Norwalk virus [21] and feline calicivirus (FCV) [38] further suggest that HEV may belong to a larger (or possibly new) family of single-stranded, polyadenylated RNA viruses that possess three ORFs and two or more 3′ subgenomic RNAs with sizes ranging between 2 and 4 Kb. More recent computer-assisted alignment studies of the nucleotide sequence that encodes the putative nonstructural polyprotein of HEV (ORF1), further suggest that HEV may belong to a "supergroup" of positive-strand RNA plant and animal viruses [24]. Computer-assisted comparison of the HEV ORF1 polyprotein to analogous regions of other positive-stranded RNA viruses (either enveloped or non-enveloped) revealed the following: (1) the RNA-dependent RNA polymerase was located at the extreme carboxy terminal portion of the ORF1 polyprotein (see above), (2) the presence of an RNA helicase, (3) the existence of a methyl transferase (found at the extreme amino terminus of the polyprotein), and, (4) presence of a virally-encoded papain-like cysteine protease (distantly related to the putative protease of rubella virus) (refer to Fig. 3). The presence of an amino terminal methyl transferase suggests that HEV is capped, in contrast to other caliciviruses. While the latter findings would appear to be at odds with the tentative classification of HEV as a "calici-like virus", it should be emphasized that HEV still shares many physico-chemical, morphological, and genetic features common to other recognized and tentatively assigned members of the *Caliciviridae*.

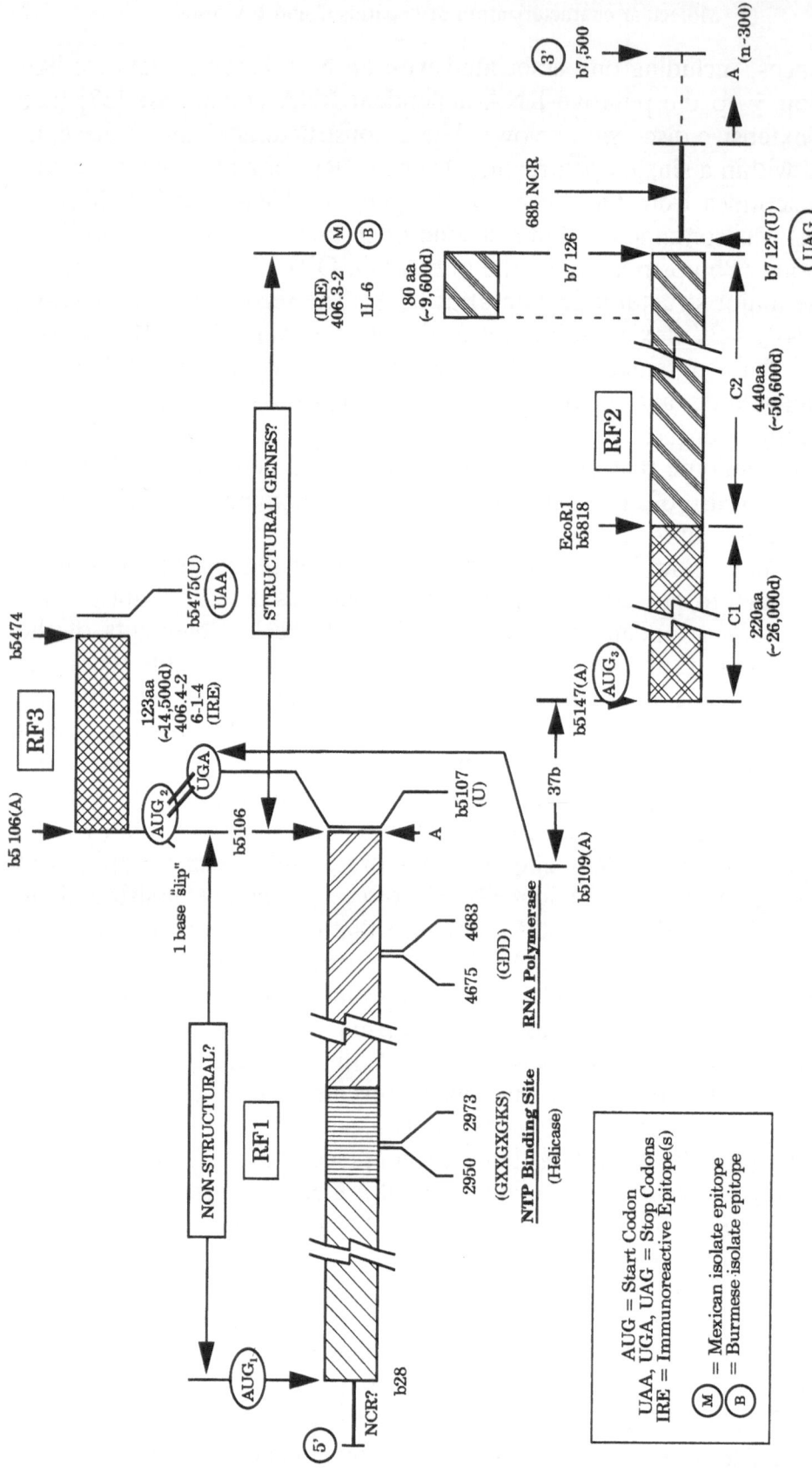

Fig. 2. General organization of the HEV genome. HEV-Hepatitis E virus genome

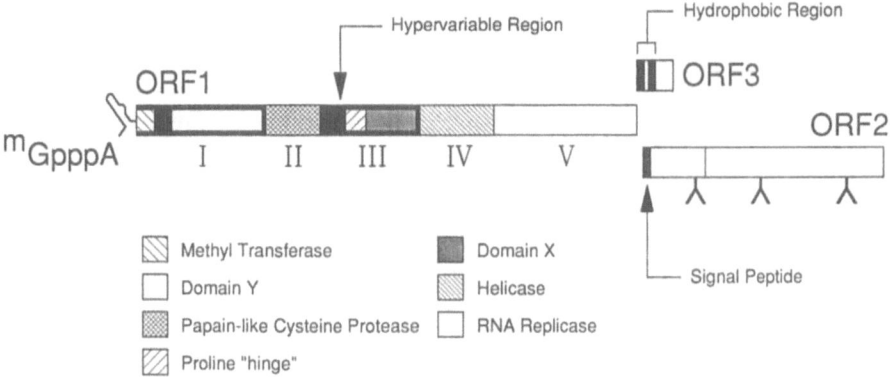

Fig. 3. Functional domains within ORF1. Genome map of the hepatitis E virus

Serologic cross-reactivity

HEV isolates from geographically distinct regions of the world have been shown by immune electron microscopy (IEM) to possess at least one major, cross-reactive epitope [8]. Acute and convalescent phase sera from well-documented cases of hepatitis E occurring in the PRC, Somalia, Burma, USSR, Mexico, Borneo, India, Pakistan, and Greece also have been shown to "block" antibody binding in a fluorescent antibody (FA) test that utilizes an FITC-labelled IgG probe (Mexico) and liver sections from a HEV-Burma infected cyno [26]. These earlier findings implied that a serodiagnostic assay for HEV-specific antibody could be developed if recombinant expressed proteins containing the immunodominant (or broadly cross reactive) epitope of HEV were utilized. In fact, a sensitive and specific western blot (WB) assay for both IgM (acute-phase) and IgG anti-HEV has been recently developed in our laboratory [32]. This assay makes use of a *trpE*-HEV fusion protein (expressed in *E. coli* as an insoluble protein [*trpE*-C2]) that contains the carboxyl two-thirds of ORF2, or the region that encodes the putative structural protein. This assay has been shown to be capable of detecting anti-HEV in both acute and convalescent phase case sera collected from outbreaks worldwide. Alternate configurations of the WB procedure that incorporate beta-galactosidase fusion proteins (representing portions of HEV ORF2 or ORF3) as antigens have shown that the Mexico-HEV isolate contains some epitopes that are different from those found on the Burma isolate [35]. Although it would appear that at least two different serotypes of HEV exist, cross challenge studies conducted in cynos have shown that animals infected with the Burma isolate are immune to re-infection by the Mexico isolate (D. Bradley, unpublished findings).

Human linear B-cell epitopes

The deduced amino acid (aa) sequence from all three ORFs of the Burma HEV isolate was used to design and synthesize a series of deca-peptides (overlapping at every fourth aa) that covered each of the polyproteins or proteins encoded by ORFs 1, 2, and 3. Solid-phase "pin" technology was used to identify the most immunoreactive peptides by standard methodologies. Each pin, coated with a unique decapeptide, was tested for its ability to specifically bind antibody from a Sudanese HEV case serum pool (n = 11 cases). Twelve immunoreactive epitopes were identified throughout ORF1, mostly within the region thought to be the virally encoded RNA-dependent RNA polymerase [23]. One epitope was identified within ORF3 at the extreme carboxy terminus, and three more were identified within ORF2. It is interesting to note, however, that no antibodies were detected against peptides representing portions of the previously identified region that encoded for the 406 3-2 epitope (see above) found at the extreme carboxy terminus of ORF2. The latter finding indicates that the highly immunoreactive beta-galactosidase fusion protein, beta-gal-406.3-2, contains conformational epitopes that cannot be mimicked by the synthetic peptide decamers used in this study. Longer peptides, or peptides consisting of other amino acid permutations (longer or shorter peptides that overlap each other by only 1–3 aa), may be needed in order to regain immunoreactivity against naturally-occurring antibodies. These findings also indicate that antibody tests could be developed that would readily distinguish between acute viral infection and prior immunization with a recombinant structural protein, since antibodies against nonstructural proteins, namely the RNA polymerase described above, should not be found in individuals who have only been immunized.

Genetic divergence of the HEV genome

Molecular cloning and sequencing of the Burma, USSR (Tashkent), and Mexico isolates of HEV have revealed an unexpectedly high degree of sequence divergence between the North American (Mexico) isolate and the Asian isolates [14]. Computer analysis of nucleotide sequences (869 bases in length) contained within the region encoding the putative RNA-dependent RNA polymerase (RDRP) showed that the Tashkent and Burma isolates shared homologies of 93.7% and 95.5% at the nt and aa levels, respectively, whereas these two isolates shared only a 77% and 87–90% homology with the Mexican isolate at the nt and aa levels, respectively. It is worth noting that the ratios of codonic changes in the first, second, or third position were nearly constant for all three isolates. The number of transitions (TI) far outnumbered the number of transversions (TV) when the Tashkent and Burma sequences were compared. By contrast, comparison of either the Tashkent or Burma sequence to that of the Mexico isolate showed that transversions were just as likely to be observed as transitions (M. Purdy, unpublished findings). These findings clearly indicate that the Mexico isolate is genetically distinct from the "Asian" isolates, a sign that these viruses probably diverged from each other in the distant past. Cloning and sequencing of other HEV isolates obtained from South America, Central Africa, and Borneo, for example, may reveal more subtle evolutionary changes at the nucleotide level and provide some insight into the specific nt sequences that are absolutely required for functionality of the respective, deduced proteins.

References

1. Alter MJ, Hadler SC, Judson FN, Mares A, Alexander WJ, Hu PY, Miller JK, Moyer LA, Fields HA, Bradley DW, Margolis HS (1990) Risk factors for acute non-A, non-B hepatitis in the United States and association with hepatitis C virus infection. JAMA 264: 2231–2235

2. Argos P (1988) A sequence motif in many polymerases. Nucleic Acids Res 16: 9909–9916

3. Beach M, Bradley D (1991) Analysis of the putative nonstructural gene region of hepatitis C virus. In: Hollinger FB, Lemon S, Margolis H (eds) Viral hepatitis and liver disease. Williams and Wilkins, Baltimore, pp 376–381

4. Beach MJ, Meeks EL, Mimms LT, Vallari D, DuCharme L, Spelbring J, Taskar S, Schleicher JB, Krawczynski K, Bradley DW (1992) Temporal relationships of hepatitis C virus RNA and antibody responses following experimental infection of chimpanzees. J Med Virol 36: 226–237

5. Bradley DW (1990) Hepatitis non-A, non-B viruses become identified as hepatitis C and E viruses. In: Melnick JL (ed) Progress in medical virology, vol 37. S. Karger AG, Basel, pp 101–135

6. Bradley DW (1990) Enterically-transmitted non-A, non-B hepatitis. Br Med Bull 46: 442–461

7. Bradley DW, Krawcsynski K, Beach MJ, Purdy MA (1991) Non-A, non-B hepatitis: Toward the discovery of hepatitis C and E viruses. Semin Liver Dis 11: 128–146

8. Bradley DW, Andjaparidze A, Cook EH, McCaustland K, Balayan M, Stetler H, Velazquez O, Robertson B, Humphrey C, Kane M, Weisfuse I (1988) Aetiological agent of enterically-transmitted non-A, non-B hepatitis. J Gen Virol 69: 731–738

9. Chambers TJ, McCourt DW, Rice CM (1989) Yellow fever virus proteins NS2A, NS2B, and NS4B: identification and partial N-terminal amino acid sequence analysis. Virology 169: 100–109

10. Choo Q-L, Kuo G, Weiner AJ, Overby LR, Bradley DW, Houghton M (1989) Isolation of a cDNA clone derived from a blood-borne non-A, non-B viral hepatitis genome. Science 244: 359–362

11. Choo Q-L, Richman KH, Han JH, Berger K, Lee C, Dong D, Gallegos C, Coit D, Medina-Selby A, Bar PJ, Weiner AJ, Bradley DW (1991) Genetic organization and diversity of the hepatitis C virus. Proc Natl Acad Sci USA 88: 2451–2455

12. Farci P, Alter HJ, Wong D, Miller RH, Shih JW, Jett B, Purcell RH (1991) A long-term study of hepatitis C virus replication in non-A, non-B hepatitis. N Engl J Med 325: 98–104

13. Fong T-L, Shindo M, Feinstone SM, Hoofnagle JH, DiBiscegli AM (1991) Detection of replicative intermediates of hepatitis C viral RNA in liver and serum of patients with chronic hepatitis C. J Clin Invest 88: 1058–1060

14. Fry KE, Tam AW, Smith MM, Kim JP, Luk K-C, Young LM, Paatak M, Feldman RA, Yun KY, Purdy MA, McCaustland KA, Bradley DW, Reyes GR (1991) Hepatitis E virus (HEV): Strain variation in the nonstructural gene region encoding motifs for an RNA-dependent RNA polymerase and an ATP/GTP binding site. Virus Genes 6: 173–185

15. Gorbalenya AE, Donchenko AP, Koonin EV, Blinov VM (1989) N-terminal domains of putative helicases of flavi- and pestiviruses may be serine proteases. Nucleic Acids Res 17: 3889–3897

16. Gorbalenya AE, Koonin EV, Donchenko AP, Blinov VM (1989) Two related superfamilies of putative helicases involved in replication, recombination, repair and expression of DNA and RNA genomes. Nucleic Acids Res 17: 4713–4730

17. Han JH, Shyamala V, Richman KH, Brauer MJ, Irvine B, Urdea MS, Tekamp-Olson P, Kuo G, Choo Q-L, Houghton M (1991) Characterization of the terminal regions of hepatitis C viral RNA: Identification of conserved sequences in the 5′ untranslated region and poly(A) tails at the 3′ end. Proc Natl Acad Sci USA 88: 1711–1715

18. Hijikata M, Kato N, Ootsuyama Y, Nakagawa M, Shimotohno K (1991) Gene mapping of the putative structural region of the hepatitis C virus genome by *in vitro* processing analysis. Proc Natl Acad Sci USA 88: 5547–5551

19. Holland JJ (1991) Defective viral genomes. In: Fields BN, Knipe DN (eds) Fundamental virology. Raven Press, New York, pp 151–165

20. Houghton M, Weiner A, Han J, Kuo G, Choo Q-L (1991) Molecular biology of the hepatitis C viruses: Implications for diagnosis, development, and control of viral disease. Hepatology 14: 381–388

21. Jiang X, Graham DY, Wang K, Estes, MK (1990) Norwalk virus genome cloning and characterization. Science 250: 1580–1583

22. Kato N, Hijikata M, Ootsuyama Y, Nakagawa M, Ohkoshi S, Sugiyama T, Shimotohno K (1990) Molecular cloning of the human hepatitis C virus genome from Japanese patients with non-A, non-B hepatitis. Proc Natl Acad Sci USA 87: 9524–9528

23. Kaur M, Hyams KC, Purdy MA, Krawczynski K, Ching WM, Fry KE, Reyes GR, Bradley DW, Carl M (1992) Human linear B-cell epitopes encoded by the hepatitis E virus include determinants in the RNA-dependent RNA polymerase. Proc Natl Acad Sci USA 89: 3855–3858

24. Koonin EV, Gorbalenya AE, Purdy MA, Rozanov MN, Reyes GR, Bradley DW (1992) Computer-assisted assignment of functional domains in the nonstructural polyprotein of hepatitis E virus: Delineation of an additional group of positive-strand RNA plant and animal viruses. Proc Natl Acad Sci USA 89: 8259–8263

25. Krawczynski K, Beach MJ, Bradley DW, Kuo G, DiBiscegli AM, Houghton M, Reyes GR, Kim JP, Choo Q-L, Alter MJ (1992) Hepatitis C virus antigen in

hepatocytes: Immunomorphologic detection and identification. Gastroenterology 103: 622–629

26. Krawczynski K, Bradley D, Ajdukiewicz A, Alter MV, Caredda F, Dilawari J, Hlady G, Innis B, Kane M, Nasidi A, Redeker A, Stetler H, Toukan A, Yoffee B (1991) Virus-associated antigen and antibody of epidemic non-A, non-B hepatitis: Serology of outbreaks and sporadic cases. In: Shikata T, Purcell RH, Uchida T (eds) Viral hepatitis C, D, and E. Elsevier Science, Amsterdam, pp 229–236

27. Lain S, Riechmann JL, Garcia JA (1990) RNA helicase: a novel activity associated with a protein encoded by a positive strand RNA virus. Nucleic Acids Res 18: 7003–7006

28. Martell M, Esteban JI, Quer J, Genesca J, Weiner A, Esteban R, Guardia J, Gomez J (1992) Hepatitis C virus (HCV) circulates as a population of different but closely related genomes: Quasispecies nature of HCV genome distribution. J Virol 66: 3225–3229

29. Miller RH, Purcell RH (1990) Hepatitis C virus shares amino acid sequence similarity with pestiviruses and flaviviruses as well as members of two plant virus supergroups. Proc Natl Acad Sci USA 87: 2057–2061

30. Preugschat F, Yao C-W, Strauss JH (1990) In vitro processing of dengue virus type 2 nonstructural proteins NS2A, NS2B, and NS3. J Virol 64: 4364–4374

31. Prince AM, Brotman B, Huima T, Pascual D, Jaffery M, Inchauspe G (1992) Immunity in hepatitis C infection. J Infect Dis 165: 438–443

32. Purdy MA, McCaustland KA, Krawczynski K, Tam A, Beach MJ, Tassopoulos NC, Reyes GR, Bradley DW (1992) Expression of a hepatitis E virus (HEV)-trpE fusion protein containing epitopes recognized by antibodies in sera from human cases and experimentally infected primates. Arch Virol 123: 335–349

33. Reyes GR, Purdy MA, Kim JP, Luk K-C, Young LM, Frey KE, Bradley DW (1990) Isolation of a cDNA from the virus responsible for enterically transmitted non-A, non-B hepatitis. Science 247: 1335–1339

34. Reyes GR, Kim JP (1991) Sequence-independent, single-primer amplification (SISPA) of complex DNA populations. Mol Cell Probes 5: 473–481

35. Reyes GR, Huang C-C, Yarbough PO, Young LM, Tam AW, Moeckli RA, Yun KY, Nguyen D, Smith MM, Fernandez J, Guerra ME, Kim JP, Luk K-C, Drawczynski K, Purdy MA, Fry KE, Bradley DW (1991) Hepatitis E virus (HEV): Epitope mapping and detection of strain variation. In: Shikata T, Purcell RH, Uchida T (eds) Viral hepatitis C, D, and E. Elsevier Science, Amsterdam, pp 237–245

36. Shimizu YK, Weiner AJ, Rosenblatt J, Wong DC, Sapiro M, Popkin T, Houghton M, Alter HJ, Purcell RH (1990) Early events in hepatitis C virus infection of chimpanzees. Proc Natl Acad Sci USA 87: 6441–6444

37. Tam AW, Smith MM, Guerra ME, Huang CC, Bradley DW, Fry KE, Reyes GR (1991) Hepatitis E virus (HEV): Molecular cloning and sequencing of the full-length viral genome. Virology 185: 120–131

38. Tohya Y, Taniguchi Y, Utagawa E, Takeda N, Miyamura K, Yamazaki S, Mikami T (1991) Sequence analysis of the 3' end of feline calicivirus genome. Virology 183: 810–814

39. Tsukiyama-Kohara K, Iizuka N, Kohara M, Nomoto A (1992) Internal ribosome entry site within hepatitis C virus RNA. J Virol 66: 1476–1483

40. Weiner AJ, Kuo G, Bradley DW, Bonino F, Saracco G, Lee C, Rosenblatt J, Choo Q-L, Houghton M (1990) Detection of hepatitis C viral sequences in non-A, non-B hepatitis. Lancet 335: 1–3

41. Weiner AJ, Brauer MJ, Rosenblatt J, Richman KH, Tung J, Crawford F, Bonino F, Saracco G, Choo Q-L, Houghton M, Han JH (1991) Variable and hypervariable domains are found in the regions of HCV corresponding to the flavivirus envelope and NS1 proteins and the pestivirus envelope glycoproteins. Virology 180: 842–848
42. Wengler G, Wengler G (1991) The carboxy-terminal part of the NS3 protein of the west nile flavivirus can be isolated as a soluble protein after proteolytic cleavage and represents an RNA-stimulated NTPase. Virology 184: 707–715

Authors' address: D.W. Bradley, Ph.D., Hepatitis Branch, Mailstop A33, DVRD, NCID, Centers for Disease Control, 1600 Clifton Road, N.E., Atlanta, GA 30333, U.S.A.

Arch Virol (1993) [Suppl] 7: 15–25

Molecular organization and replication of hepatitis E virus (HEV)

G.R. Reyes[1], **C.-C. Huang**[2], **A.W. Tam**[2], and **M.A. Purdy**[3]

[1] Triplex Pharmaceutical Corporation, The Woodlands, TX
[2] Genelabs Incorporated, Redwood City, CA
[3] National Center for Infectious Diseases, Centers for Disease Control and Prevention,
Atlanta, GA, U.S.A.

Summary. The recently characterized fecal-orally transmitted agent of hepatitis E (formerly known as enterically transmitted non-A, non-B hepatitis) has been determined to be a new type of positive strand RNA virus. The complete sequencing of four different geographic isolates of the hepatitis E virus (HEV) has confirmed a similar genetic organization not previously recognized in nonenveloped positive strand RNA viruses. The ~7.5 kb RNA genome (including polyA tail) has nonstructural genes located at the 5' end and structural genes at the 3' end. Expression of these viral genes occurs in at least 3 different forward open reading frames. The largest open reading frame begins 27 nucleotides (nt) downstream of the apparent noncoding 5' end and extends 5,079 nt. Multiple nonstructural gene motifs/domains have been recognized in this 5' ORF1 including a methyltransferase, a papain-like protease, a helicase and the RNA-dependent, RNA polymerase. The second major ORF2 begins 37 nt downstream of ORF1 and extends 1980 nt before terminating 65 nt upstream of the polyadenylation site. A third ORF of only 369 nt was identified by immunoscreening experiments as encoding an immunogenic epitope of the virus. Expression of the downstream ORF2 may occur through internal subgenomic RNA initiation at a sequence element found to have homology to internal RNA initiation sequences in Sindbis virus. This element in the HEV genome maps near the apparent 5' end of one of two identified subgenomic messages. The genomic organization and expression of HEV will be discussed and a hypothesis presented regarding the viral replication strategy.

Introduction

The recently characterized hepatitis E virus (HEV) is the etiologic agent of an epidemic form of fecal-orally transmitted hepatitis. The disease, now known as hepatitis E, was formerly classified as a form of non-A, non-B hepatitis (NANBH); a designation based on exclusionary diagnostic criteria that were used to recognize the existence of other types of viral hepatitis different than those due to the parenterally transmitted hepatitis B virus or the fecal-orally transmitted hepatitis A virus [2, 7, 20]. The NANBH designation permitted the accumulation of epidemiological and animal modeling data prior to the definitive isolation and characterization of what might be a diverse group of agents. Analysis of these data suggested the existence of at least two different agents and formed the basis for the molecular cloning of two different viruses responsible for NANBH [reviewed in 22]. In addition to HEV, the principle agent of enterically transmitted NANBH, the principle agent responsible for the parenterally transmitted NANBH, the hepatitis C virus (HCV), has also been fully elucidated [4].

The molecular cloning of HEV has led to the development of a recombinant protein based ELISA for the detection of anti-HEV antibodies elicited in the course of current or past viral infection [6, 30, 34]. Although once believed to be a disease of young adults, hepatitis E is now recognized as a cause of sporadic hepatitis in children raised in endemic areas [9, 12]. Epidemic outbreaks continue to occur in endemic areas suggesting that neutralizing antibodies, if they exist, may be transient. The recent availability of a diagnostic test has confirmed that IgM and IgG antibodies disappear in the majority of infected individuals over the course of a year [29]. The implications of these findings for life-long immunity to hepatitis E infection remains to be clarified. Our understanding of the serological response to hepatitis E will continue to evolve as testing becomes more widely used. A recent and provocative finding has been the discovery of anti-HEV antibody in individuals from regions not previously believed to be endemic [6]. Confirmed sero-prevalence rates have been reported in the range of 3% for a number of industrialized countries [6]. The significance of this high hepatitis E seroprevalence, in the apparent absence of overt disease and epidemic activity in the indigenous population, remains to be determined through comprehensive epidemiological surveillance.

Our newly acquired knowledge of HEV and HCV was obtained by the successful application of molecular biological procedures designed to characterize viruses that predominantly replicate at low titer in the infected host. This became possible after the development of a variety of advanced molecular cloning and characterization techniques such as the polymerase chain reaction (PCR) [28], sequence-independent single-primer amplification (SISPA) [25] and immunoscreening [35; for review see 23]. These techniques facilitate not only the isolation of a virus clone but also its characterization. "Molecular phenotyping" contributes important information that is critical to the taxonomic classification of a newly discovered virus. The nature and organization of the genomic nucleic acid can lead to testable hypotheses regarding the expression and replication strategy of the virus. We now review the molecular organization of HEV and present a possible replication strategy based on currently available data.

Results

The isolation of molecular clones from the Burma isolate of HEV has been previously detailed [26]. Briefly, cynomolgus macaques infected with the third passage Burma isolate were used as a source of HEV infected samples for further biochemical characterization and molecular cloning. As it had been previously hypothesized that hepatitis E was caused by an RNA virus [3], complementary DNA (cDNA) was synthesized using random primers and clones were identified from duplicate filter lifts by differential hybridization using high specific activity ^{32}P labelled cDNA probes from infected and uninfected source bile [26]. A candidate clone was verified as being derived from the HEV genome by several criteria. The cloned sequence was first demonstrated to not only be exogenous but also present in other infected tissues (e.g., liver) and specimens derived from epidemiologically similar but temporally distinct epidemics [26]. In further studies, the identification of genetic motifs (RNA-dependent, RNA polymerase; RDRP) shared by other positive strand RNA viruses of plants and animals [15, 16], and the isolation of related exogenous clones from other geographic isolates [8], firmly established the identification of a new hepatitis virus designated as HEV. The completed cloning and sequencing of the full-length viral genome, as a set of overlapping contiguous clones, reaffirmed that the HEV sequence had little similarity to any of the previously characterized positive strand viruses [31].

In quick succession there followed the molecular cloning of additional isolates from Pakistan [33], Mexico [10] and China [1]. The isolation and sequencing of these genomes confirmed the unique properties of the virus as originally recognized in HEV(Burma) (Fig. 1). These included the recognition that the virus was expressed in two major nonoverlapping open reading frames designated ORF1 and ORF2. The initial clone contained certain recognized motifs of the RDRP that were located at the 3' end of ORF1 and therefore led to the original proposition that this was the nonstructural (NS) gene encoding reading frame [31]. Separate motifs for the RNA helicase were subsequently recognized in this same reading frame and strengthened the case for ORF1 encoding the NS proteins of the virus [8].

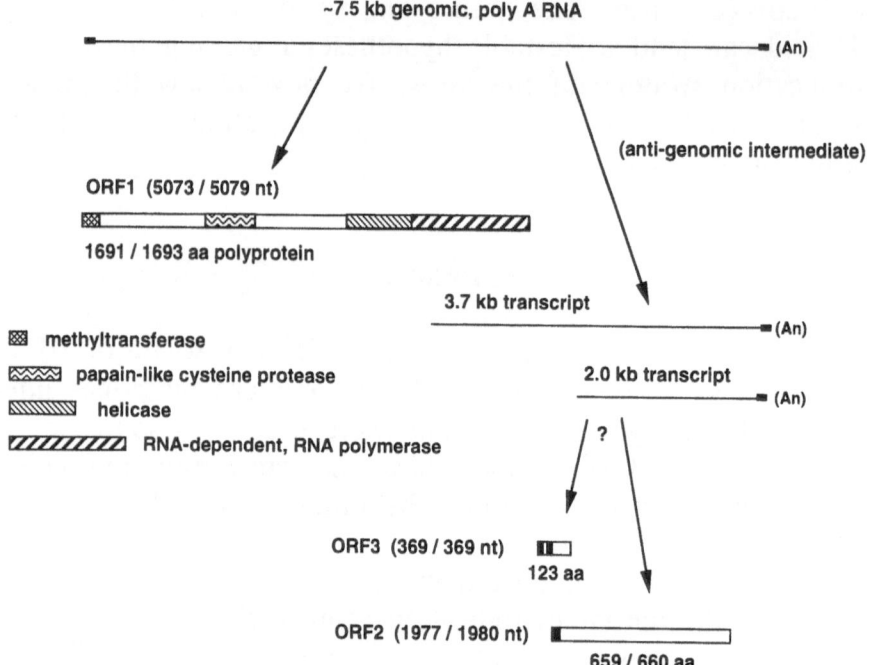

Fig. 1. Genomic organization and expression strategy of HEV. The top of the figure
illustrates the linear, single stranded, positive sense, RNA genome of HEV flanked by
small apparently untranslated sequences and a long 3' (~300 nt) poly A tail [31]. The
nonstructural polyprotein is presumed to be expressed from the genomic RNA (left
part of the diagram) and is known to vary in length between the Mexico (first number
in parentheses) and the Burma (second number in parentheses) isolates [10]. A number
of domains and motifs associated with nonstructural proteins of other positive strand
RNA viruses have been recognized [17] and their approximate locations noted in
ORF1. The right side of the figure illustrates the expression of the 3' end of the
genome (presumed structural gene[s]) via an anti-genomic (negative strand) RNA
intermediate. Two subgenomic poly A RNAs, presumed to be expressed from the antigenome, are drawn as coterminal with the 3' end [31]. The transcript(s) involved in
the expression of ORF2 or ORF3 are not known. The size of ORF2, between Mexico
and Burma, is known to vary [10]. The dark blocks near the amino termini of ORF2
and 3 are meant to indicate the presence of short stretches of hydrophobic amino
acids [10, 31]

Recently, a more detailed analysis of the entirety of ORF1 has
identified other domains with similarity to recognized NS domains from
other positive stranded viruses [17]. The limited similarities that were
recognized in the nonstructural region of the virus (see below) have
served to confirm its classification as a positive strand agent and have led
to a provocative hypothesis regarding the evolution and taxonomy of
positive sense RNA viruses [17]. This analysis has formed the basis for
the hypothesis that HEV falls within the alpha-like supergroup of viruses

based upon the many recognized motifs, their alignment, and the generation of independent gene products by co- or posttranslational polyprotein processing [21]. These characteristics are summarized in Fig. 1. In addition to the RDRP and the helicase regions, other less conserved domains include that of a methyltransferase and a cysteine-like protease. The presence of the methyltransferase implies capping of the genomic RNA [27]. The cysteine-like protease is postulated to be involved in the co- or posttranslational processing of the NS polyprotein to yield the individual NS gene products, however, the role of host derived proteases certainly cannot be excluded. The arrangement of these motifs in ORF1 is also significant in that it reinforces the evolutionary relationships between presumed members of a related group that includes rubella virus and plant furoviruses [17].

The major structural gene product is believed to be encoded by the ~2 kb ORF2. The deduced protein sequence of ORF2 contains a hydrophobic segment at the amino terminus which might imply that virion maturation occurs in association with the endoplasmic reticulum after translation [31]. ORF2 also has a high pI (pI = 10.35) in the amino terminal first half of the protein which, as discussed below, could indicate that the protein associates with the genomic RNA strand in the process of encapsidation [31]. Another feature that suggested ORF2 encoded a structural gene product was the finding of highly immunogenic epitopes selected from a cDNA library by a random immunoscreening procedure [34]. One might expect the structural proteins of a virus to elicit the strongest immune responses due to the antigenic stimulation associated with having its structural proteins exposed on the surface of the virus as opposed to the nonstructural expression that would occur principally in an intracellular compartment. A second highly immunogenic clone was isolated by the same immunoscreening protocol [34]. This clone came from yet a third forward ORF (ORF3) that overlapped both ORF1 and ORF2 (see Fig. 1). This short ORF3 of only 369 nt was not originally predicted to be used by the virus but was recognized after the isolation of an immunogenic clone [34]. ORF3 also appears to have two regions that are highly hydrophobic and might constitute transmembrane segments or a signal sequence followed by a transmembrane region [21]. The function and means of expression of the ORF3 gene product are not known.

A basic replication strategy for HEV is presented in Fig. 2. This strategy takes into account all of the elements that have been previously noted. The NS gene products are presumably expressed from the full length positive sense genome upon entering the cell. These individual gene products are then involved in the earliest stages of viral replication. The first step is the generation of negative strand RNA (anti-genomic

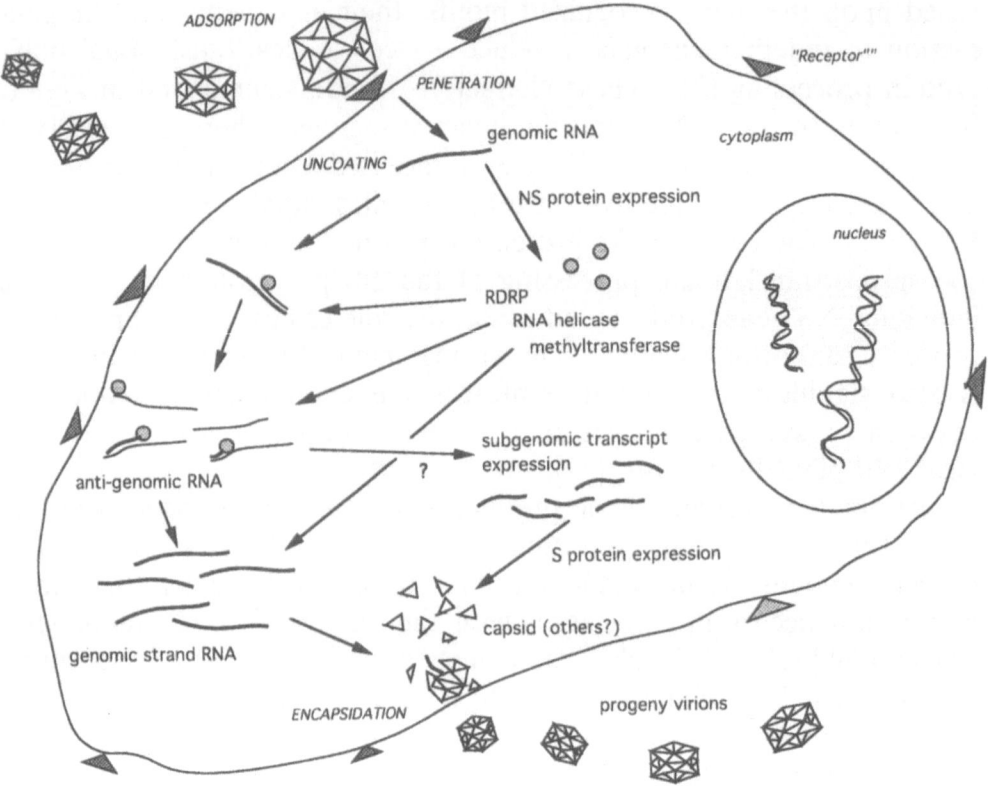

Fig. 2. Proposed replication strategy for HEV: A proposal is presented for the replication of HEV based on our current understanding of the molecular biology of the virus. After virus entry, the positive sense RNA is translated to produce the nonstructural (NS) polyprotein critical to viral replication. The NS gene products produce anti-genomic RNA which serves as template for both the full length genome as well as the subgenomic RNAs detected in infected liver [31, 34]. Structural protein(s) are presumed to be expressed from the subgenomic RNAs which can then combine with and encapsidate the genomic RNA

strand). For HEV, the RDRP may be involved in the generation of both negative strand and positive strand viral RNAs as has been shown for other positive strand RNA viruses. The available full length anti-genome RNA can then lead to production of additional full length genomic RNA and further nonstructural protein expression and accumulation. The role, if any, played by host derived proteins in this and other stages of viral replication is not known. This is especially problematic in the proposal for any model since it has already been shown that host proteins can play a role as has been shown in the 3′ end of the antigenomic strand of rubella virus [19]. The lack of a tissue culture system for HEV propagation has also hampered progress in this area.

The proposed model would have the anti-genomic RNA serve not only as the template for the production of the positive sense RNA genomic strand, but also for the production of the smaller subgenomic messages. This initiates the second phase of viral replication, the production of the viral structural protein(s). Two subgenomic messages were first detected in polyA RNA from infected cynomolgus liver using HEV hybridization probes from the 3' end of the virus [31]. The exact mechanism by which the 3.7 and 2.0 kb subgenomic messages are generated is not known. An alternative mechanism may be through the synthesis of anti-*sub*genomic messages to serve as templates for positive sense polyadenylated subgenomic message production. The subgenomic messages are then translated into the capsid protein(s). It is not known which message is used as the source for the capsid protein; nor is it clear whether more than one protein is incorporated into the virion. As noted above, the short ORF3 encodes a highly immunogenic protein. The degree of this immunogenicity might suggest that this protein has some structural role in the virion. But as already noted, the exact role of this protein and its means of expression are not known.

Production of the full-length genomic RNA is presumed to occur from the anti-genomic RNA. Encapsidation of the genomic RNA then occurs by its association with the capsid protein. The high basic amino acid content (10% arginine) of the amino terminal half of the capsid protein(s) is presumably critical to its role in genomic RNA encapsidation as postulated for other viruses [5, 31].

The expression of the individual gene products can be regulated at various stages including the expression (internal initiation) of the subgenomic RNA or perhaps the differential half-life of the individual RNA species. We have observed that the relative abundance of the individual transcripts relative to one another is nonequimolar with the smaller 2.0 kb transcript more abundant than the 3.7 kb transcript [31]. It is therefore tempting to speculate that some type of complex regulatory processing is responsible for the unequal level of transcript products. These levels might also vary according to the stage of infection, however, at this time little detail is available.

Discussion

The molecular characterization of HEV was made possible by the development of new recombinant DNA techniques and has established HEV as a distinctly different virus from the previously characterized picornavirus HAV. Although previously presumed by some to be a picornavirus variant or perhaps a different neutralization serotype, the distinctly different epidemiology did indeed indicate infection with a different virus. The occurrence of seronegative HBV due to a variant [32] might serve as a possible model for how HAV seronegative disease might occur due to a HAV genetic variant. However, the molecular isolation and characterization of HEV provided the definitive proof of the existence of an entirely new virus.

One of the immediate benefits from the molecular cloning and sequencing of HEV was the provision of specific synthetic recombinant DNA probes for use in direct hybridization analysis or as primers in the polymerase chain reaction. The probes have been used in PCR reactions since the virus titer is generally low and it's not possible to detect the genomic RNA unless in the unusual circumstance where high titer excretion has occurred, as for example, in the HEV sample that was acquired from an individual infected in the Telixtac, Mexico outbreak [24]. In the latter example, it was possible to perform a direct hybridization experiment with strand specific probes to definitively show that the virion RNA was positive sense and single stranded [24]. These probes and PCR primers have been used to further define the temporal events related to the timing of virus expression and presence in sera and excreta and hence determine the associated infectivity of these materials. These studies have confirmed that viremia occurs [18, 33], and have defined a window for virus excretion in feces [33] following inoculation, the development of biochemical hepatitis, and overt signs of infection. Studies such as these will help to determine the various pathophysiologic interrelationships in the virus life cycle.

Molecular cloning of HEV has also led to the high level recombinant expression of viral gene products useful in the serodiagnosis of disease and has led to important insights related to the newly characterized virus. These will lead to other reagents such as monoclonal antibodies that could have an impact on improving serodiagnosis but might also be used to follow virion maturation and confirm or refute some of the hypotheses presented here. It will be possible to follow the synthesis and processing of the individual gene products intracellularly to learn how they contribute to virion maturation. This is especially important in the case of the proposed structural protein or capsid which has the unusual feature of having a signal sequence at the extreme amino terminus indicating a possible association with intracellular membranes.

Cloning will lead to a better understanding of viral gene expression and replication but might also have important therapeutic implications if a neutralizing immune response can be elicited and shown to be protective. Can the expressed capsid self-associate to form empty virions as seen in parvovirus [14] and in Norwalk virus [13]? Experimental findings of this sort may ultimately prove not only critical for future vaccine development but also for defining the pathways of virion maturation.

A definitive elucidation of the replication and expression hypothesis presented here will be greatly facilitated by the availability of an established or primary cell line for studies of the in vitro propagation of the virus. Evidence now indicates that HEV can be grown in vitro [11; J. Twu et al., submitted for publication]. The utilization of cellular biological approaches will complement the wealth of molecular biological data now available for HEV.

References

1. Aye TT, Uchida T, Ma XZ, Iida F, Shikata T, Zhuang H, Win KM (1992) Complete nucleotide sequence of a hepatitis E virus isolated from the Xinjiang epidemic. Nucleic Acids Res 20: 3512
2. Bradley DW (1990) Hepatitis non-A, non-B viruses become identified as hepatitis C and E viruses. Prog Med Virol 37: 101–135
3. Bradley DW, Balayan MS (1988) Virus of enterically transmitted non-A, non-B hepatitis. Lancet i: 819
4. Choo QL, Kuo G, Weiner AJ, Overby LR, Bradley DW, Houghton M (1989) Isolation of a cDNA clone derived from a blood-borne non-A, non-B viral hepatitis genome. Science 244: 359–362
5. Dalgarno L, Rice CM, Strauss JH (1983) Ross river virus 26S RNA: Complete nucleotide sequence and deduced sequence of the encoded structural proteins. Virology 129: 170–187
6. Dawson GJ, Chau KH, Cabal CM, Yarbough PO Reyes GR, Mushahwar IK (1992) Solid-phase enzyme-linked immunoassay of hepatitis E virus IgG and IgM antibodies utilizing recombinant and synthetic peptides. J Med Virol 38: 175–186
7. Feinstone SM, Kapikian AZ, Purcell RH, Alter HJ, Holland PV (1975) Transfusion-associated hepatitis not due to viral hepatitis type A or B. N Engl J Med 292: 767–770
8. Fry KE, Tam AW, Smith MW, Kim JP, Luk K-C, Young LM, Piatak M, Feldman RA, Purdy MA, McCaustland KA, Bradley DW, Reyes GR (1992) Hepatitis E Virus (HEV): Strain variation in the nonstructural gene region encoding consensus motifs for an RNA-dependent RNA polymerase and an ATP/GTP binding site. Virus Genes 6: 173–185
9. Goldsmith R, Yarbough PO, Reyes GR, Fry KE, Gabor KA, Kamel M, Zakaria S, Amer S, Ghaffar Y (1992) Enzyme-linked immunosorbent assay for diagnosis of acute sporadic hepatitis E in Egyptian children. Lancet 339: 328–331
10. Huang C-C, Nguyen D, Fernandez J, Yun-Choe K, Tam AW, Fry KE, Bradley DW, Reyes GR (1992) Molecular cloning and sequencing of the Hepatitis E Virus Mexico Strain. Virology 191: 550–558
11. Huang RT, Li DR, Wei J, Huang XR, Yuan XT, Tian X (1992) Isolation and identification of hepatitis E virus in Xinjiang, China. J Gen Virol 73: 1143–1148
12. Hyams KC, Purdy MA, Kaur M, McCarthy MC, Krawczynski K, Bradley DW, Carl M (1992) Acute sporadic hepatitis E in Sudanese children: Analysis based on a new Western blot asay. J Infect Dis 165: 1001–1005

24 G.R. Reyes et al.

13. Jiang X, Wang M, Graham DY, Estes MK (1992) Expression, self-assembly and antigenicity of the Norwalk virus capsid protein. J Virol 66: 6527–6532
14. Kajigaya S, Fujii H, Field A, Anderson S, Rosenfeld S, Anderson LJ, Shimada T, Young NS (1991) Self-assembled B19 parvovirus capsids, produced in a baculovirus system, are antigenically and immunologically similar to native virions. Proc Natl Acad Sci USA 88: 4646–4650
15. Kamer G, Argos P (1984) Primary structural comparison of RNA-dependent RNA polymerases from plant, animal, and bacterial viruses. Nucleic Acids Res 2: 7269–7282
16. Koonin EV (1991) The phylogeny of RNA-dependent RNA polymerases of positive-strand RNA viruses. J Gen Virol 72: 2197–2206
17. Koonin EV, Gorbalenya AE, Purdy MA, Rozanov MN, Reyes GR, Bradley DW (1992) Computer-assisted assignment of functional domains in the non-structural polyprotein of hepatitis E virus: Delineation of a new group of animal and plant positive-strand RNA viruses. Proc Natl Acad Sci USA 89: 8259–8263
18. McCaustland KA, Bi S, Purdy MA, Bradley DW (1991) Application of two RNA extraction methods prior to amplification of hepatitis E virus nucleic acid by the polymerase chain reaction. J Virol Methods 35: 331–342
19. Nakahasi HL, Cao X-Q, Rouault TA, Liu T-Y (1991) Specific binding of host cell proteins to the 3' terminal stem-loop structure of Rubella virus negative-strand RNA. J Virol 65: 5961–5967
20. Prince AM, Brotman B, Grady GF, Kuhns WJ, Hazzi C, Levine RW, Millian SJ (1974) Long incubation post-transfusion hepatitis without serological evidence of exposure to hepatitis B virus. Lancet ii: 241–246
21. Purdy MA, Tam AW, Huang C-C, Yarbough PO, Reyes GR (1993) Hepatitis E virus: A nonenveloped member of the "alpha-like" RNA virus supergroup? Semin Virol (in press)
22. Reyes GR, Baroudy BM (1991) Molecular biology of non-A, non-B hepatitis agents: The hepatitis C and hepatitis E viruses. In: Maramorosch K, Murphy FA, Shatkin AJ (eds) Advances in virus research, vol 40. Academic Press, New York, pp 57–102
23. Reyes GR, Bradley DW, Lovett M (1992) New strategies for isolation of low abundance viral and host cDNAs: Application to cloning of the hepatitis E virus and analysis of tissue specific transcription. Semin Liver Dis 12: 289–300
24. Reyes GR, Huang CC, Yarbough P, Young LM, Tam AW, Moeckli RA, Yun KY, Nguyen D, Smith MM, Fernandez J, Querra ME, Kim JP, Luk K-C, Krawczynski K, Purdy MA, Fry KE, Bradley DW (1991) Hepatitis E Virus (HEV): Epitope mapping and detection of strain variation. In: Shikata T, Purcell RH, Uchida T (eds) Viral hepatitis C, D and E. Elsevier Science, Amsterdam, pp 237–245
25. Reyes GR, Kim JP (1991) Sequence-independent, single-primer amplification of complex DNA populations. Mol Cell Probes 5: 473–481
26. Reyes GR, Purdy MA, Kim JP, Luk K-C, Young LM, Fry KE, Bradley DW (1990) Molecular cloning of a cDNA from the virus responsible for enterically transmitted non-A, non-B hepatitis. Science 274: 1335–1339
27. Rozanoz MN, Koonin EV, Gorbelenya AE (1992) Conservation of the putative methyl transferase domain: A hallmark of the "sindbis-like" supergroup of positive strand RNA viruses. J Gen Virol 73: 2129–2134
28. Saiki RK, Gelfand DH, Stoffel S, Scharf SJ, Higuchi R, Horn GT, Mullis KB, Erlich HA (1988) Primer-directed enzymatic amplification of DNA with a thermostable DNA polymerase. Science 239: 487–491

29. Skidmore SJ, Yarbough PO, Gabor KA, Reyes GR (1992) Hepatitis E virus: The cause of a waterborne hepatitis outbreak. J Med Virol 37: 58–60
30. Skidmore SJ, Yarbough PO, Gabor KA, Tam AW, Reyes GR, Flower AJE (1991) Imported hepatitis E in the U.K. Lancet 337: 1541
31. Tam AW, Smith MM, Guerra ME, Huang CC, Bradley DW, Fry KE, Reyes GR (1991) Hepatitis E virus (HEV): Molecular cloning and sequencing of the full-length viral genome. Virology 185: 120–131
32. Thiers V, Nakajima E, Kremsdorf D, Mack D, Schellekens H, Driss F, Goudeau A, Wands J, Sninsky J, Tigllais P (1988) Transmission of hepatitis B from hepatitis-B seronegative subjects. Lancet ii: 1273–1276
33. Tsarev S, Emerson SU, Reyes GR, Tsareva TS, Legters JP, Malik IA, Iqbal M, Purcell RH (1992) Characterization of a prototype strain of hepatitis E virus. Proc Natl Acad Sci USA 89: 559–563
34. Yarbough PO, Tam AW, Fry KE, Krawczynski K, McCaustland KA, Bradley DW, Reyes GR (1991) Hepatitis E virus: Identification of type-common epitopes. J Virol 65: 5790–5797
35. Young RA, Davis RW (1983) Efficient isolation of genes by using antibody probes. Proc Natl Acad Sci USA 80: 1194–1196

Authors' address: Dr. G.R. Reyes, Triplex Pharmaceutical Corporation, 9391 Grogan's Mill Road, The Woodlands, TX 77380, USA.

Arch Virol (1993) [Suppl] 7: 27–39

Variability of the envelope regions of HCV in European isolates and its significance for diagnostic tools

M. Roggendorf[1], M. Lu[1], K. Fuchs[2], G. Ernst[2], M. Höhne[3], and E. Schreier[3]

[1] Institute of Virology, University of Essen, Essen
[2] Max von Pettenkofer-Institute, University of Munich, Munich
[3] Robert Koch-Institut des Bundesgesundheitsamtes, Berlin,
Federal Republic of Germany

Summary. Following the original description of HCV in 1989 a tremendous amount of sequence data is now available. Based on the 8 complete nucleotide sequences published so far at least 4 genotypes can be distinguished. Partial sequences of additional HCV isolates indicate the existence of further genotypes. A serological typing is not yet possible. For detection of virus, reverse transcription and amplification of the 5′ non coding region is most commonly performed. This region of the genome is highly conserved among all isolates. In this study we used regions of the E1 and E2 gene in order to classify HCV isolates. The nucleotide sequences of regions in E1 and E2 gene of different European isolates from Germany, Croatia, Hungaria, and Rumania were determined and compared to recently published RNA sequences of American and Japanese HCV isolates. The cDNA, obtained by reverse transcription of viral RNA extracted from sera was amplified by nested PCR, cloned and sequenced. Within 564 nucleotides (nt) of E1 we found 87–90% homology (and 89–92% homology at aa level) compared to sequences of Japanese origin and 73–74% homology (77–81% at aa level) compared to the prototype HCV sequence (ptHCV-I). In all characterized isolates the sequence of E2 (643 nucleotides) showed a homology of about 83% at the nucleotide level as compared to genotype II sequences, and a homology of about 70% to genotype I. Our results confirm the existence of two hypervariable regions in the E2 gene of genotype II sequences. Our results also indicate together with other reports from European HCV isolates that genotype II is predominant in Europe.

Introduction

The discovery of hepatitis C virus (HCV) in 1989 [1] was a milestone in the research of non-A, non-B hepatitis. To date, at least seven complete

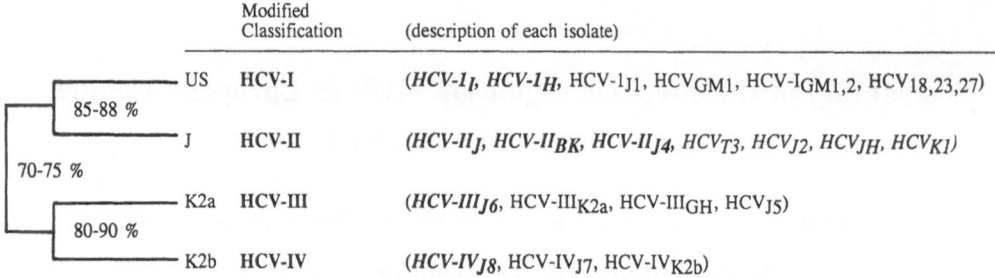

Fig. 1. Phylogenetic tree on classification of genotypes of HCV according to Okamoto et al. [7] and Mori et al. [6]

sequences [2–8] and a large number of partial sequences have been published [9–17] (Fig. 1). The HCV RNA genome consists of about 9,400 nucleotides. The 5′ non-coding region of 342 nucleotides is highly conserved within all genotypes sequenced so far. The N-terminal proteins (structural proteins) of the large polyprotein are cleaved by cellular proteases [18]. The exact cleavage sites of the core, E1 and E2 protein have been determined after in vitro translation of respective proteins [18] (Table 1). The putative E1 and E2/NS1 gene products are highly glycosylated. The non structural proteins (NS2–NS5) are most probably cleaved by viral and cellular protease(s). The 3′ end of the HCV genome consists of about 50 nucleotides and a poly-A tail has been reported for some isolates. The genomic organization of HCV summarized in Fig. 2 is very similar to flavi- and pestiviruses and might represent a unique genus in the *Flaviviridae*.

Several suggestions for phylogenetic relationships of HCV isolates have been introduced [7, 16, 19, 20]. One general consent for the definition of a HCV genotype is a divergence of at least 15–20% at the amino acid level as calculated from the total polyprotein. According to the published full length sequences HCV isolates can be divided into

Table 1. Cleavage sites in the N-terminal region of HCV polyprotein [18]

Cleavage	Site	Enzyme	Name	Size	Glycosilated
C/E1	191/192	host signalase	C	19 kD	no
E1/E2	383/384	host signalase	E1	33 kD	yes Endo H → 18 kD
E2/NS2	~750	host enzyme (in microsomal membranes host signalase?)	E2	72 kD	yes Endo H ~ 38 kD

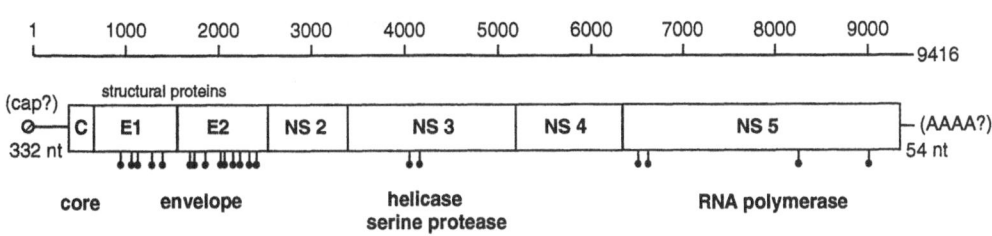

(Takamizawa et al., 1991)

Fig. 2. Genetic organization of the HCV genome

four genotypes (Fig. 1). However, from partial sequences obtained from isolates in Thailand, United States, South Africa, Great Britain, and Germany the existence of additional genotypes was postulated. The amount of sequence variation is not evenly distributed over the genome. As shown in Table 2 a high degree of variation has been reported for the E1 and E2 gene [7].

We report here about partial sequences in the E1 and E2 region of 5 isolates from Eastern European countries, i.e. two from Eastern Germany, one from Hungary, one from Rumania, one from Croatia. All of these isolates are suggested to belong to the previously established genotype II group. The advantages and limitations of HCV diagnosis using the envelope regions E1 and E2 will be discussed.

Table 2. Region-dependent sequence heterogeneity among four representative HCV isolates according to Okamoto [7]

Regions	Percentage difference in nucleotides (deduced amino acids) among four types of HCV genomes					
	I[a]/II[b]	I[a]/III[b]	I[a]/IV[d]	II[b]/III[c]	II[b]/IV[d]	III[c]/IV[d]
5'UTR	2	6	6	7	8	3
C	9 (2)	19 (10)	19 (10)	18 (9)	19 (10)	13 (6)
E1	25 (21)	38 (40)	44 (45)	41 (47)	45 (49)	31 (29)
E2	28 (22)	33 (29)	33 (26)	31 (28)	35 (30)	28 (21)
NS2	29 (23)	43 (42)	43 (44)	41 (40)	43 (44)	30 (26)
NS3	20 (9)	30 (19)	30 (19)	30 (19)	30 (20)	22 (8)
NS4	21 (13)	33 (27)	35 (28)	33 (26)	33 (28)	22 (9)
NS5	21 (16)	34 (30)	34 (31)	34 (29)	34 (29)	23 (18)
3'UTR	33	70	74	62	60	21
Total	21 (15)	32 (28)	33 (28)	32 (27)	33 (29)	23 (16)

[a] HCV-1, [b] HCV-J, [c] HC-J6, [d] HC-J8

M. Roggendorf et al.

Materials and methods

Source of HCV containing sera

Sera were obtained in 1991 from patients (Cr-1, Hu-1, Ru-1) with chronic post transfusion or sporadic HCV infection. The patients who are recent immigrants from Croatia, Hungaria and Rumania to Germany were treated at the Klinikum Großhadern, Munich, FRG. These sera were kindly supplied by Dr. Zachoval. From the clinical history of the patients it can be assumed that they were infected in their respective home countries. The two serum samples (EG-1, EG-2) from Eastern Germany were obtained in 1989 from two patients who were infected by contaminated anti D immunoglobulin in 1978 [21]. Both patients developed chronic HCV infection.

RNA extraction, amplification and sequencing of E1/E2

RNA extraction was done by a 2 h incubation at 37°C of 350 µl sera in 50 mM Tris · HCl pH 8.0/10 mM EDTA/100 mM NaCl/4% SDS/3 mg proteinase K per ml (Boehringer Mannheim, F.R.G.). The total volume was 500 µl. As a carrier, 1 µg of tRNA was added. After three times of extraction with phenol, phenol/chloroform, and chloroform, respectively, 0.3 M Na · acetate (final concentration) was added and the RNA was precipitated by three volumes of ethanol. After centrifugation and lyophilization, the RNA was resuspended in 40 µl TE buffer and reverse transcribed by 200 units of recombinant MMLV-RT (BRL Life Technologies Inc., Gaithersburg, MD) in the presence of anti-sense primer.

In order to prove viremia the 5' NCR was amplified first. Reverse transcription of the 5' end was done by using an anti-sense primer 5'-GGTGCACGGTCTACGAGACC-3', corresponding to nt 305–324 of the isolate HCJ1 [9]. Two pairs of primers were used for amplification by nested PCR [22]. The outer pair consists of the RT primer shown above and the sense primer 5'-GGCGACACTCCACCATAGAT-3' (nt 1–20) [9]. The inner primers were 5'-*GGAGGATC*CACTCCCCTGT-3' sense) (nt 21–31) and a *Bam*HI cloning site (underlined), and the primer 5'-*GGAAAGCTTGAATT*-CACCCTATCAGGCAGT-3' (anti-sense) (nt 271–286) and *Hind*III and *Eco*RI cloning sites (underlined). The numbering of nt is according to Okamoto et al. [9]. A single band of about 280 bp after the second round of PCR (total of 60 PCR cycles) was detected in all five samples.

The amplification strategy for the E1 and E2 region by nested PCR are given in Fig. 3a. The primers were selected from the ptHCV sequence [3] and JH1 [23]. Since these are sequence mismatches in the sequence of ptHCV to JH1, we also used primers which contained these nt exchanges (for details see legend to Fig. 3).

Nested PCR was done by means of a commercial kit using a standard protocol (Perkin Elmer/Cetus Corp., Norwalk, CT). Both rounds of amplification were done after an initial denaturation step (94°C, 5 min) followed by 40 cycles (1 min 94°C/2 min 37°C/3 min 72°C) and 10 min at 72°C (Weiner et al., 1990) in a thermocycler (Perkin Elmer/Cetus Corp.). The DNA fragments of the second PCR round were isolated from the gel and cloned into the *Bam*HI and *Hind*III sites of pUC8 [26].

For sequencing of the cloned amplification products the dideoxy chain-termination method was used [25] after alkaline denaturation of plasmid DNA [24] using a commercial Sequenase Kit (U.S. Biochemical Corp., Cleveland, OH).

Results

Using RT and nested PCR, the nucleotide sequences according to positions 906 to 1,443 of the HCV E1 region [23] of five European isolates from Croatia (Cr-1), Hungaria (HU-1), Rumania (Ru-1) and East Germany (EG-1 and EG-2) were cloned and analyzed. For the three strains Cr-1, Hu-1 and EG-1, a region of about 643 bp of the E2 according to position 1,492 to 2,135 were also sequenced. The amino acid sequences, deduced from the nucleotide sequence, are presented in Fig. 3.

The sequence comparison of the different genotypes (I–IV) reveals that the HCV genome is very heterogeneous. The amount of heterogeneity is not equally distributed along the entire HCV genome but appears to concentrate in some regions (Table 2) [7]. The difference between genotypes I–IV in the E1 region ranges from 21% to 49% at the amino acid level which is a higher divergence compared to other HCV genes, e.g. core region.

The five European isolates analyzed here show differences of about 9–11% at the amino acid sequence level as compared to HC-J1 (HCV-II) [2] and a significantly higher difference to other isolates (in average: 23.5% to HCV-I, 46% to HCV-III and 50% to HCV-IV). We conclude that all these European isolates belong to HCV type II.

The alignment of partial E1 sequences among HCV type II isolates and the homology to each other are given in Fig. 3b and Table 3, respectively. There is a small difference of only 4% in the E1 sequence between EG-1 and EG-2. This high degree of sequence homology between these two isolates can be explained by the same source of virus in the contaminated anti-D-Immunoglobulin preparation. Five isolates

Table 3. Comparison of partial E1 sequences of European, Type II-HCV isolates, (540 bases $\doteq$ 180 aa)

	Abbr.	% Homology (aa)					
		HCV-J1	HCV CR-1	HCV HU-1	HCV RU-1	HCV W2	HCV W5
Japan	HCV-J1	100	89	89	89	91	92
Croatia	HCV CR-1	89	100	96	88	90	90
Hungary	HCV HU-1	89	96	100	89	92	91
Rumania	HCV RU-1	89	88	89	100	92	90
Germany	HCV Eg-1	91	90	91	90	100	96
Germany	HCV Eg-2	92	90	92	92	96	100

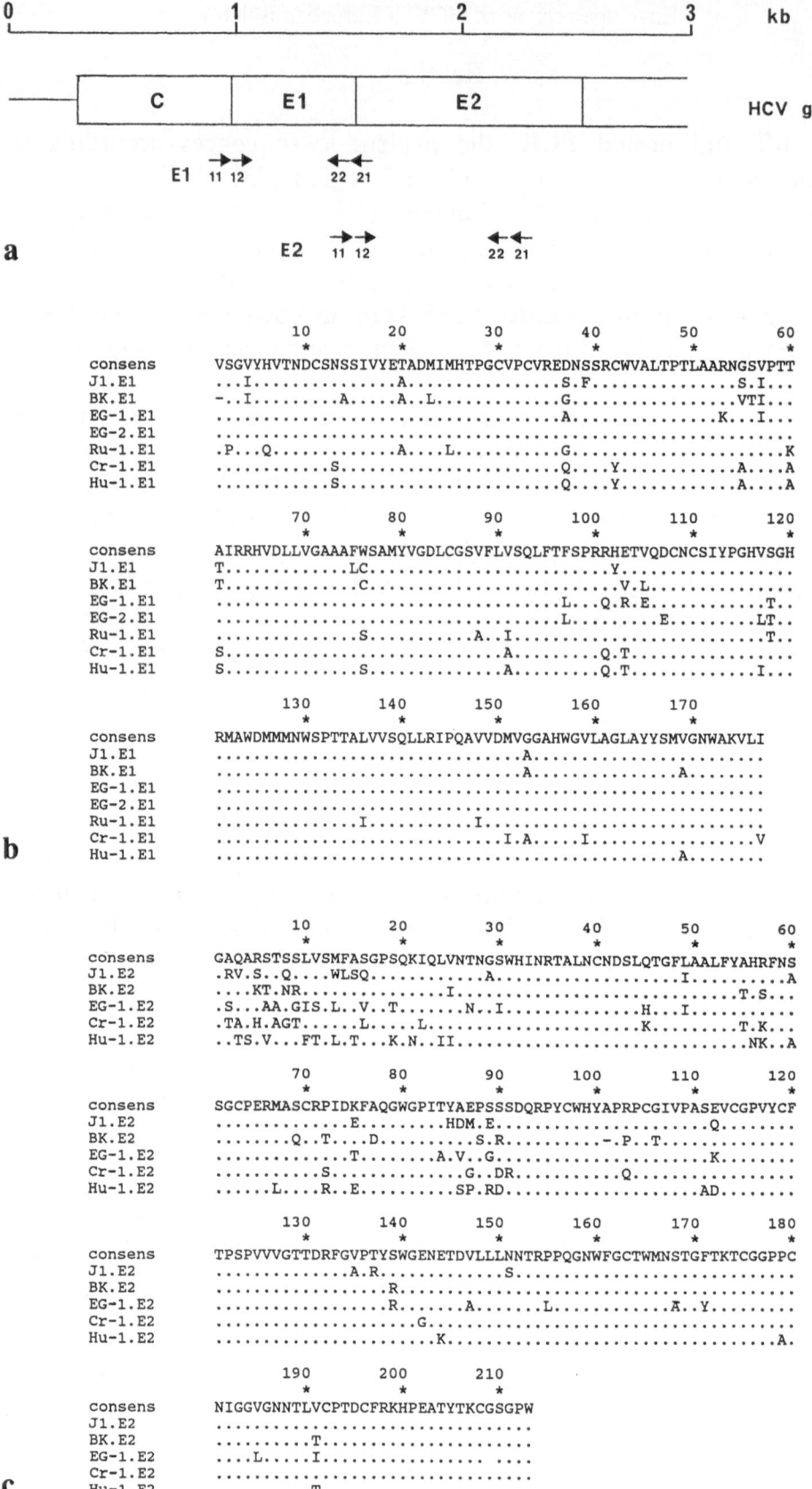

a

b

c

from other patients infected by the anti-D-immunoglobulin showed also a low diversity of 1 to 6% in the E1 sequence, reflecting their common origin (Höhne et al., unpubl. results). The Croatian Cr-1 and the Hungarian isolate Hu-1 are also closely related, comparable to the german isolates. It is not known whether the patients from Croatia and Hungaria were infected by the same source. The isolate from Rumania showed 90% homology, and seems to be more remote from all other sequences shown here.

The sequence of the E2 region of three isolates were compared with HC J1 and HC BK. A distinct feature of E2 are hypervariable regions [27, 28]. In Fig. 4, the variability was defined as the number of different

---

Fig. 3. a The strategy for amplification of HCV E1 and E2 regions. The primer sequences for E1 region are:

E1.11 5'-TGCTCTTTCTCTTATCTTCCT-3'		833–852
E1.12 5'-TTGAACTTGTGGTGATAGAA-3'		1643–1662
E1.21 5'-GGA*GGATCC*GCCTACCAAGTGCGCAA-3'		890–906
E1.22 5'-GGA*AAGCTTGAATTCTTA*GCCGGCAAATAGCAGCA-3'		1443–1459

The primer sequences for E2 region are:

E2.11 5'-GCTCACTGGGGATCCTGGCGGG-3'		1392–1414
E2.12 5'-GTACAAGGATAATGCCAAAGCCT-3'		2203–2181
E2.21 5'-AAT*GGATCC*CTCCATGGTGGGGAACTGGGC-3'		1427–1447
E2.22 5'-AGG*AAGCTTC*TAGTCGACCAGGCACCTGGGTGT-3'		2173–2159

Primers E1.11 and E1.12 which was used for RT were taken for amplification of the E1 region, primers. The sense primer E1.21 with a restriction site for BamH1 (underlined) was used for the nested PCR. Since there are sequence mismatches in the sequence of JH1 to ptHCV (Takeuchi et al. [23]) at positions 12(T), 15(T) and 16(G) of the primer, we also used the primer which contained these degeneration. The antisense primer for the nested PCR E1.22 carried a HindIII and an EcoRI site and the primer E2.22 only an EcoRI site (underlined). It is degenerated at the nt position 28 (with G) and 31 (with T). For amplification of E2 region, RT of RNA was done with an antisense primer E2.21 identical to nt 2,181 to 2,203 of the ptHCV (3) with degenerations at nt positions 3(G), 6(G), 9(G) and 18(G). The outer primer pair E2.21 and E2.11 which carried degenerations at positions 13(G) and 18(A) were chosen for the first PCR. The E2.12 of the inner primer pair was synthesized with an extention at 5′ end for a BamHI restriction site. The other one E2.22 had an 5′-extention for an HindIII site and degenerations at nt positions 21(T) and 28(A). **b** Alignment of the E1 region of seven HCV type II isolates. The amino acid sequence between positions 197 and 373 was considered in the alignment. For every position, amino acid residues common (mostly frequent) in seven sequences were taken as consensus sequence, **c** Alignment of amino acid residue 390 to 602 (numbering of amino acids according to Kato et al. (2) of five HCV type II isolates. For every position, amino acid residues that are common in all in five sequences were taken as consensus sequence. The sequences shown here have been assigned the accession Nos. X72975–X72982 in the Gen Bank Data Library (Los Alamos, NM)

M. Roggendorf et al.

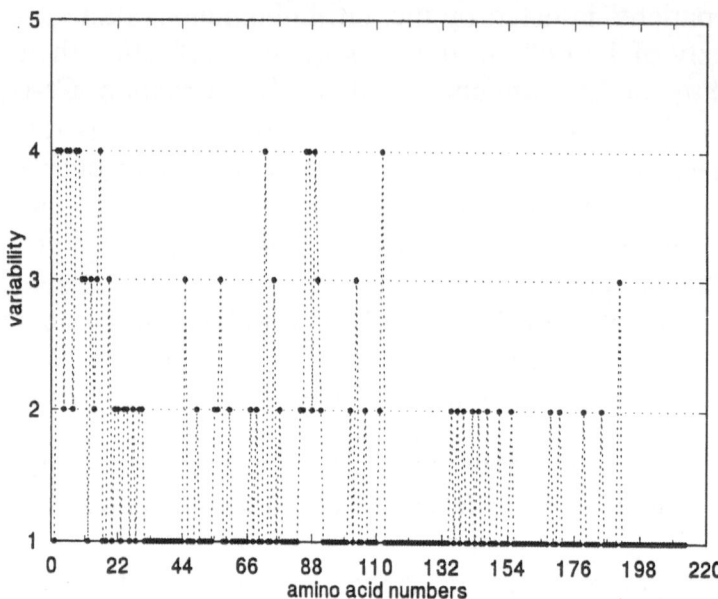

Fig. 4. Variability of HCV E2 sequence (from amino acid residue 390 to 602). The variability is defined as the number of different amino acid residue occurring at one position in the E2 sequences

amino acid residues occurring at a position. Two clusters (the first one from amino acid residue 390 to 410 and the second from 474–480) with extensive variation are easily identified in the graph. The location of the clusters correspond to those in HCV type I and type II described by Hijikata et al. [27].

Discussion

We analyzed parts of the RNA sequences of the E1 and E2 gene of different HCV isolates from Eastern and South Eastern European countries. All isolates have high homologies to genotype II of HCV which was first described in Japan. This genotype seems to be predominant in Europe, and exceeds the frequency of the American prototype HCV (ptHCV/HCV-I) [12].

In Europe genotype I and II and another genotype which has not yet definitely been classified have been found in England [20] and East Germany (Table 4). The German isolates of the latter new genotype have been found in preferentially in drug addicts and show homology to two isolates from Thailand (Ta) and UK (E-b1) classified as genotype V [16, 20]. These drug addicts showed multiple infections with several genotypes (I, V and II, V).

Table 4. Distribution of HCV genotypes (I–V) [16] within Europe

Country	I	II	III	IV	V	Ref.
Croatia (n = 1)		1				
England (n = 113)	56 (50%)	12 (10%)			45 (40%)	Simmonds et al.[a]
France (n = 119)	41 (34%)	69 (59%)			9 (7%)	Li et al.
Germany (n = 74)	9 (12%)	44 (59%)	4 (6%)		17 (23%)	Schreier et al.[a]
Hungary (n = 1)		1				
Italy (n = 5)	1	1	1	2		Cha et al.
Spain (n = 31)	9 (29%)	22 (71%)				Martell et al./ Pernas et al.[a]
Rumania (n = 1)		1				
Russia (n = 28)	8 (30%)	17 (62%)	0	1 (4%)	1 (4%)	Viazov[a]

[a] Personal communication

The homology of E1 on the amino acid level of the characterized HCV isolates is about 89–96% when compared to genotypes II. The isolates from Croatia (Cr-1) and Hungary (Hu-1) are very closely related to each other (96%). However, the isolate from Rumania (Ru-1) seems to be remote from the isolates Cr-1 and Hu-1 (88–89%).

In the sequenced E2 regions of the three isolates (EG-1, Cr-1, Hu-1) we detected two hypervariable regions discribed previously in Japanese isolates [27, 28]. The conserved stretches of amino acids in E1 and E2 are very limited (Fig. 4). Nevertheless, we found evidence for conserved epitopes in different genotypes. We tested whether patients who were infected with genotype II have antibodies to denatured and native E1/ E2-proteins (data not shown). Preliminary results (Chien et al., un-publ. results) indicate that in at least 50% of patients who completely recovered from disease antibodies are present to native proteins. In only 5% of the patients antibodies to denatured proteins have been found. In the group of patients with chronic HCV infections 100% of the patients were positive for antibodies to native proteins, and 56% to denatured proteins. This indicates the cross reactivity of antibodies against E1/E2 of isolates belonging to genotype I and II. This could reflect the close relationship between type I and type II HCV isolates [7]. It will be interesting to see whether this cross reactivity holds to the other genotypes. The immune response in patients with complete recovery seems to be mostly directed against conformational epitopes. For a detailed study of humoral immune response it will be necessary to express the appropriate proteins of each genotype and use genotype specific proteins to determine the immune response in infected patients. At present, we extend our studies on the immune response against envelope proteins with a population of patients who received comtaminated anti-D-immunoglobulin [21].

Despite the knowledge of HCV classifications that are based on nucleotide sequences we do not know whether the such defined genotypes belong to the same serotype. From hepatitis A (HAV) viruses it is known, that there exist several genotypes – at least 7 – but only one serological type. It has been shown that all genotypes of HAV can by neutralized by monoclonal antibodies directed against the major neutralizing epitope of this virus [29].

Further analysis of serotypes depend on the establishment of a tissue culture system for HCV. This would allow to test whether genotypes of HCV can also be distinguished serologically by neutralizing antibodies. To date, it is not known whether HCV infected patients are protected against infections with other genotypes. Primary results from chimpanzees [30] indicate that chimpanzees show multiple episodes of infections when inoculated with the same genotype or when inoculated with a different genotype. Under these conditions no cross neutralization was observed.

For the diagnosis of HCV as well as for epidemiological studies there is obvious need for the determination the genotype. This would also allow follow the immune response. Typing of HCV isolates is commonly performed using PCR [31–33] technologies and analyse the products either by restriction length polymorphism, by conserved primers and specific probes or mixtures of conserved and genotype specific primers.

Our approach to characterize regions of the envelope proteins might be valuable for the development of a serological diagnosis and studies of the immune response to these highly exposed proteins. Further experiments are needed to establish the relationship between mutability of the HCV genome and the immune system.

Acknowledgement

Part of this study was supported by the Fritz Bender Foundation, Munich, Germany.

References

1. Choo QL, Kuo G, Weiner AJ, Overby LR, Bradley DW, Houghton M (1989) Isolation of a cDNA clone derived from a blood-borne non-A, non-B viral hepatitis genome. Science 244: 359–362
2. Kato N, Hijikata M, Ootsuyama Y, Nakagawa M, Ohkoshi S, Sugimura T, Shimotohno K (1990) Molecular cloning of the human hepatitis C virus genome from Japanese patients with non-A, non-B hepatitis. Proc Natl Acad Sci USA 87: 9524–9528
3. Choo QL, Richman KH, Han JH, Berger K, Lee C, Dong C, Gallegos C, Coit D, Medina-Selby A, Barr PJ, Weiner AJ, Bradley DW, Kuo G, Houghton M (1991) Genetic organization and diversity of the hepatitis C virus. Proc Natl Acad Sci USA 88: 2451–2455
4. Inchauspe G, Zebedee S, Lee DH, Sugitani M, Nasoff M, Prince AM (1991) Genomic structure of the human prototype strain of hepatitis C: Comparison with American and Japanese isolates. Proc Natl Acad Sci USA 88: 10292–10296
5. Ogata N, Alter HJ, Miller RH, Purcell RH (1991) Nucleotide sequence and mutation rate of the H strain of hepatitis C virus. Proc Natl Acad Sci USA 88: 3392–3396
6. Okamoto H, Okada S, Sugiyama Y, Kurai K, Iizuka H, Machida A, Miyakawa Y, Mayumi M (1991) Nucleotide sequence of the genomic RNA of hepatitis C virus isolated from a human carrier: Comparison with reported isolates for conserved and divergent regions. J Gen Virol 72: 2697–2704
7. Okamoto H, Kurai K, Okada SI, Yamamoto K, Lizuka H, Tanaka T, Fukuda S, Tsuda F, Mishiro S (1992) Full-length sequence of a hepatitis C virus genome having poor homology to reported isolates: Comparative study of four distinct genotypes. Virology 188: 331–341
8. Takamizawa A, Mori C, Fuke I, Manabe S, Murakami S, Fujita J, Onishi E, Andoh T, Yoshida I, Okayama H (1991) Structure and organization of the hepatitis C virus genome isolated from human carriers. J Virol 65: 1105–1113
9. Okamoto H, Okada S, Sugiyama Y, Yotsumoto S, Tanaka T, Yoshizawa H, Tsuda F, Miyakawa Y, Mayumi M (1990) The 5'-terminal sequence of the hepatitis C virus genome. Jpn J Exp Med 60: 167–177
10. Enomoto N, Takada A, Nakao T, Date T (1990) There are two major types of hepatitis C virus in Japan. Biochem Biophys Res Commun 170: 1021–1025
11. Chen PJ, Lin MH, Tu SJ, Chen DS (1991) Isolation of a complementary DNA fragment of hepatitis C virus in Taiwan revealed significant sequence variations compared with other isolates. Hepatology 14: 73–78
12. Fuchs K, Motz M, Zachoval R, Deinhardt F, Roggendorf M (1991) Characterization of nucleotide sequences from European hepatitis C virus isolates. Gene 103: 163–169
13. Kremsdorf D, Porchon C, Kim JP, Reyes GR, Brechot C (1991) Partial nucleotide sequence analysis of a French hepatitis C virus: Implications for HCV genetic variability in the E2/NS1 protein. J Gen Virol 72: 2557–2561
14. Li JS, Tong SP, Vitvitski L, Lepot D, Trepo C (1991) Two French genotypes of hepatitis C virus: Homology of the predominant genotype with the prototype American strain. Gene 105: 167–172
15. Li JS, Tong SP, Vitvitski L, Lepot D, Trepo C (1991) Evidence of two major genotypes of hepatitis C virus in France and close relatedness of the predominant one with the prototype virus. J Hepatol 13: S33–S37

16. Mori A, Kato N, Yagyu A, Tanaka T, Ikeda Y, Petchclai B, Chiewsilp P, Kurimura T, Shimotohno K (1992) A new type of hepatitis C virus in patients in Thailand. Biochem Biophys Res Commun 183: 334–342
17. Schreier E, Fuchs K, Höhne M, Motz M, Zachoval R, Esteban J, Dittmann S, Deinhardt F, Roggendorf M (1992) Detection and characterization of hepatitis C virus sequence in the serum of a patient with chronic HCV infection. Arch Virol 124: 179–183
18. Hijikata M, Kato N, Ootsuyama Y, Nakagawa M, Shimotohno K (1991) Gene mapping of the putative structural region of the hepatitis C virus genome by in vitro processing analysis. Proc Natl Acad Sci USA 88: 5547–5551
19. Cha TA, Beall E, Irvine B, Kolberg J, Chien D, Kuo G, Urdea MS (1992) At least five related, but distinct, hepatitis C viral genotypes exist. Proc Natl Acad Sci USA 89: 7144–7148
20. Chan SW, McOmish F, Holmes EC, Dow B, Peutherer JF, Follett E, Yap PL, Simmonds P (1992) Analysis of a new hepatitis C virus type and its phylogenetic relationship to existing variants. J Gen Virol 73: 1131–1141
21. Dittmann S, Roggendorf M, Dürkop J, Wiese M, Lorbeer B, Deinhard F (1991) Long-term persistence of hepatitis C virus antibodies in a single source outbreak. J Hepatol 13: 323–327
22. Garson JA, Tedder RS, Briggs M, Tuke P, Glazebrook JA, Trute A, Parker D, Barbara JAJ, Contreras M, Aloysius S (1990) Detection of hepatitis C viral sequences in blood donation by "nested" polymerase chain reaction and prediction of infectivity. Lancet i: 1419–1422
23. Takeuchi K, Kubo Y, Boonmar S, Watanabe Y, Katayama T, Choo QL, Kuo G, Houghton M, Saito I, Miyamura T (1990) Nucleotide sequence of core and envelope genes of hepatitis C virus genomes derived directly from human healthy carriers. Nucleic Acids Res 18: 4626
24. Chen EJ, Seeburg HP (1985) Supercoil sequencing: a fast simple method for sequencing plasmid DNA. DNA 4: 165–170
25. Sanger F, Nicklen S, Coulson AR (1977) DNA sequencing with chain-terminating inhibitors. Proc Natl Acad Sci USA 74: 5463–5467
26. Vieira J, Messing J (1982) The pUC plasmids, an M13mp7-derived system for insertion mutagenesis and sequencing with synthetic universal primers. Gene 19: 259–268
27. Hijikata M, Kato N, Ootsuyama Y, Nakagawa M, Ohkoshi S, Shimotohno K (1991) Hypervariable regions in the putative glycoprotein of hepatitis C virus. Biochem Biophys Res Commun 175: 220–228
28. Kato N, Ootsuyama Y, Tanaka T, Nakagawa M, Nakazawa T, Muraiso K, Ohkoshi S, Hijikata M, Shimotohno K (1992) Marked sequence diversity in the putative envelope proteins of hepatitis C viruses. Virus Res 22: 107–123
29. Robertson BH, Khanna B, Naina OV, Margolis HS (1991) Epidemiologic patterns of wild-type hepatitis A virus determined by genetic variation. J Infect Dis 163: 286–292
30. Farci P, Alter HJ, Govindarajan S, Wong DC, Engle R, Lesniewski RR, Mushahwar IK, Desai SM, Miller RH, Ogata N, Purcell RH (1992) Lack of protective immunity against reinfection with hepatitis C virus. Science 258: 135–140
31. Okamoto H, Okada S, Sugiyama Y, Tanaka T, Sugai Y, Akahane Y, Machida A, Mishiro S, Yoshizawa H, Miyakawa Y, Mayumi M (1990b) Detection of hepatitis C virus RNA by a two-stage polymerase chain reaction with two pairs of primers deduced from the 5'-noncoding region. Jpn J Exp Med 60: 215–222

32. Nakao T, Enomoto N, Takada N, Takada A, Date T (1991) Typing of hepatitis C virus genomes by restriction fragment length polymorphism. J Gen Virol 72: 2105–2112

33. Okamoto H, Sugiyama Y, Okada SY, Kurai K, Akahane Y, Tanaka T, Miyakawa Y, Mayumi M (1992) Typing hepatitis C virus by polymerase chain reaction with type-specific primers: Application to clinical surveys and tracing infectious sources. J Gen Virol 73: 673–679

Authors' address: Dr. M. Roggendorf, Institute of Virology, University of Essen, Hufelandstrasse 55, D-45147 Essen, Federal Republic of Germany.

Arch Virol (1993) [Suppl] 7: 41–52

Molecular characterization of positive-strand RNA viruses: pestiviruses and the porcine reproductive and respiratory syndrome virus (PRRSV)

H.-J. Thiel[1], G. Meyers[1], R. Stark[1], N. Tautz[1], T. Rümenapf[2], G. Unger[1],
and **K.-K. Conzelmann**

[1] Federal Research Centre for Virus Diseases of Animals, Tübingen,
Federal Republic of Germany
[2] California Institute of Technology, Division of Biology, Pasadena, CA, U.S.A.

Summary. Molecular characterization has become an important tool for the analysis of viruses including their classification. The manuscript focuses on the molecular analysis of two members of the genus pestivirus (hog cholera virus, HCV and bovine viral diarrhea virus, BVDV) and of the recently discovered porcine reproductive and respiratory syndrome virus (PRRSV). The first protein encoded within the single large pestivirus ORF is a nonstructural protein with autoproteolytic activity. The cleavage site between the protease and the capsid protein p14 has been predicted previously, but recent experimental data indicate that processing occurs at a different site. The capsid protein is followed by a putative internal signal sequence and three glycoproteins which are part of the virion envelope. According to a new proposal for the nomenclature of the structural proteins of pestiviruses they are termed C, E0, E1 and E2. The genomes of BVDV pairs isolated from animals which came down with mucosal disease were analyzed. The genomes from cytopathogenic (cp) BVD viruses may contain insertions highly homologous to cellular sequences. In addition, cp BVDV may differ from its non cytopathogenic (noncp) counterpart by mere rearrangement of viral sequences. The disease PRRS, which emerged a few years ago, is caused by a single strand RNA virus; the viral genome is of positive polarity and has a size of 15 kb. Data concerning morphology, morphogenesis and virion composition suggested already that PRRSV belongs to a group of so-called arteriviruses which comprises equine arteritis virus (EAV), lactate dehydrogenase elevating virus (LDV) and simian hemorrhagic fever virus (SHFV). This conclusion has now been confirmed by analysis of genome organization, gene expression strategy and by comparison of deduced protein sequences.

Introduction

Pestiviruses are causative agents of important animal diseases such as hog cholera, bovine viral diarrhea and border disease of sheep. The pestivirus genome represents a single-stranded RNA of about 12.5 kb, which is of positive polarity. The genomic RNA comprises a single large open reading frame [8, 15]; accordingly, membership of the genus pestivirus in the family *Togaviridae* was no longer justified. Instead, pestiviruses have recently been reclassified as members of the family *Flaviviridae* which now comprises the genera flavivirus, pestivirus and the hepatitis C virus group. These viruses have certain characteristics in common like overall genome organization and strategy of gene expression. Antisera against bacterial fusion proteins and synthetic peptides as well as monoclonal antibodies were used to identify proteins encoded by pestiviruses [5, 7, 27–30]. Interestingly, the genomes of several cytopathogenic BVDV strains contain insertions highly homologous to cellular sequences [16, 17, 19, 23].

The porcine reproductive and respiratory syndrome virus (PRRSV) has been described as a small enveloped RNA virus [2, 32] with morphological and morphogenetical similarities to members of the arterivirus group, including equine arteritis virus (EAV) and lactate dehydrogenase-elevating virus of mice (LDV). In addition, relationships between PRRSV and arteriviruses are suggested by the nature of permissive cells. The arteriviruses infect particular subpopulations of macrophages [21] and PRRSV apparently grows exclusively in alveolar lung macrophages. However, serological crossreactions could so far not be demonstrated between PRRSV and any of the arteriviruses.

Results and discussion

Pestiviruses

a. Hog cholera virus

Analysis of the autoprotease HCV p23. The nonstructural protein HCV p23 represents the first protein of the pestivirus ORF. As already shown for BVDV p20 [33], HCV p23 also possesses autoproteolytic activity. A precursor molecule consisting of p23 and p14 could never be demonstrated even when very short pulse periods were used. This result was obtained after infection with HCV as well as different Vaccinia virus/HCV recombinants. Processing was also observed after in vitro translation of HCV RNA and in vitro transcription/translation experiments

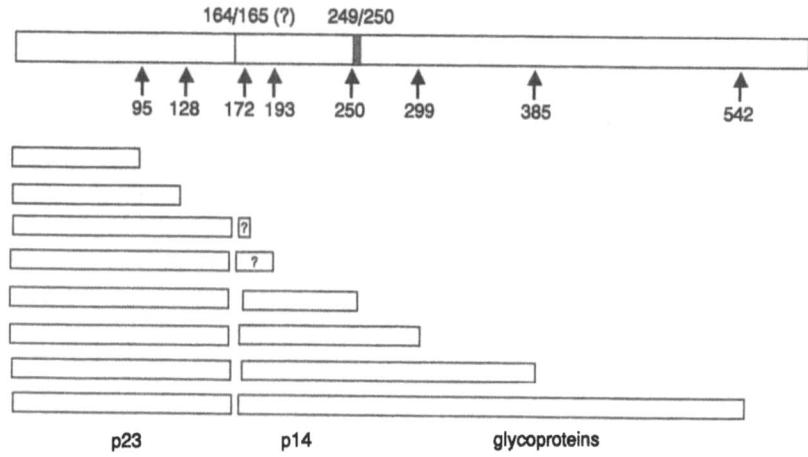

Fig. 1. Evidence that HCV p23 possesses proteolytic activity. After linearization of an HCV cDNA construct with different restriction enzymes, in vitro transcription/translation was performed. The proteins were identified after immunoprecipitation by SDS-PAGE

with cDNA constructs. The latter experiments also showed that the proteolytic activity resides in HCV p23 (Fig. 1).

For BVDV and HCV it has been proposed that cleavage occurs after Trp-164 of the ORF [28, 33] and this assumption was supported by in vitro mutagenesis studies [33]. However, there is new evidence indicating that cleavage occurs after Cys-168 (Fig. 2). Firstly, p14 could not be labeled with [35S]-cysteine which is absent from p14 only if cleavage occurs after Cys-168. Secondly, the genomic region comprising HCV p23 and p14 was expressed in bacteria. In this prokaryotic expression system the release of a 14 kD protein was observed. N-terminal sequencing of this protein revealed that the protein starts with Ser-169 (manuscript in preparation). Further efforts concern characterization of the protease and elucidation of its function.

Interestingly, the analysis of two genomes of cytopathogenic (cp) BVDV strains revealed that their genomes contain duplications encompassing the autoprotease p20. The 3'-terminal codon of the inserted p20 coding region is identical in both cp-BVDV genomes and codes for a cysteine analogous to Cys-168 of the pestiviral ORF [18].

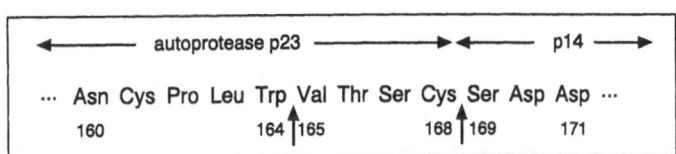

Fig. 2. Cleavage site of pestivirus autoprotease

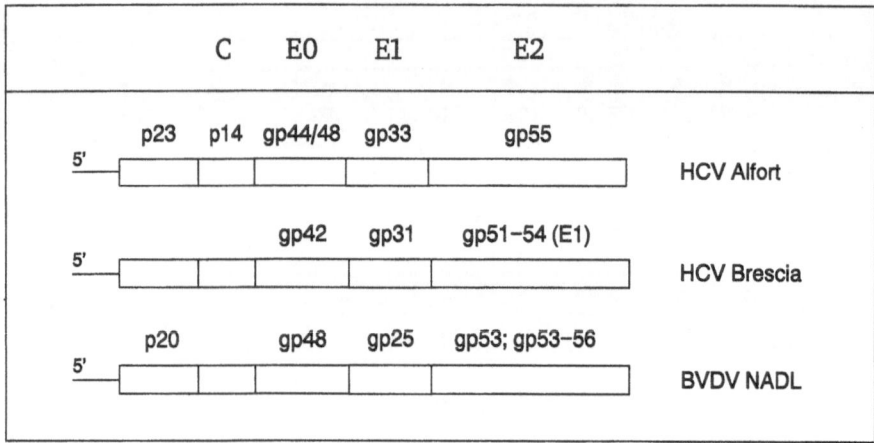

Fig. 3. Suggested nomenclature for structural proteins of pestiviruses

Structural proteins of HCV

Within the pestiviral polyprotein the autoprotease HCV p23/BVDV p20 is followed by the nucleocapsid protein p14 which has been demonstrated for HCV as well as BVDV [28]. The following internal signal sequence probably mediates translocation of the first structural glycoprotein, HCV gp44/48. Together with gp33 and gp55, the three glycoproteins constitute the virion envelope. The glycoproteins form parts of disulfide-linked dimers [28, 29]. While gp33 probably represents a transmembrane protein, the other two glycoproteins both are exposed on the surface of virions and induce virus neutralizing antibodies [30].

So far, the designation of pestivirus encoded glycoproteins varies among different laboratories because most research groups use the apparent molecular weights as a basis for nomenclature (Fig. 3). In order to obtain a common nomenclature, we suggest to use the abbreviations C, E0, E1 and E2 for nucleocapsid protein and the glycoproteins, respectively (Fig. 3). This approach allows also to have the same designations for analogous glycoproteins from the hepatitis C virus group (Fig. 4). It remains to be seen whether the (glyco)proteins from members of the genus flavivirus will be termed accordingly. At this point the designation E0 (for HCV gp44/48, BVDV gp48) is being debated because (1) it is clearly a structural protein of pestiviruses and (2) the "0" may be misleading because it is used in other virus systems for precursor molecules. It appears that E0 represents a unique glycoprotein among pestiviruses which is not only a structural protein but also secreted from infected cells (manuscript in press).

In order to determine the processing sites between C and E0 as well as between the glycoproteins, N-terminal sequencing was performed.

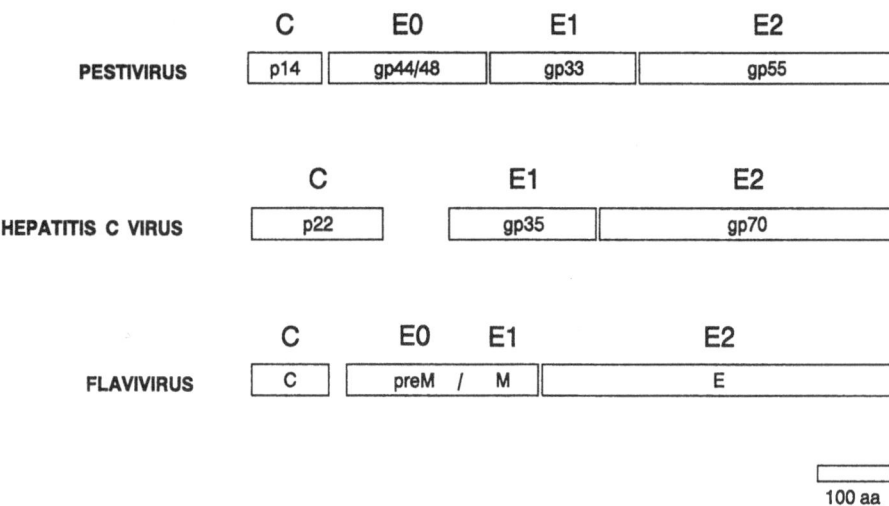

Fig. 4. Structural proteins of *Flaviviridae*

	E 0	E 1	E 2
Number of Amino Acids	227	195	373 (?)
Calculated Molecular Weight (kDa)	25.8	21.8	41.9
Apparent Molecular Weight (kDa)	48 (gp 44/48)	33 (gp33)	55 (gp55)
Number of Possible N–glycosylation Sites	9	3	6
Apparent Molecular Weight after Endoglycosidase F (kDa)	similar to the respective calculated molecular weights		

Fig. 5. Glycoproteins of hog cholera virus

For this purpose the glycoproteins were purified from virus infected cells by immunoaffinity columns using monoclonal antibodies against E0 and E2. E1 was copurified with E2 because of the linkage of the two glycoproteins by disulfide bridges. The results of N-terminal sequencing allow important conclusions concerning the three glycoproteins, namely the actual sizes of the protein backbones and the calculated contribution of carbohydrate moieties to apparent molecular weights (Fig. 5). The indicated sizes of the protein backbones are based on the assumption that only one cleavage occurs between the glycoproteins. With regard to E2 the C terminus remains to be determined.

On the basis of the data outlined above we intend to study the biosynthesis of pestivirus glycoproteins. A particularly interesting aspect of this study will be the analysis of a supposed hierarchy of cleavage events (manuscript in press). In addition, C-terminal truncations of HCV gp55 (E2) are being performed to determine its C-terminus and to identify signals important for heterodimerization and homodimerization (manuscript in preparation).

With regard to the development of vaccines against pestiviruses the role of individual glycoproteins in the induction of protective immunity will be studied. These efforts include use of (1) recombinant vaccines based upon different vectors, primarily vaccinia virus and pseudorabies virus and (2) subunit vaccines containing pestivirus glycoprotein(s) expressed for example in insect cells after infection with recombinant baculoviruses.

b. Bovine viral diarrhea virus

Pathogenesis of mucosal disease. Comparison of the genomic sequences of two BVDV strains (BVDV Osloss [26] and BVDV NADL [8]) led to the identification of small insertions located in a region coding for a nonstructural protein [6, 15, 19]. The insertion of 228 nucleotides identified in the BVDV Osloss genome encodes a complete ubiquitin-like element with only two amino acid exchanges with respect to the ubiquitin sequence conserved in all animals [16, 19]. The sequence of 270 nucleotides which is inserted in the BVDV NADL genome shows no homology to a ubiquitin gene but is almost identical with another bovine mRNA sequence (Fig. 6) [16].

In tissue culture two BVDV biotypes, cytopathogenic BVDV (cpBVDV) and noncytopathogenic BVDV (noncpBVDV), can be distinguished [1] (Table 1). Both biotypes are involved in pathogenesis of mucosal disease (MD), the most severe clinical manifestation of BVDV infections. A prerequisite for MD is a persistent infection with

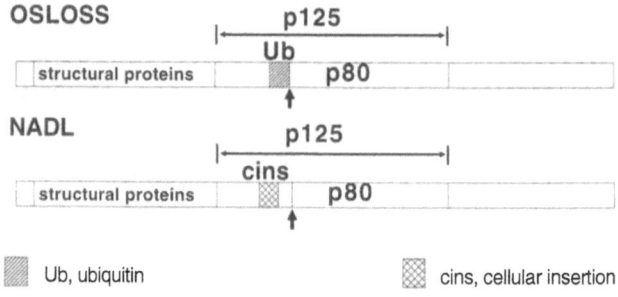

Fig. 6. Insertions identified in CP BVDV strains. *OSLOSS* 228 nucleotides; coding for a ubiquitin-like protein; equivalent to one ubiquitin monomer; 2 amino acid exchanges. *NADL* 270 nucleotides; no homology to ubiquitin; 99% identity to a bovine mRNA sequence

Table 1. Characteristics of BVDV biotypes

	Non-CP	CP
Cytopathogenicity (CP) in tissue culture cells	No	Yes
Strains (laboratory)	New York	NADL, Osloss, Singer, Oregon, Danmark
Detection of p80	No (only p125)	Yes (in addition to p125)
Pathogenicity in cattle	Yes	Yes
Occurrence of mucosal disease	Always both biotypes isolated	

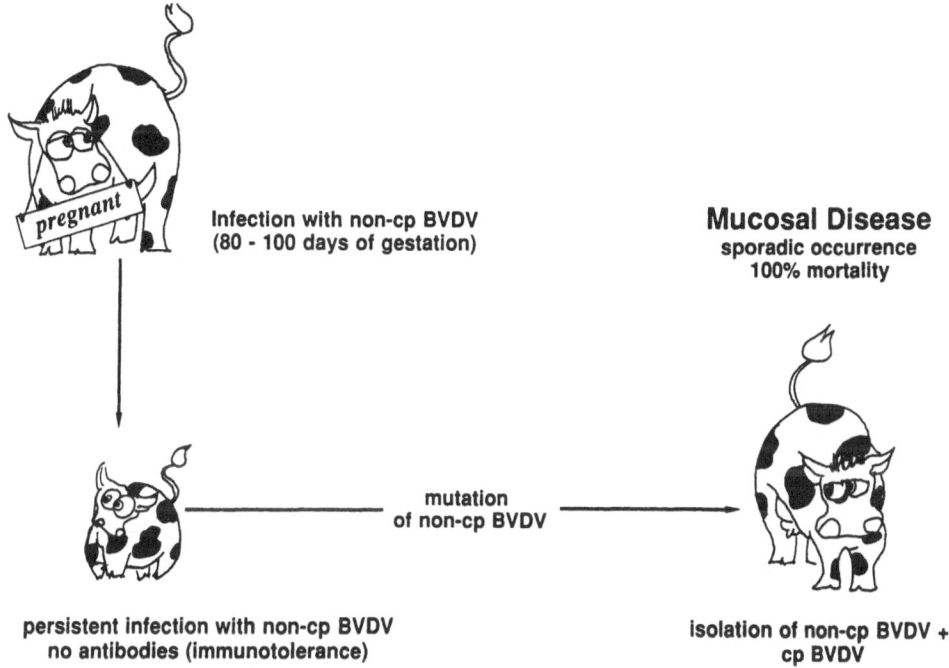

Fig. 7. Pathogenesis of mucosal disease

noncpBVDV (Fig. 7). Surprisingly, cpBVDV can always be isolated from MD animals in addition to the persisting noncp virus [3, 4]. In contrast to the described antigenic variability of BVDV field isolates the members of such a "pair" of noncpBVDV and cpBVDV are antigenically very closely related [10, 22]. This observation led to the hypothesis that during pathogenesis of MD a cpBVD virus develops from the noncp virus by acquiring some kind of mutation [10]. We proposed as a working hypothesis that recombination between viral and cellular RNA led to the formation of these cpBVDV genomes [16, 19].

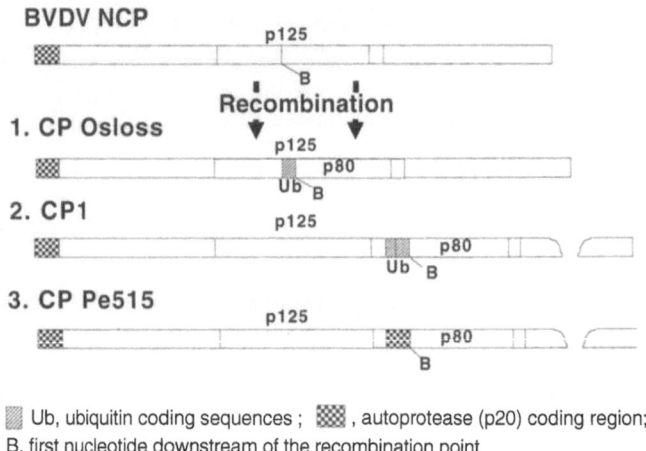

Ub, ubiquitin coding sequences ; ▓▓, autoprotease (p20) coding region;
B, first nucleotide downstream of the recombination point

Fig. 8. Comparison of BVDV genomes

To investigate directly the difference between a cytopathogenic virus and its noncytopathogenic counterpart we analyzed the genomes of a pair of cpBVDV (CP1) and noncpBVDV (NCP1) isolated from one MD animal. The RNA of CP1 was found to contain a ubiquitin-coding element which is embedded in a large duplication of viral sequence encompassing the p80-coding region (Fig. 8). In contrast, the genome of NCP 1 does not contain either insertion or duplication [17]. According to the results of these studies, one possible mutation leading to cytopathogenic BVDV is a recombination process between cellular and viral RNA.

In the case of two other cpBVDV isolates from BVDV pairs no host cell-derived insertion could be identified. However, elaborate duplication and rearrangement of viral sequences were found in both genomes (for CP Pe515 see Fig. 8). The analysis of the noncp virus (NCP Pe515) revealed that neither duplication nor rearrangement of sequences are present, and thus supported the linkage between recombination and establishment of the cytopathogenic phenotype. Accordingly, generation of cpBVDV is not restricted to recombination between cellular and viral sequences but can also be achieved by mere rearrangement of viral sequences.

One important change at the protein level distinguishing cpBVDV from noncpBVDV is the expression of p80 in cpBVDV infected cells (Fig. 6, Fig. 8). A prerequisite for generation of p80 is the presence of a cleavage site at the aminoterminus of this protein. Based on our current knowledge two basically different ways can be distinguished. One possibility is represented by the Osloss strain where the signal for cleavage is provided by a cellular ubiquitin sequence inserted into the p125 region of the viral polyprotein. Ubiquitin itself is synthesized within eukaryotic cells in form of fusion proteins which are subsequently cleaved at the carboxyterminal end of the ubiquitin moiety [12, 13, 20, 24, 25]. Accordingly, the insertion of ubiquitin into p125 introduces a signal for processing by a cellular protease (manuscript in preparation). The second way to generate p80 requires duplication of the respective coding region. Again a processing signal is necessary which is now placed in front of the duplicated sequence. In the case of CP1 ubiquitin provides again the signal for cleavage by a cellular protease (manuscript in preparation). For Pe515CP and CP6 a processing site encoded at the end of a second virus-derived duplicated element serves this purpose. This second duplicated element codes for the viral protein p20. Interestingly, insertion of the p20-coding region also transfers the protease responsible for the cleavage to its point of action. The integration of both protease and cleavage site might be necessary because of the lack of trans function of the p20 protease. This would be in accordance with the hypothesis that p20 can only act as an autoprotease [33].

Both alternatives outlined above would allow identical aminoterminal ends for p80. The expression of this protein is strictly correlated with the cytopathogenic phenotype and the development of MD. Therefore p80 represents the prime candidate for the agent responsible for killing the infected cells.

Porcine reproductive and respiratory syndrome virus (PRRSV)

Members of the arterivirus group are currently classified within the *Togaviridae* family [31], but the need for reclassification has become obvious after cloning and molecular analysis of the total EAV genome [11] and of parts of the LDV genome [14]. In contrast to togaviruses, arterivirus gene expression is characterized by transcription of multiple subgenomic mRNAs, each encoding one protein. Similar to coronaviruses, arteriviral mRNAs form a 3′ coterminal nested set and possess common 5′ terminal leader sequences which are joined to the bodies of the mRNAs during transcription. Moreover, the putative EAV polymerase gene is probably expressed by ribosomal frameshifting as in coronaviruses and possesses conserved domains also present in corona- and torovirus polymerases [11].

Using purified PRRS virions from infected macrophages as starting material molecular cDNA cloning and sequencing was performed. PRRSV specific cDNA clones spanning the 3′ terminal 5 kb of the genomic RNA were isolated, sequenced and used as probes for identification of PRRSV specific RNAs. The PRRSV genome is a positive stranded polyadenylated RNA of about 15 kb. In infected cells a 3′ coterminal nested set of six major subgenomic mRNAs could be demonstrated. Within the 3′ terminal 3.5 kb of the PRRSV genome six overlapping reading frames (ORFs) were identified, each most likely expressed by one of the subgenomic mRNAs. Amino acid sequence comparisons revealed that the most 3′ terminal ORF (ORF7) encodes the PRRSV nucleocapsid protein with a calculated molecular weight of 14 kD. It displays 44.8% amino acid identity with the capsid protein of lactate dehydrogenase-elevating virus (LDV) and 23.6% with that of equine arteritis virus (EAV). The product of ORF6, the second 3′ terminal ORF, represents a putative membrane protein and exhibits 53.2% and 27.2% amino acid identity with the corresponding LDV and EAV polypeptides. Similar to EAV, ORFs 2 through 5 might encode glycosylated viral proteins. The polypeptide deduced from the most 5′ ORF (ORF1b) contains two conserved domains common to EAV and coronavirus polymerases. Genome organization, strategy of gene expression and the sequence of deduced proteins show that PRRSV belongs to the arterivirus group of viruses [9].

Acknowledgements

This study was in part supported by the Deutsche Forschungsgemeinschaft (Th 298/3-1), the Bundesministerium für Forschung und Technologie and Intervet International BV (project 0319028A).

References

1. Baker JC (1987) Bovine viral diarrhea virus: A review. J Am Vet Med Assoc 190: 1449–1458
2. Benfield DA, Nelson E, Collins JE, Harris L, Goyal SM, Robison D, Christianson WT, Morrison RB, Gorcyca D, Chladek D (1992) Characterization of swine infertility and respiratory syndrome virus (isolate ATCC VR-2332). J Vet Diagn Invest 4: 127–133
3. Bolin SR, McClurkin AW, Cutlip RC, Coria MF (1985) Severe clinical disease induced in cattle persistently infected with noncytopathic bovine viral diarrhea virus by superinfection with cytopathic bovine viral diarrhea virus. Am J Vet Res 46: 573–576
4. Brownlie J, Clarke MC, Howard CJ (1984) Experimental production of fatal mucosal disease in cattle. Vet Rec 114: 535–536
5. Collett MS, Wiskerchen MA, Welniak E, Belzer SK (1991) Bovine viral diarrhea virus genomic organization. In: Liess B, Moennig V, Pohlenz J, Trautwein G (eds) Ruminant pestirus infections. Springer, Wien New York, pp 19–27 (Arch Virol [Suppl] 3)

6. Collett MS, Moennig V, Horzinek MC (1989) Recent advances in pestivirus research. J Gen Virol 70: 253–266

7. Collett MS, Larson R, Belzer SK, Retzel E (1988) Proteins encoded by bovine viral diarrhea virus: the genomic organization of a pestivirus. Virology 165: 200–208

8. Collett MS, Larson R, Gold C, Strinck D, Anderson DK, Purchio AF (1988) Molecular cloning and nucleotide sequence of the pestivirus bovine viral diarrhea virus. Virology 165: 191–199

9. Conzelmann K-K, Visser N, Van Woensel P, Thiel H-J (1993) Molecular characterization of porcine reproductive and respiratory syndrome virus, a member of the arterivirus group. Virology 193: 329–339

10. Corapi WV, Donis RO, Dubovi EJ (1988) Monoclonal antibody analyses of cytopathic and noncytopathic viruses from fatal bovine viral diarrhea virus infections. J Virol 62: 2823–2827

11. Den Boon JA, Snijder EJ, Chirnside ED, De Vries AAF, Horzinek MC, Spaan WJM (1991) Equine arteritis virus is not a togavirus but belongs to the coronavirus superfamily. J Virology 65: 2910–2920

12. Finley D, Bartel B, Varshavsky A (1989) The tails of ubiquitin precursors are ribosomal proteins whose fusion to ubiquitin facilitates ribosome biogenesis. Nature 338: 394–401

13. Finley D, Özkaynak E, Varshavsky A (1987) The yeast polyubiquitin gene is essential for resistance to high temperatures, starvation and other stresses. Cell 48: 1035–1046

14. Godeny EK, Speicher DW, Brinton MA (1990) Map location of lactate dehydrogenase-elevating virus (LDV) capsid protein (Vp1) gene. Virology 177: 768–771

15. Meyers G, Rümenapf T, Thiel H-J (1989a) Molecular cloning and nucleotide sequence of the genome of hog cholera virus. Virology 171: 555–567

16. Meyers G, Rümenapf T, Thiel H-J (1990) Insertion of ubiquitin-coding sequence identified in the RNA genome of a Togavirus. In: Brinton MA, Heinz FX (eds) New aspects of positive strand RNA viruses. American Society for Microbiology, Washington, D.C., pp 25–29

17. Meyers G, Tautz N, Dubovi EJ, Thiel H-J (1991) Viral cytopathogenicity correlated with integration of ubiquitin-coding sequences. Virology 180: 602–616

18. Meyers G, Tautz N, Stark R, Brownlie J, Dubovi EJ, Collett MS, Thiel H-J (1992) Rearrangement of viral sequences in cytopathogenic pestiviruses. Virology 191: 368–386

19. Meyers G, Rümenapf T, Thiel H-J (1989b) Ubiquitin in a togavirus. Nature 341: 491

20. Özkaynak E, Finley D, Solomonm MS, Varshavsky A (1987) The yeast ubiquitin genes: A family of natural gene fusions. EMBO J 6: 1429–1439

21. Plagemann PGW, Moennig V (1992) Lactate dehydrogenase-elevating virus, equine arteritis virus and simian hemorrhagic fever virus. Adv Virus Res 41: 99–192

22. Pocock DH, Howard CJ, Clarke MC, Brownlie J (1987) Variation in the intracellular polypeptide profiles from different isolates of bovine viral diarrhea virus. Arch Virol 94: 43–53

23. Qi F, Ridpath JF, Lewis T, Bolin SR, Berry ES (1992) Analysis of the bovine viral diarrhea virus genome for possible cellular insertions. Virology 189: 285–292

24. Rechsteiner M (1987) Ubiquitin-mediated pathways for intracellular proteolysis. Annu Rev Cell Biol 3: 1–30

25. Redmann KL, Rechsteiner M (1989) Identification of the long ubiquitin extension as ribosomal protein S27a. Nature 338: 438–440
26. Renard A, Dino D, Martial J (1987) Vaccines and diagnostics derived from bovine diarrhea virus. European Patent Application number 86870095.6. Publication number 0208672, 14 January 1987
27. Stark R, Rümenapf T, Meyers G, Thiel H-J (1990) Genomic localization of hog cholera virus glycoproteins. Virology 174: 286–289
28. Thiel H-J, Stark R, Weiland E, Rümenapf T, Meyers G (1991) Hog cholera virus: molecular composition of virions from a pestivirus. J Virol 65: 4705–4712
29. Weiland E, Stark R, Haas B, Rümenapf T, Meyers G, Thiel H-J (1990) Pestivirus glycoprotein which induces neutralizing antibodies forms part of a disulfide-linked heterodimer. J Virol 64: 3563–3569
30. Weiland E, Ahl R, Stark R, Weiland F, Thiel H-J (1992) A second envelope glycoprotein mediates neutralization of a pestivirus, hog cholera virus. J Virol 66: 3677–3682
31. Westaway EG, Brinton MA, Gaidamovich SYA, Horzinek MC, Igarashi A, Kääriäinen L, Lvov DK, Porterfield JS, Russel PK, Trent DW (1985) Togaviridae. Intervirology 24: 125–139
32. Wensvoort G, Terpstra C, Pol JMA, Wagenaar F (1991) Lelystad virus, the cause of porcine epidemic abortion and respiratory syndrome (mystery swine disease). In: Proc Porcine Reprod Respir Syndrome Semin, November 4–5, Brussels, Belgium, pp 27–35. European Commission VI/BII.2
33. Wiskerchen MA, Belzer SK, Collett MS (1991) Pestivirus gene expression: The first protein of the bovine viral diarrhea virus large open reading frame, p20, possesses proteolytic activity. J Virol 65: 4508–4514

Authors' address: Dr. H.-J. Thiel, Federal Research Centre for Virus Diseases of Animals, Paul-Ehrlich-Strasse 28, D-72076 Tübingen, Federal Republic of Germany.

Arch Virol (1993) [Suppl] 7: 53–62

Serological and antigenical findings indicating pestivirus in man

M. Giangaspero[1], **G. Vacirca**[1], **M. Buettner**[2], **G. Wolf**[2], **E. Vanopdenbosch**[3],
and **G. Muyldermans**[4]

[1] Institute of Special Pathology and Veterinary Medical Clinic,
Faculty of Veterinary Medicine, University of Milan, Milan, Italy
[2] Institute of Medical Microbiology, Faculty of Veterinary Medicine,
Ludwig Maximilians University, Munich, Federal Republic of Germany
[3] Department of Bovine Virology, The National Institute for Veterinary Research,
Brussels, Belgium
[4] Institute of Molecular Biology, Free University of Brussels, Sint-Genesius-Rode,
Belgium

Summary. An epidemiological survey for pestivirus was undertaken in Zambia and Europe, in view of the recent serological findings obtained by previous studies in Europe with humans. Collected sera were tested for anti-bovine viral diarrhea virus (BVDV) specific antibodies by IIF and Western Blotting. Of those individuals tested (n = 1272), 15.3% showed a seropositive reaction to the BVDV. Anti-BVDV antibody prevalence in immuno-depressed patients (e.g. HIV positive) was investigated. A higher prevalence was revealed in HIV patients suffering from chronic diarrhoea and in those having developed AIDS Related Complex (ARC). Out of 212 persons tested for pestivirus isolation, a non cytopathic virus strain was detected in 2 buffy coat samples using IIF with a specific anti-BVDV serum. The isolation could be repeated three times during 31 days in one person. The virus was identified as a pestivirus with radioimmuno-precipitation assays and IIF-flow cytometry. A doublet of 120 kD was identified only in cell lysates, indicating a non-structural protein. In order to rule out cross reactivity 30 sera from Hepatitis C seropositive patients were tested against the isolate by IIF-flow cytometry. No antigen-specific binding could be observed. These findings indicated the occurrence of a pestivirus in man and might suggest a relationship with a pestivirus of animal origin.

Introduction

The pestivirus genus, flavivirus family [5], is represented by three virus species from which Bovine Viral Diarrhoea (BVD) virus is the prototype.

The wide distribution of pestiviruses in animals, the frequent contacts between man and virus carriers as well as the human consumption of virus-contaminated animal products were the reasons to explore the prevalence in man.

Serological surveys on the occurrence of pestivirus-specific antibodies in man started in 1984 in Northern Italy [6, 17], then were extended to other European, African and Central and South American countries [9]. Specific immunoreaction (IIF, ELISA, NT and WBT) occurred especially against the BVDV Belgium non-cytopathic (ncp) A19 strain. The results also indicated that contact with susceptible animals was probably an important factor, and a transmission to man may occur. Serological tests and statistical analysis indicated that cross-reactions with Hog Cholera virus (HCV), Yellow fever virus and Rubella virus were not involved [7, 20]. Furthermore, WBT demonstrated an immune responsiveness to BVDV NADL strain 120 kD protein. Some sera reacted also with a protein of 70 kD [7].

The current study was performed with the aim to provide further evidence of a possible occurrence of pestivirus in man, in view of the recent serological findings obtained by the above mentioned studies. Since immunocompromised people are at high risk to get infected by various microorganisms, it appeared worthwhile to explore the anti-pestivirus antibody prevalence in human immunodeficiency virus (HIV) infected individuals.

An epidemiological study was performed in Africa and Europe. Among the African countries, Zambia was selected for the survey because BVDV was reported in cattle [8]. Furthermore, AIDS has an important occurrence in this country. The number of HIV-antibody positive individuals does increase rapidly and clinical evidence of disease, like Aids Related Complex (ARC) and chronic diarrhoea is high. In addition, diarrhoea of unknown etiology often causes high morbidity and mortality, especially in new-born children. The tested human serum samples were collected from HIV negative (control group) and positive persons showing different clinical syndromes: asymptomatics, with associated pathologies, with chronic diarrhoea and having developed the ARC syndrome. Our objective was to investigate a possible association between BVDV reactive antibody prevalence and the various clinical expressions of HIV infection.

In view of the serological results, in 1989 attempts for pestivirus isolation were made using buffy coat samples.

Materials and methods

Sera

Blood serum samples from 1,272 patients originating from Europe and Africa were examined of which 771 were HIV positive. The anamnestic data showed that 539 patients were asymptomatic, 81 suffered from associated diseases (Tuberculosis, Kaposi's sarcoma, pneumonia, encephalopathy, Herpes zoster and generalised lymphadenopathy), 73 were affected by chronic diarrhoea and 78 had developed the ARC syndrome. HIV negative persons (n = 501) composed the control group. The sera were obtained from Zambian (n = 1,159) and European (n = 113) people. 30 sera, positive for anti-Hepatitis C virus antibodies in a commercially available ELISA (Ortho, FRG), were selected for cross reactivity tests. The sera were stored at −20°C until examination.

Buffy coats

212 blood samples were examined. Sampling was made from brachial vein and 3 ml of blood were added in tubes containing 2 ml of Alsever medium. The samples were rapidly delivered to the laboratory and the buffy coat was obtained the same day.

Indirect Immune-Fluorescence Test (IIF)

The antibody detection by IIF was performed following the drop method as described by Wellemans and Leunen [18]. The antigen substrate was obtained by BVDV cytopathic (cp) NADL strain grown on primary foetal bovine kidney cells or BVDV Belgium ncp A19 strain grown on MDBK cells. The sera were tested following a base 3 dilution starting from 1:15, in PBS. Hyperimmunised calf serum against BVD virus C24V strain (Phylaxia) was used as the positive control. The second antibody consisted of anti-human IgG, obtained from hyperimmunised goat (GAH), FITC conjugated. The second antibody for bovine serum consisted of anti-bovine IgG, obtained from hyperimmunised goat (GAB), conjugated with FITC; the conjugate was tested for non-specific binding before and after lyophilization (National Institute for Veterinary Research NIVR, Belgium). The GAB and GAH conjugates were used at a dilution of 1:400 and 1:300 respectively, in PBS with 1:20 000 W/V Evans blue. Slides were observed under an epi-fluorescence microscope (Olympus). Titres of 1:45 were considered positive, excluding non-specific reactions.

Western Blotting Technique (WBT)

SDS-Page was performed in a discontinuous gel buffer system [10]. The stacking gel had a concentration of 4% acrylamide and 12% acrylamide in the separating gel. The BVDV cp NADL strain-infected primary foetal bovine kidney cell suspension and an uninfected cell suspension as negative control were used as antigen substrate. Electroblotting was performed as described by Towbin et al. (1979) [16]. Gels were soaked in transfer buffer and proteins were transferred to a nitrocellulose membrane. Proteins on blots were identified by antibody binding through ELISA method. Human test sera were diluted 1:10. Biotinylated GAH and streptavidin alkaline phosphatase were used at a dilution of 1:5000 and 1:1000, respectively.

Virus isolation

The routine method used for the screening of the BVDV-immunotolerant cattle from buffy coat was applied [19]. From each sample, tested in double, buffy coat cells were placed into 4 wells of a microtiter plate (Nunclon, 96 wells with flat bottom). Pestivirus free bovine testicular cells (BTC), MDBK, Vero and HeLa were added alternatively. MDBK cells were tested at regular intervals in IIF using flow cytometry (analysis of 10000 cells) and in addition by pestivirus-specific polymerase chain reaction (PCR, courtesy of Prof. Thiel, BFA, Tuebingen, FRG) for possible pestivirus contamination. The cells always were found not to contain pestivirus p125 antigen in IIF and with PCR no pestivirus-specific amplification was obtained. However, MDBK cells infected with the human isolate and BVDV infected cells as a control produced specific signals in both tests. Cellular suspensions contained 10% foetal serum (anti-BVDV Ab and BVDV free) irradiated with 1 megarad. In each plate, 4 wells were reserved for BVDV positive control and other 4 to BVDV negative control. Serum from BVDV hyperimmune cattle (Machelen, NIVR) and FITC labelled GAB were used to perform IIF. The plates were observed with IF inverted microscope (Olympus IMT2).

Characterisation of the viral isolate

The original virus isolate was replicated several times on MDBK-BVDV free cells. Assays for viral replication were performed using primary lung and spleen cell cultures of bovine origin, as well as human fetal lung cells and human diploid amnion cells. For inoculation of the cell cultures the MDBK passaged pestvirus isolate was taken. Virus characterisation was performed using IIF-flow cytometry and Radio Immuno Precipitation. Some monoclonal and polyclonal anti-sera were used for the execution of the tests (see below). An eventual cross reaction of Hepatitis C virus-specific antibodies with the isolate and the ncp BVDV strain Iowa was explored.

Radio immuno precipitation assay (RIPA)

The test was performed according to the method described by Muyldermans et al. [11].

The metabolic labelling with [^{35}S]methionine (100 µCi/ml) was done on MDBK cells infected with the ncp human isolate. The immuno-precipitation test was performed using polyclonal antibodies anti-HCV from pig (NIVR), anti-BVDV from bovine (Machelen, NIVR) and the monoclonals 1B11 (pan-pestivirus recognizing an antigenic determinant of the C-terminal cleavage product of the highly conserved non-structural p80/p125) and 8D10 (recognizing HCV p46) [2]. Purified IgG from BVDV seronegative fetal calf serum and HCV seronegative pig serum were used as negative controls. Electrophoretic analysis were made by SDS-PAGE on a 12% acrylamide gel or a 10%–18% gradient gel. Molecular weight markers were [^{14}C]methylated standard proteins (Amersham). The gel was subsequently enhanced by fluorography techniques and after drying, exposed for 5 days or 1 month to FUJI XR-films.

IIF-flow cytometry

IIF in flow cytometry was performed according to the method described by Qvist et al. [13], using the pan-pestivirus Mab C16 (specific for BVDV NADL strain p80/p125), on permeabilized non infected MDBK cells and infected ones with the isolated virus of human origin. A bovine foetal spleen cell culture infected with an equal dose of the same virus was also tested. Furthermore, cross reaction of anti-Hepatitis C virus antibodies was explored, taking into account the recent classification of BVDV as *Flaviviridae* [5]. Thirty Hepatitis C virus ELISA antibody-positive sera were tested using cells infected either with BVDV ncp Iowa strain or the human isolate. Secondary FITC labelled antibodies rabbit anti-human and sheep anti-mouse (Sigma) were used separately as negative control.

Results

The results of IIF test performed with 1,272 human sera originating from Zambia and Europe for the search of anti-BVDV antibodies are reported in Table 1. 15.3% of the sera were positive for anti-BVDV antibodies using BVDV cp NADL strain. In the WBT 52 sera showed a specific immune responsiveness to NADL viral strain protein of 120 kD. Furthermore, some sera also reacted to a protein of 33 kD.

Comparison between different serological and clinical aspects of HIV infection and seropositiveness to BVDV is reported in Table 2.

Comparison of serological results in Zambian and European samples did not show relevant differences. From Zambia, 14.7% of HIV negatives (n = 477), 15.2% from HIV positive without clinical symptoms (n = 480), 15.8% from HIV positive with associated pathologies (n = 70), 19.2% from HIV positive with chronic diarrhoea (n = 73) and 20.3% from HIV positive with ARC syndrome (n = 59) were anti-BVDV antibody positive. From the 113 persons of European origin, seropositiveness to BVD virus was found as follows: HIV negatives (n = 24) 4.2%, HIV positive without clinical symptoms (n = 59) 8.5%, HIV

Table 1. Variation of Ab titres against BVD NADL strain (IIF method) detected in 1272 human sera originated from Zambia and Europe

IFI Titres	Number of Sera	%
Neg.	1077	84.7
1:45	131	10.3
1:135	59	4.6
1:405	5	0.4

Table 2. Comparison among different serological and clinical aspects of HIV infection with the BVD virus seropositiveness in 1272 persons (IIF method)

Serological and clinical aspects of HIV infection	Number of Sera	% of BVDV antibody positiveness
HIV negative	501	14.2
HIV positive asymptomatics	539	14.5
HIV positive with associated diseases	81	17.3
HIV positive with chronic diarrhoea	73	19.2
HIV positive with ARC syndrome	78	23.3

Table 3. Viraemia and serology from a patient infected with a non-cytopatic (ncp) virus strain

Date of sampling	Viraemia	Anti-BVD antibody titres	
		strain ncp A19	strain cp NADL
day 0	pos.	1:45	neg.
day 24	pos.	1:405	1:45
day 31	pos.	1:135	neg.
day 37	neg.	1:135	1:45

positive with associated pathologies (n = 11) 27.3%, HIV positive with ARC syndrome (n = 19) 31.6%. This last group was re-tested with BVDV ncp A19 strain and a higher positiveness was detected: 13 sera were positive (68.4%), 4 showing an antibody titre of 1:135 and 9 with 1:45.

The viral isolation assay from 212 human buffy coats revealed a viraemia in two persons. In both cases, a ncp viral strain, growing on MDBK cells only, was detected using specific anti-BVDV serum. These two samples originated from clinically healthy persons. From the buffy coat cells of a 30 years old woman, the isolation was repeated three times during 31 days (Table 3) and negative results were obtained on the subsequent five attempts from the 37th to the 103rd day. At the first isolation, MDBK cell layers were completely infected by the virus; only limited numbers of infected cells were found in the further isolations.

The serological tests carried out in parallel with the viral isolation showed an immune responsiveness against BVDV cp NADL and ncp A19 strains, with a major affinity to the latter (Table 3).

Results of tests performed with RIPA on the original virus isolate were evident after 5 days of exposure. Mab 1B11, polyclonals anti-HCV

MW 1 2   3   4 5 6 7   8   9 10 11 12

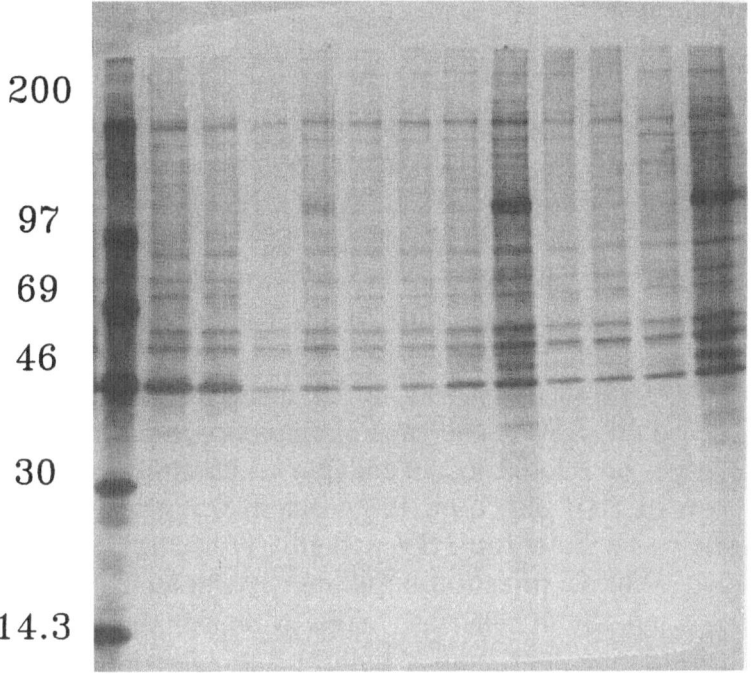

Fig. 1. Radio Immuno Precipitation Assay. Evaluation of the human virus isolate [35S]methionine labelled MDBK infected cell lysate (lanes *2, 4, 6, 8, 10* and *12*) and mock-infected (lanes *1, 3, 5, 7, 9* and *11*) with Mab 8D10 (lanes *1* and *2*), Mab 1B11 (lanes *3* and *4*), polyclonal anti-HCV (lanes *7* and *8*) and polyclonal anti-BVDV (lanes *11* and *12*). Negative control: HCV seronegative pig serum (lanes *5* and *6*) and BVDV seronegative fetal calf serum (lanes *9* and *10*). *MW* [14C]methylated molecular weight standards (values expressed in kD) (left lane) (FUJI XR-film)

and anti-BVDV precipitated a virus specific doublet of 120 kD but only in cell lysates, indicating a non-structural protein (Fig. 1). Both polyclonals also recognized a protein of 50.2 kD.

Furthermore, IIF-flow cytometry using pan-pestivirus Mab C16 demonstrated specific fluorescence signals in MDBK cells infected with the virus of human origin. A lower level of fluorescence was found in bovine foetal spleen cells infected with an equal dose of the same virus. The attempts to multiply the virus in human amnion and fetal lung cells did not give any positive result.

Analysis of 30 human hepatitis C antibody positive sera for reactivity with MDBK cells infected with BVDV ncp strain Iowa or the human isolate using IIF in flow cytometry did not show specific binding.

Discussion

The observation in human population showed a prevalence of specific anti-BVDV antibodies in 15.3% of the tested sera. This percentage probably could vary widely depending on the pestivirus used as antigen. In fact, the prevalence increased in a group of HIV ARC from Europe, when re-tested with the BVDV ncp A19 strain. In addition, previous screening indicated a higher affinity against BVDV ncp strains [9].

The hypothesis that diseased HIV positive people might be more susceptible to pestivirus take was confirmed by the detection of antibodies. Seropositiveness to BVD virus did not differ between asymptomatic HIV positive patients and those who were serologically negative. A higher percentage of seroreactivity with BVDV was found in HIV diseased patients, especially ARC. This was similar in Zambian and European people. It might be related to the collapse of immune defence at certain clinical stages of HIV infection. Pestivirus invasion and probably replication might be facilitated in HIV patients. However, it is questionable whether such a double infection implies any pathological meaning or the viruses may even play a synergistic role in the etiopathology of clinical features, considering that, under natural and experimental conditions, the "helper" role of BVDV and HIV to some viral and bacterial pathogens is well known in man and animals [12, 13–15].

Taking into account the limited number of samples of European origin, the comparison between the results of the sera from Zambia and Europe did not show any significant differences. The epidemiological aspects were similar in the two populations investigated.

Other studies demonstrated the presence of anti-BVDV antibodies in human population. Wilks et al. [20] reported seropositiveness to BVD virus in persons from New Zealand. The authors' conclusions indicated the possibility of a presently unknown Togavirus.

The isolation of a ncp strain supported the previous serological findings, especially the high affinity of antibodies against BVDV ncp strains, and explained the persistence of antibody titres for several years reported in some persons [7]. The human "pestivirus-like" isolate may support the suggestion that the infection might be of animal origin. Genome analysis of the isolate in comparison with animal pestiviruses may reveal relationship or differences. High risk of BVDV transmission from immunotolerant, pestivirus ncp strain persistently infected animals to man should be considered [1].

The time of viraemia in one of the two positive persons was rather long when compared with that occurring in animals. Following acute forms of the infection in cattle, viraemia is demonstrable from 1 to 2 weeks. Nevertheless, it may be longer, up to 2 months in certain postnatal forms and in some recovering subjects. However, the immunotolerants show a lifelong viraemia. In pigs, the virus is isolable up to 6 days after infection; then isolation becomes sporadic and impossible after 20 days. In experimentally infected rabbits, viraemia does not exceed 14 days [3].

The presence of a doublet of 120 kD, demonstrated using radioisotopes, and the test performed with the IIF-flow cytometry, using Mab C16, supported the correlation of the isolated human strain with the pestivirus genus. The testing of anti-Hepatitis C virus ELISA antibody positive sera using BVDV ncp strain Iowa and the isolate demonstrated no reactivity with these antigens. The results concerning the virus isolation were corroborated by the research undertaken by Yolken et al. [21]. Using a monoclonal antibody-based solid phase enzyme immuno assay with anti-BVDV cp Singer strain Mab, they reported the presence of a 48 kD viral glycoprotein, from faeces of children under two years affected by gastroenteritis often associated with respiratory illness, where no pathogenic agent could be identified. This glycoprotein has also been detected in BVDV strains isolated from cattle [4]. The authors excluded any correlation of the detected virus with other related RNA viruses of human and animal origin (e.g. Venezuelan equine encephalitis virus, dengue viruses, Rubella virus, West Nile virus).

These findings indicate the occurrence of a pestivirus in man but it is still unclear whether this is of human origin or might be an adaptation of animal strains; nevertheless, serological and antigenical findings in man suggest a relationship with the animal pestivirus. In addition, possible heterogeneity in pestivirus populations, implying virus particles with higher affinity to human tissues remains to be investigated using many different cells of human origin.

The results merit further attention in order to confirm the finding and to examine the full extent of the observations in human beings. The support of molecular biology (e.g. characterization of the genome of the isolate and comparison with related RNA viruses and possibly with the pestiviruses isolated in man in North America) and the improvement of antibody/antigen detection techniques will provide an essential contribution for the future research on this topic.

References

1. Bolin SR, McClurkin AW, Coria MF (1985) Frequency of persistent BVDV infection in selected cattle herds. Am J Vet Res 46: 2378–2385
2. Caij A, Muyldermans G, De Smet A, Hamers R, Koenen E (1993) Production and characterization of monoclonal antibodies against hog cholera virus (Alfort 187 strain). Arch Virol 131: 185–192

3. Castrucci G (1978) Togaviridae. In: Infezione da virus degli animali domestici. Edagricolae, Milan 3: 114–201
4. Corapi WV, Donis R, Dubovi EJ (1988) Monoclonal antibody analyses of cytopathic and noncytopathic viruses from fatal bovine viral diarrhoea virus infections. J Virol 62: 2823–2827
5. Franki RIB, Fauquet CM, Knudson DL, Braun F (eds) (1991) Classification and nomenclature of viruses. Fifth Report of the International Committee on Taxonomy of Viruses. Springer, Wien New York (Arch Virol [Suppl] 2)
6. Giangaspero M (1986) Infezione da virus BVD/Recettivit umana al BVD. Thesis of Doctorat. University of Milan, Italy
7. Giangaspero M (1989) La Diarrhee Virale des Bovins – BVD. Detection d'anticorps specifiques chez l'homme. Thesis of MSc 8 IMT Antwerp Belgium
8. Gianpaspero M, Blondeel H, Bear J, Alders R, Morgan D (1991) Epidemiological survey on Bovine Viral Diarrhoea virus in Zambian cattle. Zimbabwe Vet J 22: 57–63
9. Giangaspero M, Wellemans G, Vanopdenbosch E, Belloli A, Verhulst A (1988) Bovine viral diarrhoea. Lancet ii: 110
10. Laemmli UK (1970) Cleavage of structural proteins during the assembly of the head of bacteriophage T4 Nature 227: 680–686
11. Muyldermans G, Caij A, De Smet A, Koenen F, Hamers R (1993) Characterization of structural and non-structural proteins of Hog Cholera virus by means of monoclonal antibodies. Arch Virol 131: 405–417
12. Potgieter LND, McCracken MD, Hopkins JM (1984) Effect of BVDV infection on the distribution of infectious bovine rhinotracheitis virus in cows. Am J Vet Res 45: 687
13. Qvist P, Houe H, Aasted B, Meyling A (1991) Comparison of flow cytometry and virus isolation in cell culture for identification of cattle persistently infected with bovine viral diarrhea virus. J Clin Microbiol 29: 660–661
14. Reggiardo C, Kaeberle ML (1981) Detection of bacteraemia in cattle inoculated with BVDV. Am J Vet Res 42: 218
15. Stoeber M (1982) Streptotrichose der Haut (Dermatophytose) bei Mastbullen mit BVD. Tieraerztl Umschau 37: 629–630
16. Towbin H, Staehelin T, Gordon J (1979) Electrophoretic transfer of proteins from polyacrylamide gels to nitrocellulose sheets: procedure and some applications. Proc Natl Acad Sci 76: 4350–4354
17. Vacirca G, Giangaspero M, Belloli A, Bertoli G, Citrino A (1988) Ricerca di Anticorpi anti-BVD in siero di sangue umano. Meeting 8: 37–39
18. Wellemans G, Leunen J (1973) La rhinotracheite infectieuse des bovins (IBR) et sa serologie. Ann Med Vet 117: 507–511
19. Wellemans G, Vanopdenbosch E (1990) Depistage des bovins BVD-immunotolerants (BVD-IT) a partir des lymphocytes: un apercu de 22 mois de controle de la population bovine Belge. Ann Med Vet 134: 175–178
20. Wilks CR, Abraham G, Blackmore DK (1989) Bovine pestivirus and human infection. Lancet i: 107
21. Yolken R, Leister F, Almeido-Hill J, Dubovi E, Reid R, Santosham M (1989) Infantile gastroenteritis associated with excretion of pestivirus antigens. Lancet i: 517–519

Authors' address: Dr. H. Giangaspero, Institute of Special Pathology and Veterinary Medical Clinic, Faculty of Veterinary Medicine, University of Milan, Via Celoria 10, I-20133 Milan, Italy.

Arch Virol (1993) [Suppl] 7: 63–74

Molecular analysis of the human coronavirus
(strain 229E) genome

J. Herold, T. Raabe, and **S. Siddell**

Institute of Virology, University of Würzburg, Würzburg,
Federal Republic of Germany

Summary. The nucleotide sequence of the human coronavirus strain 229E (HCV 229E) has been determined. This article describes the organization of the virus genome, the predicted viral gene products and the mechanisms which regulate viral gene expression. This information provides a basis to investigate the biology and pathogenesis of HCV.

Introduction

Human coronaviruses are one of the causative agents of the common cold. HCV infections are generally mild, last only a few days and are seldom associated with severe symptoms such as headache, fever or diarrhea. Nevertheless, the economic consequence of respiratory disease caused by HCVs is significant [10]. HCVs can be divided into 2 major serological groups, represented by the prototypes, HCV 229E and HCV OC43. The two groups are about equally responsible for human respiratory infections [11]. Both virus types are difficult to isolate and both grow poorly in tissue culture. HCV 229E can, however, be adapted to grow in human lung fibroblasts, but even so, there is only a limited amount of information on the protein components of the virion, viral RNA and protein synthesis, the regulation of viral gene expression and the function of the viral gene products [1, 12, 21].

As a basis for our studies on the biology and pathogenesis of HCV 229E, we have undertaken a sequence analysis of the viral genome. With this information it is possible to

1. describe the organization of the HCV 229E genome
2. identify structural features which may play a role in the regulation of viral gene expression
3. predict the entire complement of viral gene products and
4. make predictions concerning the possible structure-function relationships of the viral proteins.

Materials and methods

Virus

The HCV 229E isolate used in these studies was obtained from a volunteer at the MRC Common Cold Unit, Salisbury, U.K. The virus was adapted to culture in C16 cells [16], titrated to limiting dilution and the supernatant from a well with one focus of infection was used to prepare a virus stock. C16 cells were infected with HCV 229E at an m.o.i. of 3, incubated at 33°C and cytoplasmic RNA was isolated 48 h.p.i. using standard procedures. Poly-A RNA was obtained by poly-U-Sepharose chromatography.

cDNA cloning

cDNA libraries were prepared essentially by the method of Gubler and Hoffman [9] using random hexanucleotide or virus specific oligonucleotide primers. The cDNA was size fractionated and cloned into the Bluescript vector pKS II⁺ (Stratagene). Recombinant clones were identified by colony hybridization with HCV 229E specific oligonucleotides. Plasmid purification, agarose gel electrophoresis, colony hybridizations and standard recombinant DNA procedures were done as described by Sambrook et al. [20].

PCR amplification

PCR was performed using a GeneAmp/RNA PCR kit (Perkin Elmer Cetus) according to the manufacture's procedures using biotinylated and non-biotinylated, HCV specific primers. The resulting cDNA strands were separated with streptavidin coupled magnetic beads (Dynal) and the nucleotide sequence of both strands was determined.

Sequence analysis

Sequencing was done on double and single strand DNA templates (cDNA and PCR amplification products) using the chain termination method and M13, T7, T3 and HCV 229E specific primers. To generate sequencing templates, cDNAs were subcloned by restriction enzyme digestion and overlapping deletions were introduced by exonuclease III. Sequence data were assembled by the programmes of Staden [23] and analysed by the programmes of the University of Wisconsin Genetics Computer Group [8].

Results

The genomic RNA

The genomic RNA of HCV 229E is comprised of 27,277 nucleotides and a 3′ poly-A tract of not less than 50 residues. By analogy to murine hepatitis virus (MHV), it is assumed that the 5′ end of the genome is

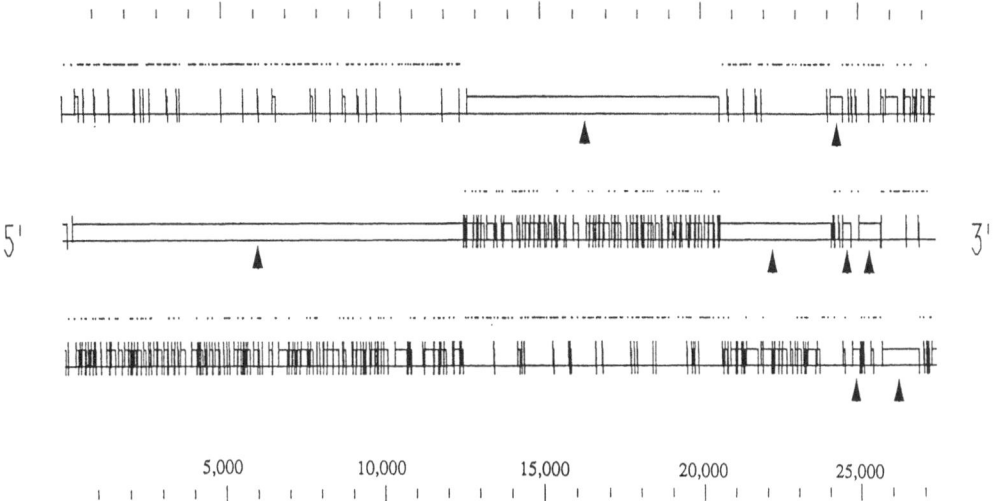

Fig. 1. Open reading frame and codon usage analysis of the HCV 229E genomic RNA. The analysis was performed using the UWGCG programmes FRAMES and CODONPREFERENCE

linked to a "cap" structure but this has not been directly demonstrated for HCV. Computer assisted analysis of the HCV 229E genome sequence reveals 8 non-redundant open reading frames (nrORFs) of more than 50 codons. There are also smaller nrORFs at the 5' and 3' ends of the genome and many redundant open reading frames (rORFs) located within the coding regions of the larger nrORFS. If a codon frequency table is deduced on the basis of the 8 nrORFs and applied to the entire genome (Fig. 1) there is no indication that the small nrORFs or the rORFs are expressed.

This analysis would not, however, be sensitive enough to predict viral gene products generated, for example, by mechanisms such as RNA editing or frameshift mutation. Also, it does not exclude the translation of an 11 codon nrORF located at the 5' end of the genome, in frame with and preceding the first large nrORF. Indeed, this small ORF is conserved in the genomes of HCV 229E, MHV and IBV [3, 13] and it may be speculated to have a role in the regulation of the initiation of protein synthesis from genomic RNA.

By analogy to the MHV and IBV genomes and their gene products, it is possible to assign the products of the HCV 229E genome to the 8 nrORFs described above. In some cases, these assignments can be strengthened by comparing the properties of predicted gene products with those of the known HCV proteins. This assignment is shown in Fig. 2.

　　　　　　　　　　　　　　　J. Herold et al.

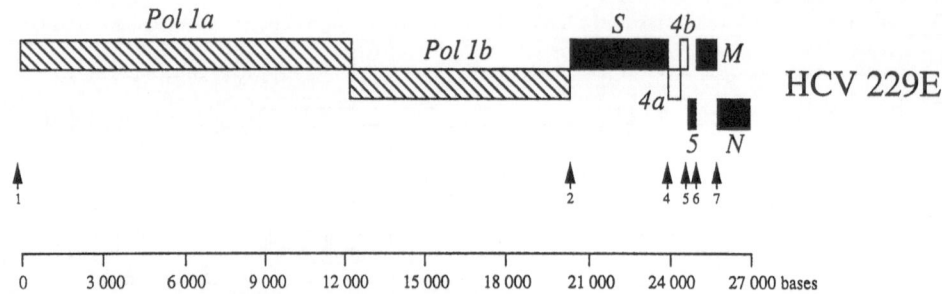

Fig. 2. Organization of the HCV 229E genome. The diagram is drawn to scale and the non-redundant ORFs are drawn in the correct reading frames. The structural protein genes are shown as black boxes, the non-structural protein genes as open (unknown function) or hatched (RNA polymerase) boxes

The genes of HCV 229E

The genes of HCV 229E are listed in Table 1 together with the predicted sizes of their assigned gene products.

The nucleocapsid gene

The nucleocapsid protein gene lies at the 3′ end of the genome. It encodes a polypeptide of M_r 43,500 which is in agreement with the apparent molecular weight of the HCV 229E N protein in SDS-PAGE [14]. In common with other coronavirus nucleocapsid proteins, the HCV

Table 1. The gene products of HCV 229E

Gene	Bases	Codons	Protein	Molecular mass
5′ UTR	293	–	–	–
ORF 1a	12,258	4,086	[polymerase]	454,200
ORF 1ab	20,277	6,759	[polymerase]	754,200
S	3,522	1,174	surface	128,600
ORF 4a	402	134	unknown	15,300
ORF 4b	267	89	unknown	10,200
ORF 5	234	78	[small membrane]	9,100
M	678	226	membrane	26,000
N	1,170	390	nucleocapsid	43,500
3′ UTR	422 +poly A	–	–	–

229E N protein is a serine-rich, basic protein (net charge +16 at neutral pH) and the protein is most probably phosphorylated. The distribution of basic and acidic residues is compatible with a three domain structure, as proposed for the MHV N protein by Parker et al. [15].

The membrane glycoprotein gene

The membrane glycoprotein gene is located adjacent to the N protein gene and encodes a polypeptide of 225 amino acids with an M_r of 26,000. The HCV M protein has several features which are characteristic of a coronavirus membrane protein. First, there are 3 potential N-linked glycosylation sites, one of which is near the amino terminus. It has been shown that the HCV 229E M protein is N-glycosylated [12]. Second, the polypeptide displays three internal hydrophobic domains within the amino terminal half and a relatively hydrophilic carboxy terminus. Third, the polypeptide is slightly basic with a net charge of +4 at neutral pH. These data suggest that the membrane topology of the HCV 229E M protein is very similar to that proposed by Rottier et al. [19] for the MHV M protein.

The surface glycoprotein gene

The HCV 229E surface glycoprotein gene encodes a polypeptide of 1,173 amino acids with an M_r of 128,600. The polypeptide has 30 potential N-glycosylation sites. The difference in the predicted M_r of the S protein and its apparent molecular weight in SDS-PAGE (180,000; [21]) suggests that the majority of these sites are used. A number of structural features typical of coronavirus S proteins can be recognized in the HCV 229E S protein gene product. These include an amino terminal signal sequence, a carboxy terminal membrane anchor, heptad repeat structures and a carboxy terminal cysteine cluster. In contrast to the S proteins of MHV and IBV, the HCV 229E S protein does not contain a basic region with the motif RRXRR or RRAHR (where X is F, S, H or A) which have been identified as the sites at which the MHV and IBV S proteins are proteolytically cleaved. Apparently, the HCV S protein is not post-translationally cleaved.

A detailed, computer-assisted comparison of the HCV 229E S protein with the published S protein sequences of other coronaviruses has also revealed a number of highly conserved cysteine residues (excluding those within the cysteine cluster) which are probably important in determining the three-dimensional structure of the protein. Clearly, however, the

```
              (-)                              (+)    cysteine-rich
IBV  3c   MMNLLNKSLEENGSFLTALYIFVGFLALYLLGRALQAFVQAADACCLFW  -  60aa.  12,4K

TGEV 4    MTFPRALTVIDDNGMVISIIFWFLLIIILILLSIALLNIIKLCMVCCNLG  -  32aa.  10,2K

MHV  5b     MFNLFLTDTVWYVGQIIFIVAVCLMVTIIVVAFLASIKRCIQLCGLC  -  41aa.  10,0K

BCV  6     MFMADAYFADTVWYVGQIIFIVAICLLVIIVVVAFLATFKLCIQLCGMC  -  35aa.   9,5K

HCV  5      MFLKLVDDHALVVNVLLWCVVNIVILLVCITIIRLIKLCFTCHMFC  -  31aa.   9,1K
```

Fig. 3. Sequence similarity of different coronavirus SM proteins. The protein sequences are taken from the literature and have been aligned to emphasize the amino proximal hydrophobic domain, the flanking charged residues and the cysteine cluster

relevance of these features will only become apparent when a detailed structural analysis of the HCV 229E S protein is made.

The ORF 5 gene

On the basis of its structural similarity to the SM proteins of IBV and TGEV (Fig. 3), the HCV ORF 5 gene product is most probably a structural protein of the virus. The predicted M_r of the HCV SM protein is 9,100. To date, no protein of this size has been identified in HCV 229E virions, however, this may be due to the low amount of protein which is incorporated into the virus particle. In addition to its similarity to other coronavirus SM proteins, there is a striking resemblance with the M2 protein of influenza virus. This protein has recently been shown to have an associated ion channel activity selective for monovalent ions [17] but the possible relevance of this activity for the coronavirus infection process is unknown.

The coronavirus non-structural protein genes fall into two categories. On the one hand, a single, large gene located at the 5′ end of the genome encodes the viral RNA polymerase, whilst, a variable number of small genes encoding proteins of as yet unknown function are interspersed between the N, M and S protein structural genes. In the case of HCV 229E this pattern is repeated.

The RNA polymerase gene

The RNA polymerase gene is comprised of two overlapping ORFs of 4,086 and 6,759 codons respectively. Together these ORFs have the potential to encode polypeptides with a total M_r of 754,200. As described below, there is evidence from studies on MHV and IBV that the coronavirus polymerase gene is polycistronic and its expression appears to involve mechanisms such as ribosomal frame shifting and autoproteolytic processing. The size of the gene leads one to suspect that this region of the genome may also encode functions which are not related to viral RNA synthesis.

A computer-assisted analysis of the HCV 229E polymerase gene reveals a number of sequence motifs which have been associated with RNA replicative functions (an RNA polymerase module, a helicase motif and a metal binding domain) as well as protease motifs (two papain-like and one 3C-like motif) which may encode activities involved in the post-translational proteolytic processing of the RNA polymerase (and perhaps other) gene products. These motifs have also been reported for the MHV and IBV RNA polymerase genes [13]. Additionally, a number of highly conserved regions can be identified in the HCV 229E, MHV and IBV polymerase genes, although it is not yet possible to ascribe them to any particular function.

The ORF 4a and ORF 4b gene

The remaining two HCV 229E nrORFs are ORF 4a and ORF 4b. The proteins encoded by these genes have predicted M_rs of 15,300 and 10,200 respectively. To date, these proteins have not been identified in the infected cell or in virions and they are provisionally considered to be non-structural proteins of unknown function.

Gene regulation

Transcriptional regulation

The expression of coronavirus genes is mediated by a set of 3' coterminal subgenomic mRNAs. In the case of HCV 229E, 5 subgenomic mRNAs have been identified [18]. With the exception of the smallest mRNA each of these subgenomic RNAs is structurally polycistronic but, in general, they are believed to be functionally monocistronic. This principle has been established, at least for MHV and IBV. As one mRNA encodes only one gene product, the amount of each subgenomic mRNA will largely determine the amount of the individual proteins synthesized.

The coronavirus mRNAs are synthesized in non-equimolar amounts but in a constant ratio throughout the infection cycle. Studies on MHV [22] have shown that the generation of coronavirus mRNAs involves a process of discontinuous transcription. A specific sequence, related to motif UCUAAAC, is found at those positions in the genome which define the 5' end of the unique region of each mRNA, i.e. the region not

Table 2. The nucleotide sequence of HCV 229E intergenic regions

Region	Sequence
5' UTR / Polymerase	G U C U A C U U U U C U C A A C U A A A C G A A A $[N_{214}]$ <u>A U G</u>
Polymerase / S	A U C A U U U A G U C U C A A C U A A A U A A A A <u>U G</u>
S / ORF 4	U G U G A A U C A A C U A A A C U U C C U U U U A $[N_{30}]$ <u>A U G</u>
ORF 4 / ORF 5	U U U C U U A U U U C U C A A C U A A C G A C U U $[N_{146}]$ <u>A U G</u>
ORF 5 / M	U U A U U G A U U U C U A A A C U A A A C G A C A A <u>U G</u>
M / N	U U C A U U U U U U C U A A A C U G A A C G A A A A G <u>A U G</u>

found in the next smallest RNA. This motif is thought to have an important function in mRNA synthesis, however, its precise role is still uncertain. On the one hand, it may be involved in determining the frequency of leader primed initiation on genome length negative strands or it could also act as a termination signal for negative strand RNA synthesis. At the present time, it is not possible to present a definitive model for the genesis of coronavirus subgenomic mRNAs. Because of the exceedingly low amounts of viral RNA in the earliest stages of infection, it will be difficult to distinguish between alternative models. In Table 2, the sequence of the HCV 229E intergenic regions which precede each of the subgenomic mRNA body sequences are shown.

Translational regulation

The basic premise of coronavirus translation is that only the information encoded in the 5′ unique region of each mRNA is expressed as protein. (Fig. 4). For the majority of HCV 229E mRNAs, the 5′ unique region encompasses only a single ORF and it can be assumed that these ORFs are functionally monocistronic. There are 2 mRNAs (mRNA 1 and mRNA 4) in which the 5′ unique region contains more than 1 ORF, and in these cases translational strategies appear to regulate gene expression.

mRNA 1

The HCV 229E RNA polymerase gene (or perhaps more correctly, RNA polymerase locus) is comprised of two large overlapping ORFs,

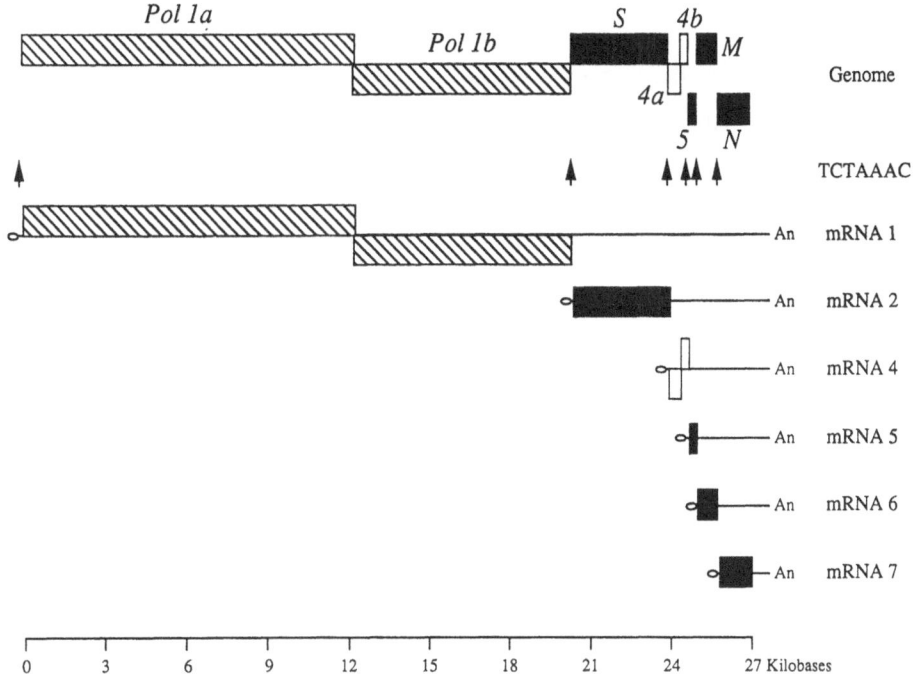

Fig. 4. Expression of the HCV 229E genome. The genomic organization and the 3'coterminal set of subgenomic mRNAs are illustrated. The structural and non-structural protein genes are shown as in Fig. 1

ORF 1a and ORF 1b. There is now substantial evidence that ORF 1b is expressed by a (-1) ribosomal frame shifting event mediated by a specific structure, the RNA pseudoknot, which is located at the ORF 1a/ORF 1b junction. This element has been identified for IBV, MHV and HCV 229E and its function in vitro and in vivo has been demonstrated by transcription/translation experiments [4, 5, 13]. In the absence of any further processing, the HCV 229E polymerase gene could encode an ORF 1a product (454,200 M_r) or an ORF 1a/b product (754,200 ORF M_r).

In addition to the flexibility offered by ribosomal frame shifting, there is both genetic and biochemical evidence which suggests that the coronavirus polymerase gene contains 5 or 6 complementation groups [2]. Almost certainly, the functionally separate gene products are generated by post-translational proteolytic processing. This conclusion is supported by the presence within the polymerase locus of protease motifs characteristic of both papain ($\times2$) and 3C-like proteases. These protease "domains" are located in ORF 1a. Recently, the first experiment on proteolytic processing of the MHV gene products in vivo have been reported, [6, 7] but it is too early to propose a processing pathway.

mRNA 4

The second HCV 229E mRNA which contains more than one nrORF in its 5′ unique region is mRNA 4. At the present time it is premature to speculate on the mechanisms used to express the downstream ORF (ORF 4b) because neither gene product has been identified in vivo or in vitro.

Conclusions

The molecular analysis of the HCV 229E genome reported here forms the basis for a detailed study of the biology and pathogenesis of this infectious agent. Should these studies reveal that HCV 229E infection is relatively benign, it may also open up the possibility of using this agent as a live virus vaccine for respiratory pathogens.

Acknowledgements

We thank Barbara Schelle-Prinz and Atiye Toksoy for excellent technical help and Andrea Feyrer for typing the manuscript.

References

1. Arpin N, Talbot JP (1990) Molecular characterization of the 229E strain of human coronavirus. In: Cavanagh D, Brown TDK (eds) Coronaviruses and their diseases. Advances in experimental biology and medicine, vol 276. Plenum Press, New York, pp 73–80
2. Baric RS, Fu KS, Schaad MC, Stohlman SA (1990) Establishing a genetic recombination map for MHV-A59 complementation groups. Virology 177: 646–656
3. Boursnell MEG, Brown TDK, Foulds IJ, Green PF, Tomley FM, Binns MM (1987) Completion of the sequence of the genome of the coronavirus avian infectious bronchitis virus. J Gen Virol 68: 57–77
4. Breedenbeek PJ, Pachuk CJ, Noten AFH, Charite A, Luytjes W, Weiss SR, Spaan WJM (1990) The primary structure and expression of the second open reading frame of the polymerase gene of the coronavirus MHV-A59: a highly conserved polymerase is expressed by an efficient ribosomal frameshifting mechanism. Nucleic Acids Res 18: 1825–1832
5. Brierley I, Digard P, Inglis SC (1989) Characterization of an efficient ribosomal frameshifting sequence: requirement for an RNA pseudoknot. Cell 57: 537–547
6. Denison MR, Zoltick PW, Hughes SA, Giangreco B, Olsen AL, Perlman S, Leibowitz JL, Weiss SR (1992) Intracellular processing of the N-terminal ORF 1a

proteins of the coronavirus MHV requires multiple proteolytic events. Virology 189: 274–284

7. Denison MR, Zoltick PW, Leibowitz JL, Pachuk CJ, Weiss SR (1991) Identification of polypeptides encoded in open reading frame 1b of the putative polymerase gene of the murine coronavirus mouse hepatitis virus A59. J Virol 65: 3076–3082

8. Devereux J, Haeberli P, Smithies O (1984) A comprehensive set of sequence analysis programs for the VAX. Nucleic Acids Res 12: 387–395

9. Gubler U, Hoffman BJ (1983) A simple and very efficient method for generating cDNA libraries. Gene 25: 263–269

10. Hierholzer JC, Tannock GA (1988) Coronaviridae: the coronaviruses. In: Lenette EH, Halonen P, Murphy FA (eds) Viral, rickettsial, and chlamydial diseases. Laboratory diagnosis of infectious disease-principles and practice, vol 2. Springer, Berlin Heidelberg New York Tokyo, pp 451–483

11. Isaacs D, Flowers D, Clarke JR, Valman B, Macnaughton MR (1983) Epidemiology of coronavirus respiratory infections. Arch Dis Child 38: 500–503

12. Kemp MC, Hierholzer JC, Harrison A, Burks JS (1984) Characterization of viral proteins synthesized in 229E infected cells and effect(s) of inhibition of glycosylation and glycoprotein transport. In: Rottier PJM, van dere Zeijst BAM, Spaan WJM, Horzinek MC (eds) Molecular biology and pathogenesis of coronavirus. Advances in experimental biology and medicine, vol 173. Plenum Press, New York, pp 65–79

13. Lee H-J, Shieh C-K, Gorbalenya AE, Koonin EV, LaMonica N, Tuler J, Bagdzhadzhyan A, Lai MMC (1991) The complete sequence (22 kilobases) of murine coronavirus gene 1 encoding the putative proteases and RNA polymerase. Virology 180: 567–582

14. Myint S, Harmsen D, Raabe T, Siddell S (1990) Characterization of a nucleic acid probe for the diagnosis of human coronavirus 229E infections. J Med Virol 31: 165–172

15. Parker MM, Masters PS (1990) Sequence comparison of the N genes of five strains of mouse hepatitis virus suggest a three domain structure for the nucleocapsid protein. Virology 179: 463–468

16. Phillpots JR (1983) Clones of MRC-C cells may be superior to the parent line for the culture of 229E-like strains of human respiratory coronaviruses. J Virol Methods 6: 267–269

17. Pinto LH, Holsinger LJ, Lamb RA (1992) Influenza virus M2 protein has ion channel activity. Cell 69: 517–528

18. Raabe T, Schelle-Prinz B, Siddell SG (1990) Nucleotide sequence of the gene encoding the spike glycoprotein of human coronavirus HCV 229E. J Gen Virol 71: 1065–1073

19. Rottier PJM, Welling GW, Welling-Wester S, Niesters HGM, Lenstra JA, van der Zeijst BAM (1986) Predicted membrane topology of the coronavirus protein E1. Biochemistry 25: 1335–1339

20. Sambrook J, Fritsch EF, Maniatis T (1989) Molecular cloning: a laboratory manual, 2nd edn. Cold Spring Harbor Laboratory, Cold Spring Harbor

21. Schmidt OW, Kenny GE (1982) Polypeptides and functions of antigens from human coronaviruses 229E and OC43. Infect Immun 35: 515–522

22. Spaan W, Rottier P, Smeekens S, van der Zeijst BAM, Delius H, Armstrong J, Skinner M, Siddell SG (1983) Coronavirus mRNA synthesis involves fusion of non-contigious sequences. EMBO J 2: 1839–1844

23. Staden R (1982) Automation of the computer handling of gel reading data produced by the shotgun method of DNA sequencing. Nucleic Acids Res 10: 4731–4751

Authors' address: Dr. S.G. Siddell, Institute of Virology, University of Würzburg, Versbacher Strasse 7, D-97078 Würzburg, Federal Republic of Germany.

Arch Virol (1993) [Suppl] 7: 75–80

Toroviruses – members of the coronavirus superfamily?

M.C. Horzinek

Divison of Virology, Department of Infectious Diseases and Immunology,
Veterinary Faculty, University of Utrecht, Utrecht, The Netherlands

Introduction

During the last twenty years in Utrecht we have focused on the characterization of enveloped positive-stranded RNA viruses, above all on the well-characterized family of coronaviruses, with the mouse hepatitis and feline infectious peritonitis viruses receiving most attention; in addition, less common and not properly classified agents e.g. equine arteritis virus (EAV) and Berne virus (BEV) have been analyzed. It was quite unexpected when nucleotide sequence comparisons suggested that EAV and BEV are phylogenetically related to each other and to coronaviruses; the properties of the latter have been reviewed [16, 17]. The present paper summarizes some data on toroviruses and gives the arguments why these are considered as evolutionarily related with corona- and arteriviruses.

Berne virus – the torovirus prototype

Berne virus (BEV) has become the prototype of a new taxon in virology [4]. Toroviruses represent a group of enveloped, positive-stranded RNA viruses with a unique and altogether protean morphology. An elongated, bacilliform core with two rounded ends is surrounded by a membrane which may either tightly adhere or "shrink-wrap" it; in the first instance straight or curved rods are formed, in the latter case a biconcave disk results [for reviews see 10, 11]. Virion pleomorphism in negatively stained preparation is probably the reason why these ubiquitous viruses have not been discovered earlier; in ultrathin sections, however, the characteristic "hollow" tubular capsids are conspicuous, especially when forming twin circular structures in transversal sections [22].

The history of torovirology is brief – it started with a study on the cause of calf diarrhoea. Breda virus (BRV) was discovered in 1979

during investigations in a dairy herd in Breda (Iowa), in which severe
neonatal calf diarrhea had been a problem for 3 subsequent years.
Faecal material was found to contain pleomorphic particles which carried
club-shaped projections similar to coronaviruses, but which were antigenically unrelated to them [23]. Despite repeated attempts, BRV had at
that time not been adapted to growth in cell or tissue culture which has
hampered its biochemical, biophysical and molecular characterization. In
the meanwhile, the problem of in-vitro propagation has been solved:
respiratory strains of bovine toroviruses were grown with cytopathic
effect in Madin-Darby bovine cell culture, and subsequently in a range
of other cells from different animal species [19].

Breda virus (BRV) had many properties in common with a chance
isolate from a horse presented at the Veterinary School in Berne,
Switzerland; the isolation had been made already in 1972, and early
characterization of BEV resulted from a collaboration between laboratories at Berne and Utrecht [21]. Toroviruses occur in rodents, carni-
vores, ungulates and man. Most infections are enteric, but antigen has
also been found in respiratory lesions of calves and in aborted bovine
fetuses [20].

Toroviruses and coronaviruses are related by divergence of their
polymerase and envelope proteins from common ancestors. In addition,
their genome organization and expression strategy, which involves the
synthesis of a 3′-coterminal nested set of mRNAs, are comparable.
Nucleotide sequence analysis of the genome of BEV has revealed the
results of two independent non-homologous RNA recombinations during
torovirus evolution. Berne virus open reading frame 4 encodes a protein
with significant sequence similarity (30–35% identical residues) to a part
of the hemagglutinin esterase proteins of coronaviruses and influenza
virus C. The sequence of the C-terminal part of the predicted BEV
polymerase open reading frame 1a product contains 31–36% identical
amino acids when compared with the sequence of a non-structural 30/
32K coronavirus protein. The cluster of coronaviruses which contains
this non-structural gene does not express it as a part of their polymerase,
but by synthesizing an additional subgenomic mRNA [14, 15].

Arteriviruses

Another virus which has occupied my groups in Tübingen and Utrecht
is EAV. As a result of early attempts at characterization, it became
the prototype of the genus arterivirus in the family *Togaviridae*, with
lactic dehydrogenase virus (LDV) as another possible member [4]. The

spherical enveloped arterivirion has a diameter of 50–70 nm and contains an isometric, probably icosahedral nucleocapsid 35 nm in diameter, as determined for EAV [8] and LDV [6]. The envelope carries ring-like surface structures – no peplomeric subunits like coronaviruses [7]. The similarity in molecular weights of the structural proteins was another trait that motivated their grouping in one cluster ("lactiviruses": [6]). The genome is a single positive-stranded RNA molecule, less than half the size of coronaviral genomes, which contains multiple ORFs. During EAV replication again a 3'-coterminal nested set of subgenomic RNAs appears. They are composed of leader and body sequences which are not contiguous on the EAV genome and perhaps are formed by alternative splicing. The leader sequence is derived from the extreme 5' end of the EAV genome [2]. Leader sequences are also present in coronaviruses, but have not been found in toroviruses.

We have recently shown that the EAV genome contains seven open reading frames (ORFs) and presented data on the structural proteins and the assignment of their respective genes. Virions are composed of a 14 kDa nucleocapsid protein (N) and 3 membrane proteins designated M, G_S, and G_L. M is an unglycosylated protein of 16 kDa, G_S and G_L are N-glycosylated proteins of 25 kDa and 30–42 kDa, respectively. Using monospecific antisera and expression of individual ORFs, the genes for the structural proteins were identified: ORF 7 codes for N, ORF 6 for M, ORF 5 for G_L, and ORF 2 for G_S. With the exception of G_S, the proteins are about equally abundant in EAV virions, being present at a molar ratio of 3(N):2(M):3(G_L). The G_S protein which is expressed at a level similar to that of M in infected cells, is strikingly under-represented (1–2%) in virus particles [3].

Study of arteriviruses at the molecular level has received a boost during the last two years, and genomic properties similar to those described above were identified for LDV [12] and Lelystad virus (LV), the cause of "mystery swine disease" or swine infertility and respiratory syndrome. A sequence of 15,088 nucleotides with eight ORFs was determined; ORFs 1a and 1b contain sequence elements that are conserved in the RNA polymerase genes of BEV, EAV and LDV, coronaviruses and of other positive-stranded RNA viruses [13]. We had shown before that in corona- and toroviruses the polymerase is translated from ORF1 by a process of ribosomal frame-shifting, for which a pseudoknot and a preceding "slippery" nucleotide sequence are preconditions [1, 14]; the same arrangement has now been found also for LV. Comparison of the amino acid sequences has indicated that LV is more closely related to LDV than to EAV [13]. The genome of Simian Haemorrhagic Fever virus, a fourth representative of this cluster, is presently being sequenced (Brian Mahy 1992, pers. comm.).

The "superfamily" concept

During the last decade, various "replicase modules" have been recognized among positive-stranded RNA viruses. They form the basis for "superfamilies" of plant and animal viruses, the two largest of which are those of the picornavirus- and alphavirus-like "superfamilies" [18]. More recently, we have proposed a third "superfamily" [17] which would comprise the corona-, toro- and arteriviruses. The importance of at least three conserved domains in corona- and toroviral polymerases [14] is underlined by their presence in the putative polymerase of the only distantly related EAV. Two domains are believed to possess polymerase and helicase activities and are common in positive-stranded RNA viruses. However, the conservation, both in sequence and in relative position, of a third domain suggests that it also plays a role in viral replication – a hypothesis supported by the fact that it also occurs in the putative polymerase of LDV. It is tempting to speculate that this domain performs a function which is specific for this cluster of viruses, e.g. the synthesis of multiple subgenomic mRNAs.

On the basis of the similarities in polymerase expression and amino acid sequence we postulate that the polymerase genes of arteri-, corona- and toroviruses have descended from a common ancestor. In contrast to the helical nucleocapsid structure of corona- and toroviruses, the nucleocapsids of EAV, like those of togaviruses, possess an isometric, probably icosahedral architecture. Also, the EAV envelope does not bear elongated peplomers. The coupling of different sets of structural genes to the same replicase has been explained by recombination of complete genes or gene sets (modules). Together with divergence from a common ancestor, this modular evolution can account for the diverse composition of viral genomes [18]. In arteriviruses, the coupling of a coronavirus-like replicase module to a set of structural genes – which confer togaviral morphology to the virion – might be another example of modular evolution.

Concluding remarks

There is another line of evolution that should be mentioned here: that of viral taxonomy. While virion details have dominated classification in the early days, genome organization and replication strategy are now en vogue. Indeed the traits shared between corona- and toroviruses have resulted in their classification as genera of the family Coronaviridae, as agreed upon at the interim meeting of the ICTV (Oxford, April 1992). However, taxonomy should not only be intellectually appealing but also practical, and structural features must not be abolished altogether. Lumping divergent viruses together in families only because their replication contains similar elements does a disservice to the community of virologists at large, which includes teachers and diagnosticians. Arteriviruses are structurally unrelated to the Coronaviridae and should be given separate status – a feeling unanimously expressed during the 5th International Coronavirus Symposium (Chantilly, September 1992). The hierarchical category of the order would be appropriate to replace the unofficial category of "superfamily".

Acknowledgements

The author should like to thank Miss Janneke Meulenberg, Central Veterinary Institute Lelystad, and Dr. E.Vanopdenbosch, The National Institute for Veterinary Research Brussels, for sharing unpublished information. My former collaborators Willy Spaan, Eric Snijder, Johan den Boon (Leiden, The Netherlands) and Marion Koopmans (Atlanta, USA) have contributed most of the original data reviewed above, and their continued effort and interest is gratefully acknowledged.

References

1. Bredenbeek PJ, Pachuk CJ, Noten JFH, Charité J, Luytjes W, Weiss SR, Spaan WJM (1990) The primary structure and expression of the second open reading frame of the polymerase gene of coronavirus MHV-A59. Nucleic Acids Res 18: 1825–1832
2. de Vries AAF, Chirnside ED, Bredenbeek PJ, Gravestein LA, Horzinek MC, Spaan WJM (1990) All subgenomic mRNAs of equine arteritis virus contain a common leader sequence. Nucleic Acids Res 18: 3241–3247
3. de Vries AAF, Chirnside ED, Horzinek MC, Rottier PJM (1992) The structural proteins of equine arteritis virus. J Virol 66: 6294–6303
4. Francki RIB, Fauquet CM, Knudson DL, Brown F (1991) Classification and nomenclature of viruses. 5th Report of the International Committee on Taxonary of Viruses. Springer, Wien New York (Arch Virol [Suppl] 2)
5. Godeny EK, Speicher DW, Brinton MA (1990) Map location of lactate dehydrogenase-elevating virus (LDV) capsid protein (Vp1) gene. Virology 177: 768–771
6. Horzinek MC (1975) The structure of togaviruses and bunyaviruses. Med Biol 53: 406–411
7. Horzinek MC (1981) Non-arthropod-borne togaviruses. Academic Press, London
8. Horzinek M, Maess J, Laufs R (1971) Studies on the substructure of togaviruses. II. Analysis of equine arteritis, rubella, bovine viral diarrhoea, and hog cholera viruses. Arch Ges Virusforsch 33: 306–318
9. Horzinek MC, Van Wielink PS, Ellens DJ (1975) Purification and electron microscopy of lactic dehydrogenase virus of mice. J Gen Virol 26: 217–226
10. Horzinek MC, Flewett TH, Saif LF, Spaan WJM, Weiss M, Woode GN (1987a) A new family of vertebrate viruses: Toroviridae. Intervirology 27: 17–24

11. Horzinek MC, Weiss M, Ederveen J (1987b) Toroviridae: a proposed new family of enveloped RNA viruses. In: Bock G, Whelan J (eds) Novel diarrhoea viruses. CIBA Foundation Symp 128. Wiley, Chichester, pp 162–174
12. Kuo L, Chen Z, Rowland RRR, Faaberg KS, Plagemann PGW (1992) Lactate dehydrogenase-elevating virus (LDV): subgenomic mRNAs, mRNA leader and comparison of 3'-terminal sequences of two LDV isolates. Virus Res 23: 55–72
13. Meulenberg JJM, Hulst MM, de Meijer EJ, Moonen PLJM, den Besten A, de Kluyver EP, Wensfoort G, Moorman RJM (1992) Lelystad virus, the causative agent of porcine epidemic abortion and respiratory syndrome (PEARS), is related to LDV and EAV. Virology 192: 62–72
14. Snijder EJ, den Boon JA, Bredenbeek PJ, Horzinek MC, Rijnbrand R, Spaan WJM (1990) The carboxy-terminal part of the putative Berne virus polymerase is expressed by ribosomal frameshifting and contains sequence motifs which indicate that toro- and coronaviruses are evolutionarily related. Nucleic Acids Res 18: 4535–4542
15. Snijder EJ, den Boon JA, Horzinek MC, Spaan WJM (1991) Comparison of the genome organization of toro- and coronaviruses: evidence for two non-homologous recombination events during Berne virus evolution. Virology 180: 448–452
16. Spaan W, Cavanagh D, Horzinek MC (1988) Coronaviruses: structure and genome expression. J Gen Virol 69: 2939–2952
17. Spaan W, Cavanagh D, Horzinek MC (1990) Coronaviruses. In: van Regenmortel MHV, Neurath AR (eds) Immunochemistry of viruses, vol 2. The basis for sero-diagnosis and vaccines. Elsevier Science Publishers, Amsterdam, pp 359–375
18. Strauss JH, Strauss EG (1988) Evolution of RNA viruses. Annu Rev Microbiol 42: 657–683
19. Vanopdenbosch E, Wellemans G, Charlier G, Petroff K (1992a) Bovine torovirus: cell culture propagation of a respiratory isolate and some epidemiological data. Valams Tijschr Diergeneesk 61: 45–49
20. Vanopdenbosch E, Wellemans G, Oudewater J, Petroff K (1992b) Prevalence of torovirus infections in Belgian cattle and their role in respiratory, digestive and reproductive disorders. Vlaams Tijschr Diergeneesk 61: 187–191
21. Weiss M, Steck F, Horzinek MC (1983) Purification and partial characterization of a new enveloped RNA virus (Berne virus). J Gen Virol 64: 1849–1858
22. Weiss M, Horzinek MC (1987) The proposed family Toroviridae: agents of enteric infections. Arch Virol 92: 1–15
23. Woode GN, Reed DE, Runnels PL, Herrig MA, Hill HT (1982) Studies with an unclassified virus isolated from diarrhoeal calves. Vet Microbiol 7: 221–240

Author's address: Dr. M.C. Horzinek, Division of Virology, Dept. Infectious Diseases and Immunology, Veterinary Faculty, University of Utrecht, P.O. Box 80.165, Yalelaan 1, 3508 TD Utrecht, The Netherlands.

Arch Virol (1993) [Suppl] 7: 81–100

Molecular biology and evolution of filoviruses

H. Feldmann[1,2], **H.-D. Klenk**[1], and **A. Sanchez**[2]

[1] Institut für Virologie, Philipps-Universität, Marburg, Federal Republic of Germany
[2] Special Pathogens Branch, Division of Viral and Rickettsial Diseases, National Center for Infectious Diseases, Centers for Disease Control, Atlanta, GA, U.S.A.

Summary. The family *Filoviridae* contains extremely pathogenic human viruses causing a fulminating, febrile hemorrhagic disease. Filoviruses are enveloped, filamentous particles with a nonsegmented negative-strand RNA genome showing the gene arrangement 3'-NP-VP35-VP40-GP-VP30-VP24-L-5'. Genes are flanked by highly conserved transcriptional signals and are generally separated by variable intergenic regions. They are transcribed into monocistronic polyadenylated messenger RNAs which contain relatively long 5' and 3' untranslated regions. Seven structural proteins are encoded by the genome of which four form the helical nucleocapsid (NP-VP35-VP30-L), two are membrane-associated (VP40-VP24), and one is a transmembrane glycoprotein (GP). Comparison of filovirus genomes with those of other nonsegmented negative-strand RNA viruses suggest comparable mechanisms of transcription and replication and a common evolutionary lineage for all these viruses. Sequence analyses of single genes, however, showed that filoviruses are more closely related to paramyxoviruses, particularly human respiratory syncytial virus. These data support the concept of the taxonomic order *Mononegavirales* for all nonsegmented negative-strand RNA viruses and the classification of Marburg virus, Ebola virus, and Reston virus in the family *Filoviridae*, separate from the families *Paramyxoviridae* and *Rhabdoviridae*.

Classification

The family *Filoviridae* [23] consists of Marburg virus (MBG), two related subtypes of Ebola virus (EBO), and a recently isolated EBO-like virus, called Reston virus (RES). Filoviruses are filamentous, enveloped, nonsegmented negative-strand (NNS) RNA viruses. They are grouped together with the two other families of viruses with linear undivided negative-sense genomes, *Paramyxoviridae* and *Rhabdoviridae*, to comprise the order *Mononegavirales* [17]. MBG and EBO, prototypes of this family, are extremely pathogenic for human and nonhuman primates and often cause a fulminating hemorrhagic disease with a severe shock syndrome and high mortality [29, 47].

History

MBG was the first filovirus to be discovered, and was isolated from laboratory workers who had been exposed to tissues and blood from African green monkeys (Cercopithecus aethiops) imported from Uganda into Europe. Twentyfive primary cases and six secondary infections were reported in the outbreak, seven of the primary cases died [30, 45]. African green monkeys experimentally inoculated with the virus all died. Since then, sporadic cases of MBG disease in man have occurred in various parts of Africa (Table 1): Zimbabwe/South Africa 1975 [14], Kenya 1980 [48], and Kenya 1987 [24].

EBO was first recognized in 1976 when major outbreaks occurred simultaneously in Zaire (318 cases and 290 deaths) [21, 53] and Sudan

Table 1. Outbreaks of hemorrhagic disease caused by filoviruses and documented by virus isolation

Location	Year	Species affected	Virus	Source	Human cases (deaths)
West Germany/ Yugoslavia[1]	1967	human/monkey	MBG	monkey (Uganda)	31 (7)
Zimbabwe/ South Africa[2]*	1975	human	MBG	?	3 (1)
Zaire[3]	1976	human	EBO	?	318 (290)
Sudan[4]	1976	human	EBO	?	284 (150)
Zaire[5]	1977	human	EBO	?	1 (1)
Sudan[6]	1979	human	EBO	?	34 (22)
Kenya[7]	1980	human	MBG	?	2 (1)
Kenya[8]	1987	human	MBG	?	1 (1)
U.S.A. (VA/PA)[9]	1989	monkey	RES	monkey (Philippines)	4 (0)
Italy[10]	1992	monkey	RES	monkey (Philippines)	0 (0)

Besides the well documented episodes listed in this table two more suspected fatal and two nonfatal cases of EBO hemorrhagic disease were reported (22, 39, 49). [1](30), [2](14), [3](53), [4](54), [5](16), [6](1, 55), [7](48), [8](24), [9](19), and [10](56). * The index case was apparently infected in Zimbabwe but became ill in South Africa. *VA* Virginia; *PA* Pennsylvania

(284 cases and 150 deaths) [54], and again in 1979 in Sudan (34 cases and 22 deaths) [1, 55]. Another single case was confirmed by virus isolation in Zaire 1977 [16]. Two subtypes were isolated (EBO-Zaire; EBO-Sudan) which differ in pathogenicity, antigenicity and genomic composition. They were morphologically identical with but serologically distinct from MBG [3, 6]. Besides these episodes documented by virus isolation, two more fatal and two nonfatal cases have been reported [22, 39, 49]. No association with monkeys could be attributed to any of the EBO outbreaks (Table 1).

Another filovirus, called RES, which is serologically related to EBO, was isolated in 1989 from cynomolgus monkeys (Macaca fascicularis) imported from the Philippines into the United States [19] (Table 1). Pathogenicity of RES in these animals was uncertain due to the concurrent simian hemorrhagic fever virus infection known as a severe pathogen for macaques [38]. However, studies on experimentally infected monkeys showed that RES is less pathogenic for primates than MBG and both subtypes of EBO [13]. During the 1989 epizootic four animal caretakers with high exposure to the infected monkeys seroconverted, and virus was isolated from one case. None of the four had a febrile illness. These cases indicate that RES is infectious for humans but it appears to have lower pathogenicity causing no serious human disease [18, 31, 33, 34]. Recently (March, 1992), a filovirus was isolated from cynomolgus monkeys imported from the Philippines into Italy. These monkeys were obtained from the same exporter that shipped RES-infected monkeys into the United States in the 1989 outbreak (Table 1). Human infections were not observed during this outbreak [56].

The natural reservoirs of filoviruses are unknown. MBG and EBO appear to be indigenous to the African continent. The origin of the original EBO-like virus (RES) infecting macaques in the Philippines has not been definitively established, but this epizootic suggests the possibility that Asian filoviruses exist. Serological studies suggest that EBO or related viruses may be endemic in Zaire, Sudan, the Central African Republic, Gabon, Nigeria, Ivory Coast, Liberia, Cameroon, and Kenya. This habitat may extend to other African countries where seroprevalences have not been examined [18, 20, 46, 50]. RES and the recently isolated RES-like filovirus, however, were isolated from animals originating from the Philippines. Thus it appears that filoviruses are not restricted to the African continent. Recent serological investigations on sera of humans with varying levels of exposure to monkeys performed in the USA [32] and in Germany [2] suggest that subclinical infections with filoviruses or related agents may also occur in these countries. These observations may indicate that hitherto unknown filoviruses with varying pathogenic potential may be found in many parts of the world.

Morphology

Filovirus particles are pleomorphic, appearing as long filamentous, sometimes branched forms, or as "U"-shaped, "6"-shaped, or circular forms. They are morphologically similar to rhabdovirus particles but much longer. Virions vary greatly in length but have a uniform diameter of approximately 80 nm. Viral particles purified by rate zonal gradient centrifugation are bacilliform and have an average length of 665 nm for MBG and 805 nm for EBO; these monomorphic virion structures are associated with peak infectivity (Fig. 1a). Except for the difference in length, filoviruses seem to be very similar in morphology. Particle cores are formed by a nucleocapsid consisting of a dark central space (20 nm in

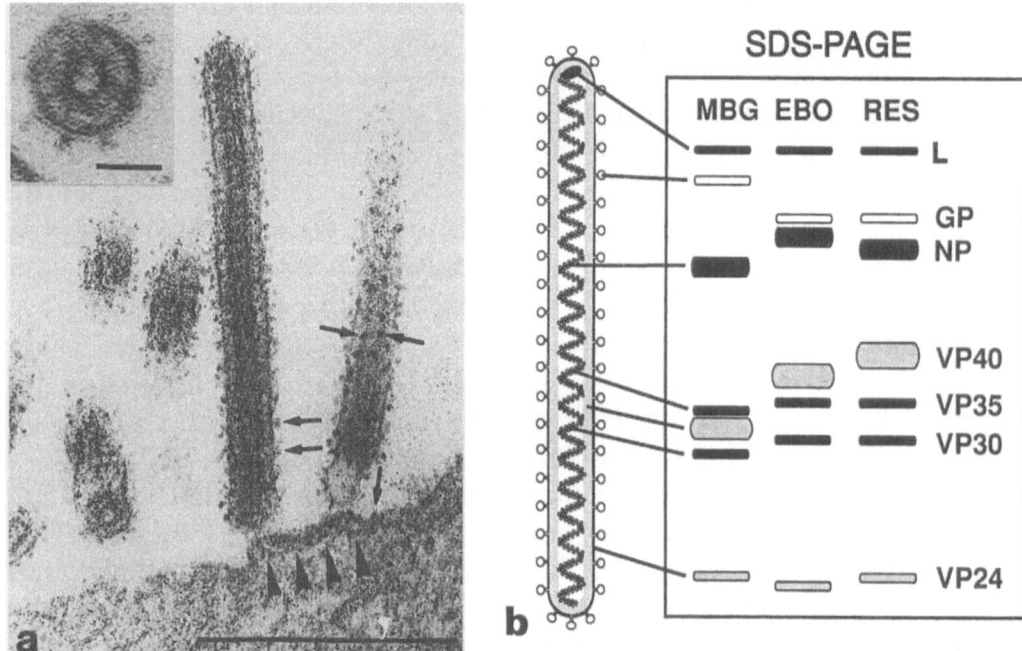

Fig. 1. Structure of filovirus particles. **a** Electron micrograph of Marburg virus particles. Ultrathin sections obtained from primary cultures of human endothelial cells three days after infection have been analyzed by transmission electron microscopy. Particles consist of a nucleocapsid surrounded by a membrane in which spikes are inserted (arrows). The nucleocapsid contains a central channel (inset). The plasma membrane of infected cells is often thickened at locations were budding occurs (arrowheads). Bar: 0.5 μm; bar inset: 50 nm). **b** Filovirus structural proteins. The nonsegmented negative-strand RNA genome is encapsidated by the nucleoprotein (*NP*). Associated with the ribonucleocapsid complex are the viral polymerase (*L*) and the viral structural proteins (*VP*) 35 and 30. VP40 and VP24 are membrane-associated proteins, and the spikes are formed by the glycoprotein (*GP*). Differences in the electrophoretic mobility patterns (*SDS-PAGE*) of filovirus structural proteins are schematically illustrated

diameter) surrounded by a helical capsid (50 nm in diameter) bearing cross-striations with a periodicity of about 5 nm. Within the nucleocapsid an axial channel of 10–15 nm is located (Fig. 1a; inset). The helical nucleocapsid is surrounded by a lipid envelope derived from the host cell plasma membrane. Spikes of approximately 7 nm in length, spaced at about 10 nm intervals are exposed on the virion surface [23, 36, 40] (Fig. 1a).

Proteins

Virion particles contain at least seven proteins with presumed identical functions for the different viruses (Table 2). The electrophoretic mobility patterns (SDS-PAGE) of the structural proteins are characteristic for MBG isolates on one hand and EBO and EBO-like viruses on the other. Differences in these patterns are most prominent in the migration of the glycoprotein (GP) and the viral structural proteins (VP) 40, 35, and 30 (Fig. 1b). Four proteins are associated with the viral ribonucleocapsid complex (RNP) (the nucleoprotein (NP), VP 35 and 30, and the large (L) protein), two proteins are associated with the viral membrane (VP40

Table 2. Filovirus structural proteins and their proposed functions

Designation	MW (MBG) A	B	MW (EBO) A	B	Encoded by gene	Proposed function
L	267.2 K	180 K	–	180 K	7	RNA-dependent RNA polymerase: transcription and replication
GP	74.8 K	170 K	74.5 K	125 K	4	glycoprotein: forms viral spikes, mediates virus entry
NP	77.9 K	96 K	83.3 K	104 K	1	major nucleoprotein: encapsidation
VP40	31.7 K	38 K	35.3 K	40 K	3	matrix protein: membrane-associated
VP35	31.0 K	32 K	38.8 K	35 K	2	transcriptase component: P protein analogue
VP30	31.5 K	28 K	29.7 K	30 K	5	minor nucleoprotein: ribonucleoprotein-associated
VP24	28.8 K	24 K	28.3 K	24 K	6	second matrixprotein: membrane-associated

MW/A Molecular weight calculated from the deduced amino acid sequences of the open reading frames of the seven filovirus genes; *MW/B* molecular weight estimated from SDS-PAGE analysis

and VP24), and one is inserted in the lipid envelope (GP) (Fig. 1b; Table 2). Metabolic labeling using different [^{3}H]-labeled carbohydrates demonstrated the presence of only one glycoprotein in mature particles. Labeling with [^{32}P]-orthophosphate revealed in two phosphorylated proteins, the NP (major phosphoprotein of the virion) and a second, weakly phosphorylated protein. This second phosphoprotein was identified as the VP30 for EBO [9]. Nonstructural proteins have not yet been identified.

L protein

The L protein functions as the viral RNA-dependent RNA polymerase. For MBG (Musoke strain) it has a predicted molecular weight (MW) of 267 kDa (Table 2). The MBG L protein shows a high content of leucine and isoleucine residues, a large positive net charge (+56), clusters of basic amino acids (putative RNA binding domains), and several putative ATP binding domains [35]. These features were also found with other NNS RNA virus L proteins. Sequencing data on other filovirus L genes are not yet available. Computer-assisted comparisons of L proteins of NNS RNA viruses revealed significant homologies, primarily in the amino-terminal halves of the proteins which seem to be the functional domains of these polymerases. The variable carboxy-terminal halves of the proteins on the other hand seem to carry virus-specific functions [35].

Glycoprotein

The GP (MBG 170 kDa [11]; EBO 125 kDa [9]; RES 125 kDa) (Table 2) is an integral membrane protein and forms the surface projections of the virion particle [24]. Since the GP is the only membrane protein exposed on the viral surface, it is reasonable to assume that it is responsible for receptor binding and membrane fusion. The GP of MBG has the structural features of a type I transmembrane protein [52] and is inserted in the lipid membrane as a homotrimer [11]. The carbohydrate structures of the highly glycosylated MBG GP account for more than 50% of the MW of the mature protein [52]. The structures include oligomannosidic and hybrid type N-glycans as well as bi-, tri-, and tetraantennary complex species, and high amounts of neutral mucin-type O-glycans (type-1 and type-2 core structures). Sialic acid residues are absent on both N- and O-glycans [11, 15] (Fig. 2).

Detailed structural analyses of the carbohydrates of EBO and RES GP are not available. However, lectin binding studies also showed

	GNA		DSA				PNA				SNA		MAA					
	Endo H		Endo F/PNGase F		O-Glyc.		O-Glyc.		NA		NA		NA		Endo F/PNGase F		O-Glyc.	
	-	+	-	+	-	+	-	+	-	+	-	+	-	+	-	+	-	+
MBG	+	-	+	-	+	+	+	+	+	+	-	-	-	-	-	-	-	-
EBO	+	-	+	-	+	+	+	+	+	++	-	+	-	+	+	+	+	+
RES	+	-	+	-	+	+	+	+	+	++	-	+	-	+	+	+	+	+

Fig. 2. Analysis of carbohydrate structures of filovirus glycoproteins using a lectin binding assay. Viral proteins were subjected to SDS-PAGE (10%) with or without prior treatment of glycohydrolases, blotted onto nitrocellulose membranes, incubated with digoxigenin-labeled lectins, and detected using an anti-digoxigenin antibody [for details see 11]. *MBG* Marburg virus; *EBO* Ebola virus; *RES* Reston virus; *GNA* galanthus nivalis agglutinin; *DSA* datura stramonium agglutinin; *PNA* peanut agglutinin; *SNA* sambucus nigra agglutinin; *MAA* maackia amurensis agglutinin; *Endo H* endoglycosidase H; *Endo F/PNGase F* endoglycosidase F/N-glycosidase F; *O-Glyc.* O-glycosidase (endo-α-N-acetylgalactosaminidase; *NA* neuraminidase (vibrio cholerae). Lectin specificities: GNA – Manα(1–3)Man (α1–3 > α1–6 > α1–2); DSA – Galβ(1–4)GlcNAc / GlcNAc-Ser/Thr; PNA – Galβ(1–3)GalNAc; SNA – NeuNAcα(2–6)Gal/GalNAc; MAA – NeuNAcα(2–3)Gal

that these glycoproteins contain N- and O-linked oligosaccharides with similar structures as found on the MBG GP (Fig. 2). In contrast to MBG GP, the glycoproteins of EBO and RES are substituted with terminal α(2–3)-linked sialic acid residues on N- as well as O-glycans (Fig. 2). The difference in sialylation appears not to be due to an intrinsic neuraminidase activity of MBG particles (C. Will, unpubl. data).

Amino acid sequence comparison of MBG, EBO, and RES GPs showed conservation at the N- and C-terminal ends of the proteins (Fig. 3D). Hydrophobic regions were found in these conserved areas, which correspond to the signal sequence at the N-terminus and the transmembrane region at the C-terminus [42, 51, 52]. The middle part is variable, extremely hydrophilic and shows a high antigenicity index (Fig. 3D). This part carries the bulk of the glycosylation sites for N- and O-linked carbohydrates. The comparison also demonstrated the close relationship between EBO and RES (data not shown). Significant homologies among filovirus GPs and envelope proteins of other NNS RNA viruses do not exist. However, a region of 26 amino acids in the external domain closely located to the transmembrane region shows significant homology to a domain observed in the envelope proteins of several retroviruses [51, 52]. This domain is purported to be responsible for certain immunosuppressive properties of these retroviruses [4].

Nucleoprotein

The NP is the major component of the viral nucleocapsid. NP molecules of the different filoviruses differ slightly in their electrophoretic mobility patterns (Fig. 1b). The MW calculated from the deduced amino acid sequences of the NP genes of MBG and EBO are 78 kDa [44] and 83 kDa [43] (Table 2), respectively. The NP proteins are highly acidic with net charges of -28 for MBG and -30 for EBO. They can be divided into a hydrophobic N-terminal half which may play a role in either folding and/or RNA binding, and a hydrophilic and acidic C-terminal half which may interact with the matrix protein and/or other viral structural proteins [43, 44; A. Randolf, unpubl. data]. Comparisons of filovirus NPs revealed weak nucleotide sequence similarity, but the predicted sequence of the first 400 amino acids showed a high degree of homology. A small region in the middle of filovirus NP sequences was found to contain a significant amino acid homology with NPs of para- myxoviruses and to a lesser extent with rhabdoviruses. This region contains sequences previously identified as highly conserved within the family *Paramyxoviridae* [8].

Viral structural proteins 40, 35, 30, and 24

Little is known about the functions of the remaining four structural proteins (Fig. 1b; Table 2). VP30 (MBG 28 kDa [24]; EBO 30 kDa [9]) may represent a minor nucleoprotein (Fig. 1b; Table 2). In vitro protein/ protein interaction studies have shown that the MBG VP30 binds to the NP protein (A. Randolf, unpubl. data). Both proteins, NP and VP30, seem to be intimately associated with the virion RNP [9; A. Randolf, unpubl. data].

VP35 (MBG 32 kDa [24]; EBO 35 kDa [9]) may be a component of the transcriptase complex, loosely associated with the RNP [9, 24] (Fig. 1b; Table 2). It shows a weak binding to the NP protein in vitro as was found for MBG (A. Randolf, unpubl. data). Since VP35 is encoded by the second gene and is removed from the RNP with increasing salt conditions, it may be functionally analogous to the P proteins of paramyxo- and rhabdoviruses (Fig. 4) [9, 10, 24, 42].

VP40 (MBG 38 kDa [24]; EBO 40 kDa [9]) and VP24 (MBG 24 kDa [24]; EBO 24 kDa [9]) are membrane-associated proteins (Fig. 1b; Table 2). They are removed from the RNP under even isotonic conditions demonstrating that they are not components of the RNP [9, 12]. VP40 is the most prominent viral structural protein, and its weak association with the RNP suggests a role as the matrix protein of filoviruses (Fig. 4).

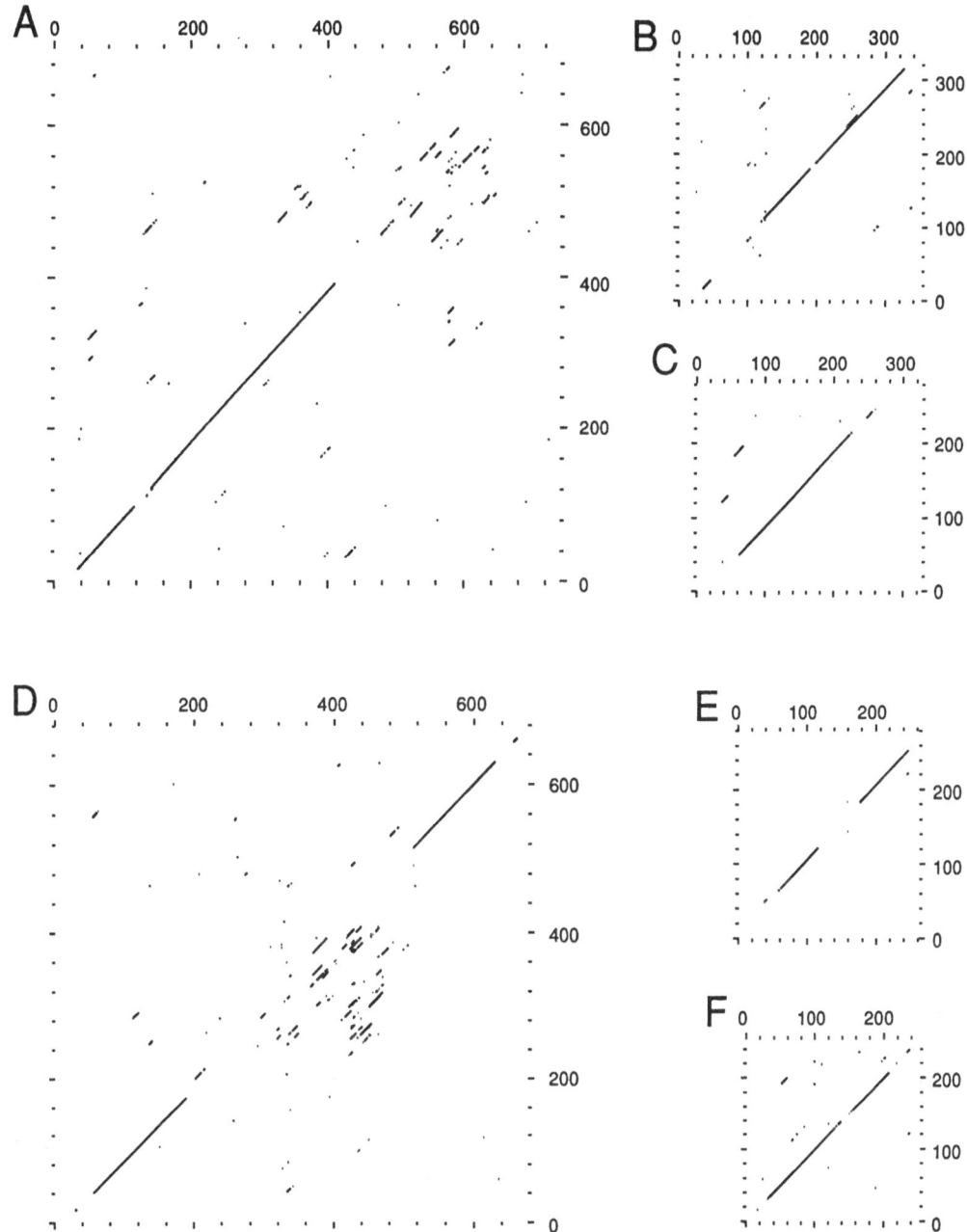

Fig. 3. Sequence homologies of Marburg and Ebola virus proteins. The DotPlot program of the GCG Sequence Analysis Software Package (7) was used with a window setting of 30 residues and a stringency setting of 15 matches. The Ebola virus proteins are shown on the horizontal and the Marburg virus porteins on the vertical axis.
A NP, **B** VP35, **C** VP40, **D** GP, **E** VP30, and **F** VP24

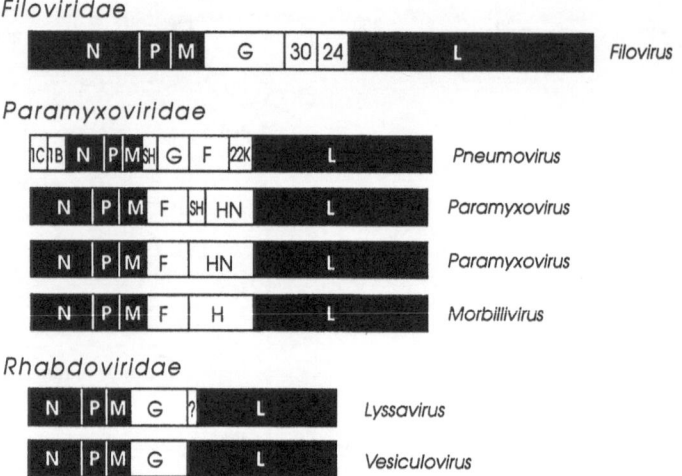

Fig. 4. Conservation in genome organization of viruses in the order *Mononegavirales*. The genome organizations of different genera in the order *Mononegavirales* and the genome sizes are schematically illustrated. Conserved parts present in all genomes are indicated by black boxes. Variable parts are indicated by white boxes. *1C* and *1B* Genes of unknown function; *N* nucleoprotein gene; *P* phosphoprotein gene; *M* nonglycosylated membrane protein gene; *G/H/HN/F* glycosylated membrane protein genes; *SH* small hydrophobic protein gene; *22K* nonglycoslyated membrane protein gene; *30* (VP30) viral structural protein gene 30 (ribonucleoprotein-associated protein); *24* (VP24) viral structural protein gene (membrane-associated protein); *?* pseudogene in the G-L intergenic region; *L* RNA-dependent RNA polymerase gene

Genome

The genomes of filoviruses consist of a single negative-strand RNA molecule which is noninfectious, not polyadenylated, and complementary to viral messenger RNA (mRNA) [10, 23, 41]. The first complete sequence of a filovirus genome was determined using the Musoke strain [48] of MBG [EMBL Data Library, emnew:MVREPCYC (accession number Z12132)] [10]. Partial sequencing analyses of the genome of the Mayinga strain of EBO (subtype Zaire) have also been performed [42, 43, 51]. Filovirus genomes are approximately 19 kb in length and very rich in adenosine and uridine residues. They are larger than the published sequences of other negative-strand RNA viruses (e.g. influenza A virus 13.5 kb, bunyavirus 12.3 kb, rhabdovirus 11.2 kb, paramyxovirus 15.5 kb).

Filovirus genomes show a linear gene arrangement in the order 3'-NP(N)-VP35(P)-VP40(M)-GP(G)-VP30-VP24-L-5' [10, 42] (Fig. 4; Table 2). Genes are generally separated by intergenic regions which all vary in length and nucleotide composition. This corresponds to the variability in intergenic regions of some paramyxo- and rhabdoviruses whereas it is in contrast to the conservations of others. Gene overlaps are present in filovirus genomes, EBO shows three (VP35/VP40; GP/VP30; VP24/L) and MBG one (VP30/VP24). Overlapping genes have also been found with respiratory syncytial virus (22K/L) [5] but not with any other NNS RNA virus.

At the 3'ends of all filovirus genes highly conserved transcriptional start signals are present which are fourteen nucleotides in length and show the consensus sequence: 3'-NNCUNCNUN*UAAUU*-5' (Fig. 5) (MBG: five out of fourteen are variable; EBO: three out of fourteen are variable). The last eleven (twelve) nucleotides of the 5'ends of all filovirus genes are nearly identical (3'-*UAAUU*CUUUUU(U)-5') (Fig. 5). Exceptions are the VP40 of MBG [C at position 2 instead of an A; (−) sense] and some EBO genes which show a run of six uridine residues at the 5'end instead of five. These sequences serve as the transcription termination signals for the viral polymerase. Conserved transcription signals are a feature common to all NNS RNA viruses. Comparative analyses based on the consensus sequences of those signals revealed a limited similarity in the start sites but an obvious conservation in the termination sites. The conservation is primarily found at the 5'end of the termination signals always containing a run of uridine residues (four to seven). This is assumed to be the site where stuttering of the viral polymerase occurs in the polyadenylation of transcripts. A filovirus-specific feature is the presence of the highly conserved pentamer 3'-*UAAUU*-5' at the 5'ends and the 3'ends of all transcription start and termination signals, respectively (Fig. 5) [10, 35, 42–44]. The function of this pentamer is unknown, but it could serve as the recognition site for the polymerase, whereas the surrounding semiconserved regions may direct the exact initiation of transcription and the termination/polyadenylation event.

Extragenic sequences are present at the 3' and 5'ends of filovirus genomes [10, 25, 35, 43, 44]. These sequences are presumed to correspond to those found in the genomes of all NNS RNA viruses and are known as (+) and (−) leader RNAs (Fig. 5). It has been proposed for other NNS RNA viruses, that transcription first begins with the synthesis of a (+) leader RNA [26–28]. However, neither (+) nor (−) leader RNAs have been detected in filovirus-infected cells. Comparison of these regions to those of other NNS RNA viruses showed two conserved regions which could serve as functional domains involved in transcription and/or replication, namely an encapsidation signal and an entry signal for the polymerase complex (Fig. 5) [10]. Filovirus genomes are transcribed to yield seven monocistronic subgenomic RNA (mRNA) species which are complementary to the viral genomic RNA (Fig. 5) [10, 24, 41, 42]. The 5'ends of the mRNAs start at the transcription start signal

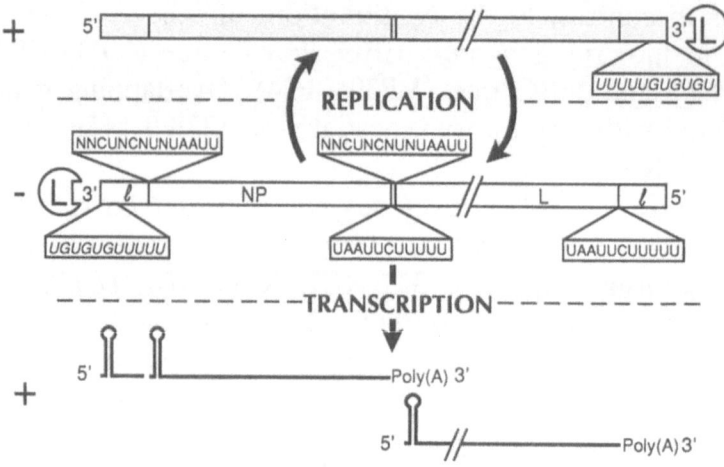

Fig. 5. Model for filovirus transcription and replication. The model illustrates the mode of filovirus transcription and replication based on the data available to date. In the middle part the nonsegmented (−) sense genome is shown with the transcriptional start signals (3′-NNCUNCNUN*UAAUU*-5′) indicated above and the termination signals (3′-*UAAUU*CUUUUU-5′) underneath. For replication a full-length (+) sense antigenome is synthesized which serves as the template for the synthesis of progeny (−) sense RNA anticomplementary to the parental RNA (upper part). The genome extremities are complementary as indicated by the sequence 3′-*UGUGUGUUUUU*-5′ at the 3′ends of the (−) sense genome and the (+) sense antigenome. This domain could provide the encapsidation site and/or the entry site for the viral polymerase (*L*). Transcription starts at the 3′end of the (−) sense genome and leads to polyadenylated mRNA transcripts (lower part). All transcripts show strong secondary structures at their 5′ends. *l* 3′ and 5′untranslated region of the genome; *NP* nucleoprotein; *L* viral RNA-dependent RNA polymerase

sequences and the 3′ends carry a poly(A) tail generated by the polymerase at uridine residues located at the 5′ends of all transcription termination signals (Fig. 5). Filoviruses contain unusually long untranslated regions at the 5′ and/or 3′ends of many of their mRNAs, which is generally not the case with other NNS RNA viruses [10, 35, 42–44]. The role of these long noncoding regions is unknown. Six mRNA species have been identified from filovirus-infected cells by in vivo labeling or Northern blot hybridization. In vitro translation of these mRNAs resulted in products comigrating with the structural proteins NP, VP40, VP35, VP30, VP24, and the unglycosylated form of the GP. A mRNA specific for the L protein has not been detected, presumably due to a low copy number in infected cells [10, 24, 41].

Transcription and replication of filoviruses take place in the cytoplasm of infected cells (Fig. 5). From the data available it is unclear, whether the strategies of filovirus transcription and replication are similar to other members of the order *Mononegavirales*. The 3'leader region of the genome probably provides the encapsidation site for the NP and the entry site for the polymerase (Fig. 5). Transcription of monocistronic mRNAs initiates at highly conserved start signals and ends with polyadenylation at a run of uridine residues located at the 5'end of the termination signals. During this process, the pentamer *3'-UAAUU-5'* might serve as a recognition signal for the polymerase wheras an upstream (start signal) or downstream (termination signal) semiconserved regions might initiate or terminate transcription, respectively. Transcription efficiency might be influenced by the following observations: (1) gene order, (2) formation of secondary structures at the 3'ends of the genes, (3) secondary structure formation within intergenic sequences, (4) overlapping genes, and (5) presence of two termination sites as found for the VP24 of EBO [10, 42]. Replication of the genome is mediated by the synthesis of a full-length complementary antigenome [(+) sense] which serves as a template for the synthesis of progeny negative-strand RNA anticomplementary to the parental template RNA (Fig. 5). The fact that the extremities of the genomes are complementary suggest a single identical encapsidation site on the genome and antigenome and an identical entry signal for the polymerase for both transcription as well as replication mode [10, 35, 43, 44].

Evolutionary relationship

Sequence analyses of the genomes of MBG and EBO clearly demonstrate that filoviruses and other NNS RNA viruses are closely related genetically, which supports the classification of the families *Filoviridae*, *Paramyxoviridae*, and *Rhabdoviridae* in the order *Mononegavirales* [17]. The genome organizations of NNS RNA viruses show a similar linear arrangement of genes according to functions (Fig. 4). This arrangement is as follows: 3'leader region – core protein genes – envelope protein genes – RNA-dependent RNA polymerase (L) gene – 5'leader region. These genomes can be viewed as containing conserved regions at the 3' and 5'ends, which encode the core proteins and the L protein, respectively, and a variable part in the middle encoding the envelope protein (Fig. 4). The complexity of the filovirus genome (seven genes) suggests an evolutionary position between the simpler organized genomes of the vesiculoviruses (five genes) and the more complex genomes of the pneumoviruses (ten genes). This may organizationally align filoviruses more closely to the paramyxo- and morbillivirus genera of the *Paramyxoviridae* (six/seven genes). However, the presence of overlapping genes, the variable length of intergenic regions, and the presence of seven linearly arranged genes suggest a progression of filovirus genomes towards the greater complexity of the pneumovirus genome [10, 42].

Besides a similar genome organization, NNS RNA virus genomes share several features in their mechanisms of transcription and replication, indicating a common ancestoral lineage. These features are: (1) complementarity of the genome extremities, (2) homologous regions in the 3'leader sequences, (3) conserved transcriptional start and termination signals, (4) interruption of genes by intergenic sequences not present in the mRNA transcripts, (5) possession of a virion-associated RNA-dependent RNA polymerase, (6) helical nucleocapsid as the functional template for synthesis of replicative and mRNA, (7) transcription of mRNAs by sequential interrupted synthesis from a single promotor, (8) replication by synthesis of a full-length (+) sense antigenome, (9) transcription and replication in the cytoplasm, and (10) maturation by envelopment of independently assembled nucleocapsids at membrane sites containing inserted viral proteins [10, 42]. All known L proteins of NNS RNA viruses show three conserved boxes which seem to represent the functional domains of these RNA-dependent RNA polymerases [35]. These observations further indicate similarities in the modes of transcription and replication for filo-, paramyxo-, and rhabdoviruses.

Comparisons of the deduced amino acid sequences of the L and NP genes with those of different NNS RNA viruses [dotplot analyses [35, 44] and dendrograms (Fig. 6)], demonstrate that filoviruses are more closely related to paramyxoviruses than to rhabdoviruses. In addition, the dendrograms suggest that filoviruses are even closer to the genera paramyxovirus and morbillivirus of the family *Paramyxoviridae* than the genus pneumovirus. This observation correlates with the genome arrangement as discussed before, and may support arguments for the creation of a separate family for pneumoviruses.

Despite similarities to paramyxo- and rhabdoviruses, comparative analyses have clearly demonstrated that filoviruses are unique, supporting their classification in a separate family within the order *Mononegavirales*. All filoviruses are morphologically similar with minor differences in the particle length (Fig. 1). They all show the same basic structural organization. Computer assisted dotplot comparisons of the deduced amino acid sequences of the first six genes of the EBO and MBG genomes demonstrate the close relationship within the family (Fig. 3). Filoviral unique features are: (1) virion morphology (long filamentous particles) (Fig. 1), (2) pathogenic potential (hemorrhagic fever), (3) genome size (about 19 kb) and coding capacity (seven open reading frames) (Fig. 4), (4) the pentamer 3'-*UAAUU*-5' present in all transcriptional signals (Fig. 5), (5) long noncoding regions at the 5' and 3'ends of all mRNA transcripts, (6) large molecular weight nucleoprotein (Fig.

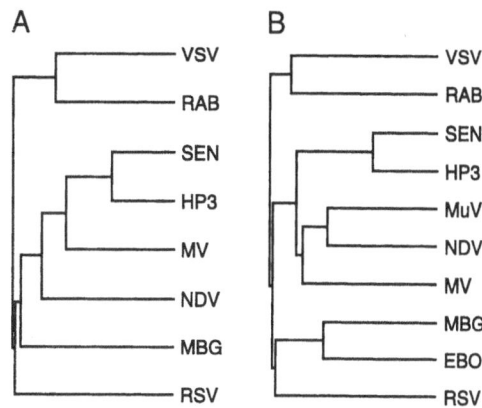

Fig. 6. Dendrograms showing the relationship among NNS RNA viruses. **A** L protein (viral RNA-dependent RNA polymerase. **B** Nucleoprotein. Plots were generated using a multiple sequence alignment program (PileUp) that employs a modification of the method of Needleman and Wunsch [37] to calculate pairwise alignments of sequence clusters [7]. *Rhabdoviridae*: *VSV* Vesicular stomatitis virus; *RAB* rabies virus. *Paramyxoviridae*: *SEN* Sendai virus; *HP3* human parainfluenza 3 virus; *MV* measles virus; *NDV* Newcastle disease virus; *MuV* mumps virus; *RSV* human respiratory syncytial virus *Filoviridae*: *MBG* Marburg virus; *EBO* Ebola virus

1B), and (7) highly N- and O-glycosylated transmembrane protein (Fig. 2).

A classification within the new family *Filoviridae* has not yet been proposed. All MBG isolates are serologically distinct from EBO and RES with no cross-reactivity, whereas the two EBO subtypes and RES do cross-react serologically. In addition, there are other features which distinguish filoviruses. Based on the knowledge to date a separation into two groups can be proposed: (A) EBO and EBO-like viruses with EBO, subtype Zaire, as the prototype virus and (B) MBG and MBG-like viruses with MBG, strain Musoke, as the prototype virus. This separation is clearly demonstrated by the dendrogram of the deduced amino acid sequences of filovirus glycoproteins shown in Fig. 7. The following criteria could be used to classify a new isolate within the family *Filoviridae*: (1) serologic cross-reactivity with the group prototype viruses (convalescent sera, monospecific polyclonal sera, monospecific monoclonal antibodies), (2) electrophoretic mobility pattern of virion structural proteins, primarily the glycoprotein (MBG-group: large GP approximately 170 kDa/EBO-group: smaller GP approximately 125 kDa) (Fig. 1b), (3) terminal linked sialic acids on carbohydrate structures (MBG-group: absent/EBO-group: present (Fig. 2), and (4) sequence analysis of genome parts [e.g.: length of 3′noncoding region of the

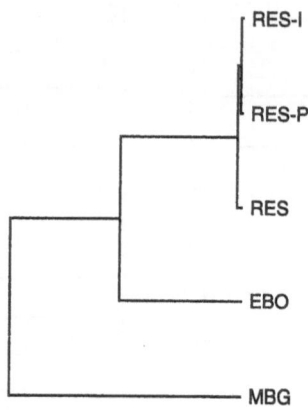

Fig. 7. Dendrogram of filovirus glycoproteins demonstrating the relationship within the family. Plots were generated as described in the legend to Fig. 6. *MBG* Marburg virus (strain Musoke) [52]; *EBO*, Ebola virus (subtype Zaire, strain Mayinga) [42]; *RES* Reston virus (A. Sanchez, unpubl. data); *RES-I* filovirus, isolated from a monkey during the 1992 outbreak in Italy (Sienna) (A. Sanchez, unpubl. data); RES-P filovirus, isolated from a monkey which died in a facility in the Philippines (1992) (A. Sanchez, unpubl. data)

NP gene (EBO-group: extremely long compared to MBG-group), and location of overlapping genes].

The different MBG isolates are antigenically closely related. Oligonucleotide and peptide mapping studies [24] as well as preliminary sequencing data (S. Netesov, pers. comm.; H. Feldmann, unpubl. data) showed that geographically and temporally distinct isolates seem to be very similar to each other but genetically distinguishable. EBO exists in two recognized subtypes (EBO-Zaire and EBO-Sudan) which differ in pathogenicity, antigenicity, and genomic composition, but show serologic cross-reactivity [3, 6, 23]. RES, based on serology and the analyses of sequencing data of the nucleoprotein (data not shown) and glycoprotein (Fig. 7), is a member of the EBO group. However, it is not known whether it is a distinct filovirus or a third subtype of EBO. This question might be answered in part by sequence analysis of an EBO-Sudan isolate. Sequencing of GP genes of new isolates from the last filovirus outbreak in Italy showed only slight variations from RES isolated in the 1989 outbreak (A. Sanchez, unpubl. data). This finding indicates that the monkeys imported into Italy were infected with RES or a filovirus very similar to RES as demonstrated by the dendrogram in Fig. 7. The question still remains, whether the variations suggest a genetic drift for RES, or an introduction of a RES-like virus from another source in the Philippines.

Acknowledgements

The authors greatly appreciate Dr. Clarence J. Peters for his helpful discussion. A part of the data summarized in this paper were obtained by studies which were supported by the Deutsche Forschungsgemeinschaft (Kl238/1-1 and SFB 286, Teilprojekt A6).

References

1. Baron RC, McCormick JB, Zubeir OA (1983) Ebola hemorrhagic fever in southern Sudan: hospital dissemination and intrafamilial spread. Bull World Health Organ 6: 997–1003
2. Becker S, Feldmann H, Will C, Slenczka W (1992) Evidence for occurrence of filovirus antibodies in humans and imported monkeys: do subclinical filovirus infections occur worldwide? Med Microbiol Immunol 181: 43–55
3. Buchmeier MJ, DeFries RU, McCormick JB, Kiley MP (1983) Comparative analysis of the structural polypeptides of Ebola virus from Sudan and Zaire. J Infect Dis 147: 276–281
4. Cianciolo GJ, Copeland TD, Oroszlan S, Snyderman R (1985) Inhibition of lymphocyte proliferation by a synthetic peptide homologous to retroviral envelope proteins. Science 230: 453–455
5. Collins PL, Olmsted RA, Spriggs MK, Johnson PR, Buckler-White AJ (1987) Gene overlap and site-specific attenuation of transcription of the viral polymerase L gene of human respiratory syncytial virus. Proc Natl Acad Sci USA 84: 5134–5138
6. Cox NJ, McCormick JB, Johnson KM, Kiley MP (1983) Evidence for two subtypes of Ebola virus based on oligonucleotide mapping of RNA. J Infect Dis 147: 272–275
7. Devereux J, Haeberli P, Smithies O (1984) A comprehensive set of sequence anlysis programs for the VAX. Nucleic Acids Res 12: 387–395
8. Elango N (1989) The mumps virus nucleocapsid mRNA sequence and homology among the paramyxoviridae proteins. Virus Res 12: 77–86
9. Elliott LH, Kiley MP, McCormick JB (1985) Descriptive analysis of Ebola virus proteins. Virology 147: 169–176
10. Feldmann H, Mühlberger E, Randolf A, Will C, Kiley MP, Sanchez A, Klenk H-D (1992) Marburg virus, a filovirus: messenger RNAs, gene order, and regulatory elements of the replication cycle. Virus Res 24: 1–19
11. Feldmann H, Will C, Schikore M, Slenczka W, Klenk H-D (1991) Glycosylation and oligomerization of the spike protein of Marburg virus. Virology 182: 353–356
12. Feldmann H, Wunder H, Huppertz S, Randolf A, Mahner F, Klenk H-D (1993) Characterization and expression of two membrane-associated virion structural proteins (VP40 and VP24) of Marburg virus. Virology (submitted)
13. Fisher-Hoch SP, Brammer L, Trappier SG, Hutwagner LC, Farrar BB, Ruo SL, Brown BG, Hermann LM, Perez-Oronoz GI, Goldsmith CS, Hanes MA, McCormick JB (1992) Pathogenic potential of filoviruses: role of geographic origin of primate host and virus strain. J Infect Dis 166: 753–763
14. Gear JSS, Cassel GA, Gear AJ, Trappler B, Clausen L, Meyers AM, Kew MC, Bothwell TH, Sher R, Miller GB, Schneider J, Koornhoff HJ, Comperts ED, Isaäcson M, Gear JHS (1975) Outbreak of Marburg virus disease in Johannesburg. Br Med J 4: 489–493

15. Geyer H, Will C, Feldmann H, Klenk H-D, Geyer R (1992) Carbohydrate structure of Marburg virus glycoprotein. Glycobiology 2: 299–312
16. Heymann DL, Weisfeld JS, Webb PA, Johnson KM, Cairns T, Berquist H (1980) Ebola hemorrhagic fever: Tandala Zaire, 1977–78. J Infect Dis 142: 373–376
17. ICTV (1991) The order Mononegavirales. Paramyxovirus Study Group of the Vertebrate Subcommittee. Arch Virol 117: 137–140
18. Jahrling PB (1991) Filoviruses and arenaviruses. In: Balows A (ed) Manual of clinical microbiology. American Society for Microbiology, Washington, D.C., pp 984–997
19. Jahrling PB, Geisbert TW, Galgard DW, Johnson ED, Ksiazek TG, Hall WC, Peters CJ (1990) Preliminary report: isolation of Ebola virus from monkeys imported to USA. Lancet 335: 502–505
20. Johnson BK, Ochen D, Oogo S, Gitau LG, Wambui C, Gichogo A, Libondo D, Tukei PM, Johnson ED (1986) Seasonal variation in antibodies against Ebola virus in Kenyan fever patients. Lancet i: 1160
21. Johnson KM, Lange JV, Webb PA, Murphy FA (1977) Isolation and partial characterization of a new virus causing acute hemorrhagic fever in Zaire. Lancet i: 569–571
22. Johnson KM, Scribner CL, McCormick JB (1981) Ecology of Ebola virus: a first clue? J Infect Dis 143: 749–751
23. Kiley MP, Bowen ETW, Eddy GA, Isaäcson M, Johnson KM, McCormick JB, Murphy FA, Pattyn SR, Peters D, Prozesky OW, Regnery RL, Simpson DIH, Slenczka W, Sureau P, van der Groen G, Webb PA, Wulff H (1982) Filoviridae: a taxonomic home for Marburg and Ebola viruses? Intervirology 18: 24–32
24. Kiley MP, Cox NJ, Elliott LH, Sanchez A, DeFries R, Buchmeier MJ, Richman DD, McCormick JB (1988) Physicochemical properties of Marburg virus: evidence for three distinct virus strains and their relationship to Ebola virus. J Gen Virol 69: 1957–1967
25. Kiley MP, Wilusz J, McCormick JB, Keene JD (1986) Conservation of the 3′ terminal nucleotide sequence of Ebola and Marburg virus. Virology 149: 251–254
26. Kurilla MG, Stone HO, Keene JD (1985) RNA sequence and transcriptional properties of the 3′end of the Newcastle disease virus genome. Virology 145: 203–212
27. Leppert M, Kolakofsky D (1980) Effect of defective interfering particles on plus- and minus-strand leader RNAs in vesicular stomatitis virus infected cells. J Virol 35: 704–709
28. Leppert M, Rittenhouse L, Perrault J, Summers DF, Kolakofsky D (1979) Plus and minus strand leader RNAs in negative-strand virus-infected cells. Cell 18: 735–747
29. Martini GA (1971) Clinical syndrom. In: Martini GA, Siegert R (eds) Marburg virus disease, 1st edn. Springer, Berlin Heidelberg New York, pp 1–9
30. Martini GA, Siegert R (1971) Marburg virus disease, 1st edn. Springer, Berlin Heidelberg New York
31. MMWR (1990) Update: filovirus infection in animal handlers. MMWR 39: 221
32. MMWR (1990) Update: filovirus infections among persons with occupational exposure to nonhuman primates. MMWR 39: 266–267
33. MMWR (1990) Update: evidence for filovirus infection in an animal caretaker in a research/service facility. MMWR 39: 296–297
34. MMWR (1990) Update: filovirus infection associated with contact with nonhuman primates or their tissues. MMWR 39: 404–405

35. Mühlberger E, Sanchez A, Randolf A, Will C, Kiley MP, Klenk H-D, Feldmann H (1992) The nucleotide sequence of the L gene of Marburg virus, a filovirus: homologies to paramyxoviruses and rhabdoviruses. Virology 187: 534–547

36. Murphy FA, van der Groen G, Whitfield SG, Lange JV (1978) Ebola and Marburg virus morphology and taxonomy. In: Pattyn SR (ed) Ebola virus hemorrhagic fever, 1st edn. Elsevier/North-Holland, Amsterdam, pp 61–84

37. Needleman SB, Wunsch CD (1970) A general method applicable to the search for similarities in the amino acid sequence of two proteins. J Mol Biol 48: 443–453

38. Palmer AE, Allen AM, Tauraso NM, Shlokov A (1968) Simian hemorrhagic fever. I. Clinical and epizootiologic aspects of an outbreak among quarantine monkeys. Am J Trop Med Hyg 17: 404–412

39. Pattyn SR (1978) Ebola virus hemorrhagic fever, 1st edn. Elsevier/North-Holland, Amsterdam, pp 1–436

40. Peters D, Müller G, Slenczka W (1971) Morphology, development, and classification of Marburg virus. In: Martini GA, Siegert R (eds) Marburg virus disease, 1st edn. Springer, Berlin Heidelberg New York, pp 68–83

41. Sanchez A, Kiley MP (1987) Identification and analysis of Ebola virus messenger RNA. Virology 157: 414–420

42. Sanchez A, Kiley MP, Holloway BP, Auperin DD (1993) Sequence analysis of the Ebola virus genome: Organization, genetic elements, and comparison with the genome of Marburg virus. Virus Res (in press)

43. Sanchez A, Kiley MP, Holloway BP, McCormick JB, Auperin DD (1989) The nucleoprotein gene of Ebola virus: cloning, sequencing, and in vitro expression. Virology 170: 81–91

44. Sanchez A, Kiley MP, Klenk H-D, Feldmann H (1992) Sequence analysis of the Marburg virus nucleoprotein gene: comparison to Ebola virus and other non-segmented negative-strand RNA viruses. J Gen Virol 73: 347–357

45. Siegert R, Shu H-L, Slenczka W, Peters D, Müller G (1967) Zur Äthiologie einer unbekannten von Affen ausgegangenen Infektionskrankheit. Dtsch Med Wochenschr 92: 2341–2343

46. Slenczka W, Rietschel M, Hoffmann C, Sixl W (1984) Seroepidemiologische Untersuchungen über das Vorkommen von Antikörpern gegen Marburg- und Ebola-Virus in Afrika. Mitt Oesterr Ges Tropenmed Parasitol 6: 53–60

47. Smith CEG, Simpson DIH, Bowen ETW (1967) Fatal human disease from vervet monkeys. Lancet ii: 1119–1121

48. Smith DH, Johnson BK, Isaäcson M, Swanapoel R, Johnson KM, Kiley MP, Bagshawe A, Siongok T, Keruga WK (1982) Marburg-virus disease in Kenya. Lancet i: 816–820

49. Teepe RGC, Johnson BK, Ocheng D, Gichogo A, Langatt A, Ngindu A, Kiley M, Johnson KM, McCormick JB (1983) A probable case of Ebola virus hemorrhagic fever in Kenya. East Afr Med J 60: 718–722

50. van der Waals FJ, Pomerov KL, Goudsmit J, Asher DM, Gajdusek DC (1986) Hemorrhagic fever virus infection in an isolated rainforest area of central Liberia. Limitations of the indirect immunofluorescence slide test for antibody screening in Africa. Trop Geogr Med 38: 209–214

51. Volchkov VE, Blinov VM, Netesov SV (1992) The envelope glycoprotein of Ebola virus contains an immunosuppressive-like domain similar to oncogenic retroviruses. FEBS Lett 305: 181–184

52. Will C, Mühlberger E, Linder D, Slenczka W, Klenk H-D, Feldmann H (1993) Marburg virus gene four encodes the virion membrane protein, a type I transmembrane glycoprotein. J Virol 67: 1203–1210

53. World Health Organization (1978a) Ebola hemorrhagic fever in Zaire, 1976. Bull World Health Organ 56: 271–293
54. World Health Organization (1978b) Ebola hemorrhagic fever in Sudan, 1976. Bull World Health Organ 56: 247–270
55. World Health Organization (1979) Viral hemorrhagic fever surveillance. WER 54: 342–343
56. World Health Organization (1992) Viral hemorrhagic fever in imported monkeys. WER 67: 142

Authors' address: Dr. H. Feldmann, Institute for Virology, Philipps-University, Robert-Koch-Strasse 17, D-35037 Marburg, Federal Republic of Germany.

Arch Virol (1993) [Suppl] 7: 101–109

Borna disease virus: nature of the etiologic agent and significance of infection in man

J.A. Richt[1], **S. Herzog**[1], **J. Pyper**[2], **J.E. Clements**[2], **O. Narayan**[2], **K. Bechter**[3], and **R. Rott**[1]

[1] Institut für Virologie, Justus-Liebig-Universität Giessen, Giessen, Federal Republic of Germany
[2] Division of Comparative Medicine, The Johns Hopkins University, Baltimore, Maryland, U.S.A.
[3] Bezirkskrankenhaus Günzburg, Günzburg, Federal Republic of Germany

Summary. This review presents data on the characterization of Borna disease virus (BDV) and its potential as a possible causative agent in humans. The isolation of (i) BDV-specific cDNA clones that encode various BDV-specific proteins and (ii) partially purified virus particles led to the conclusion that the viral genome consists of negative-sense, single-stranded RNA. The organization of the BDV-specific RNA species appears to be a nested set of overlapping subgenomic RNA transcripts. Furthermore, evidence is presented that BDV can infect humans and may cause certain psychiatric and neurological disorders. This concept is supported by (i) the finding of virus-specific antibodies in sera of patients with neuropsychiatric diseases and (ii) results obtained during attempts to isolate BDV or a BDV-related agent from the cerebrospinal fluid of seropositive patients.

Introduction

Borna disease (BD) is an infectious, immunopathological disease of the central nervous system (CNS), characterized by a disseminated meningoencephalomyelitis [9]. BD occurs as a natural infection in horses and sheep and probably other species and has been transmitted experimentally to birds, rodents, ruminants and non-human primates. BD occurs only sporadically and has not been recognized in countries other than Germany and Switzerland. In naturally infected animals the disease characteristically results in paralysis and death [9]; occasionally recovery occurs, but the infected animals may still exhibit motoric and behavioral alterations. Recent seroepidemiological surveys of BD in horses have shown that BDV-specific antibodies are present in many horses [13], surprisingly also in sera of horses from Africa and the USA (Herzog, unpubl. results), without clinical signs of the disease.

Most laboratory studies on the pathogenesis of BDV have involved experimentally inoculated Lewis rats. Under these conditions, BDV infection occurs only after mandatory replication in nervous tissues, and disseminates via neural pathways from the central nervous system (CNS) to the peripheral nervous system (PNS) or vice versa [12, 17]. In immunocompetent rats no infection is found in the extraneural tissues, whereas in newborn or Cyclosporin A-treated rats the virus also spreads to nonneural tissues in the vicinity of nerve endings [10, 27]. The clinical outcome of a BDV-infection in rats can vary greatly and depends on the passage history of the virus used for inoculation. Thus, a variety of BDV-variants exist including strains which induce long-lasting behavioral changes, obesity syndrome with fertility disturbances, paralysis with a high percentage of mortality, or inapparent infections using a MDCK-cell-adapted virus variant (Herzog, unpubl. results). In all cases, intra-cerebral (i.c.) or intransal (i.n.) inoculation resulted in productive replication in the nervous system. Therefore, it can be assumed, that the virus can tolerate various mutations and the virus inoculum seems to be genetically heterogeneous and may be selected by various host factors.

Borna disease virus

Borna disease can be induced by an infectious agent, the Borna disease virus (BDV) [31], but until recently little was known about the nature of the agent. BDV is present in relatively high concentrations in the brain tissue of affected animals and it can be readily induced in susceptible animals by inoculation of infectious brain material. The infectivity titers in infected tissue and cells were reduced or eliminated by exposure to UV light, detergents [5, 7, 9] or incubation with cycloheximide or actimomycin D [5, 7]. Therefore, it was proposed that BDV contains nucleic acid as genetic material, is likely to be enveloped and to have physical and biological properties of a conventional virus [7, 16, 22]. BDV particles have never been visualized in infectious material, but BDV-infection is associated with the expression of at least three virus-specific proteins with molecular weights of 14, 24 and 38/39 kd, respectively [8, 25, 28].

BDV-specific cDNAs were recently isolated in three laboratories from infected rat brain and tissue culture cells using subtractive cloning methods [6, 14, 21, 29, 30]. The establishment of subtractive cDNA libraries from infected material resulted in the identification of cDNAs coding either for the 14 kd and 24 kd (p14, p24) [6, 14, 21, 29, 30] or the 38/39 kd protein (p38) [19]. Database searches of nucleotide and deduced amino acid sequences for the p14, p24 and p38 proteins of BDV have shown only limited regions of homology. The p38 protein showed small regions of homology with some viral polymerases and matrix proteins; this may suggest that the p38 BDV-protein associates with RNA or ribonucleoprotein [19]. Analysis of the deduced amino acid sequence for the p24 and p38 proteins revealed nuclear targetting motifs in the p24 and p38 proteins [19, 29]. Southern blot hybridization experiments using digested genomic and episomal DNA from persistently infected tissues failed to show positive hybridization signals [6, 14, 21, 30]. Therefore, BDV is unlikely to be a DNA virus or retrovirus and the cDNA clones represent no host encoded genes.

In Northern hybridization experiments, cDNA clones coding for the p24 and p38 protein detected four BDV-specific RNAs of 10.5 (supposably the viral genome), 3.6, 2.1 and 1.40–0.85 kb in extracts from infected rat brain and infected culture cells; the p24 cDNA clone hybridizes to the 0.85 kb RNA, whereas the p38 cDNA clone hybridizes to the 1.40 kb RNA [19]. All of these RNAs seem to be enriched by polyadenylate [poly (A)] selection [30]. In contrast, Lipkin et al. [14] reported that the largest RNA species is 8.5 kb and not polyadenylated. The BDV-specific RNAs were sensitive to digestion with pancreatic RNase [6]. These findings are in accordance with the view that BDV is a single-stranded RNA virus.

The cDNA clones isolated in our laboratories and coding for the p24 protein contained two open reading frames (ORF) with 217 and 77 amino acids (aa) [21, 30], the clone coding for the p38 protein contained a single ORF with 357 aa [19]. In vitro transcription and translation of the p24 cDNA clone produced two major proteins of ~14 and 24 kd (Fig. 1); in vitro transcription and translation of the p38 clone resulted in a 38 kd protein, recognized by both monoclonal and polyclonal antibodies to BDV. Recently we have amplified and cloned p24 and p38 BDV-specific sequences from horse brain RNA using the polymerase chain reaction (PCR) and oligonucleotides derived from the BDV-specific cDNA clones encoding for the p24 or p38 BDV-proteins (Richt et al., unpubl. results). The horse brain RNA was derived from a seropositive animal with clinical BD, which was found to be positive for BDV-infectivity on fetal rabbit brain cells (FRB) and BDV-antigens by immunohistological methods.

The genetic organization and polarity of the BDV-specific RNA-species was determined using strand-specific oligonucleotide probes. Oligonucleotides with negative polarity from coding and noncoding regions of the p24 cDNA clone all hybridized to the same four positive-stranded BDV-RNAs (Fig. 2; 10.5, 3.6, 2.1 and 0.85 kb) whereas com-

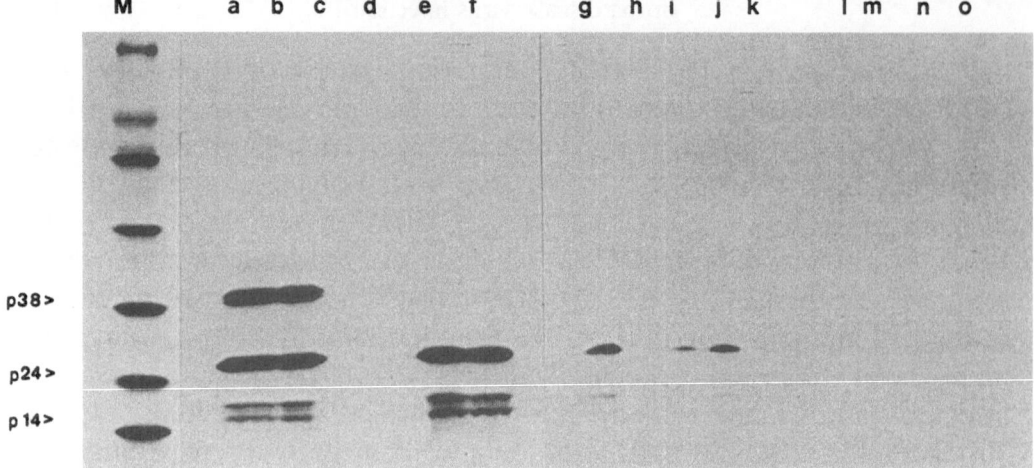

Fig. 1. Immunoprecipitations of BDV-specific proteins translated from the cDNA clone B8. Poly(A)-selected RNA from BDV-infected rat brain was translated in vitro and immunoprecipitated with polyclonal anti-BDV sera from rat (*a*) and rabbit (*b*). (−)-strand RNA (*c, d*) from cDNA clone B8 and (+)-strand RNA (*e–o*) from cDNA clone P4 (B8 subcloned in reverse orientation) were synthesized and translated in vitro and immunoprecipitated with polyclonal anti-BDV sera from rat (*c, e*) and rabbit (*d, f*) or human sera (*g–o*). Normal human sera (*l–o*) and various human sera, all positive for BDV-antigens in indirect immunofluorescence assays were analysed: human anti-p24 (*g*); human anti-p38 (*h*); patient 112 (*i*); patient 114 (*j*); patient 115 (*k*). Lane M contains the molecular weight markers from top to bottom: 200, 92.5, 69, 46, 30, 21.5, 14.3 kd proteins

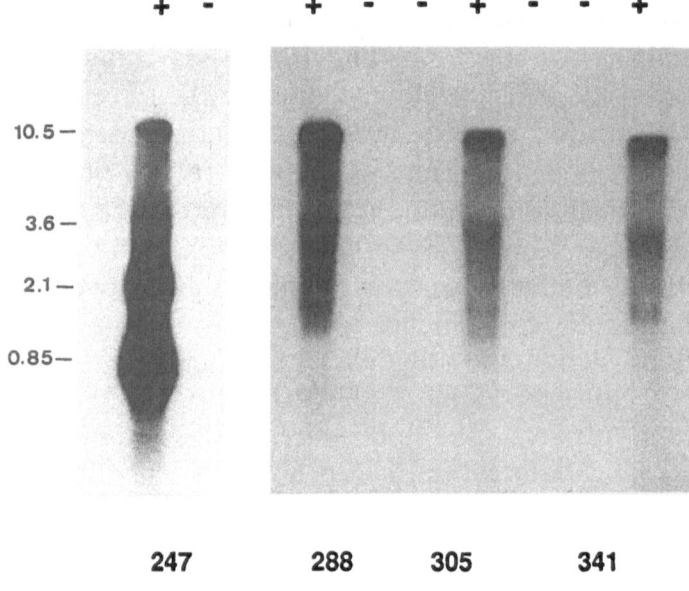

Fig. 2. Nothern blot analysis of RNA from rat brain. 10 µg total RNA from uninfected rat brain (−) and BDV-infected rat brain (+) were hybridized with a negative sense ^{32}P-labelled oligonucleotide (#247), or with positive sense ^{32}P-labelled oligonucleotides (#288, #305, #341) from different regions of cDNA clone B8 as previously described

[21]

plementary positive-stranded oligonucleotides all hybridized to three BDV-specific RNAs of 10.0, 3.5 and 1.7 kb (Fig. 2) in size [21]. This suggests that the organization of the BDV-specific RNAs is a nested set of overlapping negative- and positive-stranded subgenomic RNA-species [21], similar to members of the coronavirus superfamily [11].

The question of genome polarity has been ambiguous, because purification procedures for BDV particles were not available. The relative abundance of the negative-stranded genomic RNA over the complementary positive-stranded RNA species provided a basis for the suggestion that BDV is likely to be a negative-stranded RNA virus [14], whereas the presence of a nested set of cross-hybridizing mRNAs indicates a striking similarity of BDV to the positive-stranded coronaviruses [21]. To identify the polarity of genomic BDV-RNA, the isolation of intact viral particles is required. We have recently partially purified BDV from infected tissues with the lipid solvent Freon-113 [20]. This allowed us to examine BDV-specific RNA species and proteins after extraction in a two-phase Freon-gradient. This treatment resulted in infectious particles with a bouyant density of 1.16–1.22 g/ml. All three virus-specific proteins regularly detected in BDV-infected tissue homogenates could be demonstrated after Freon treatment. Both positive- and negative-stranded RNA species found in BDV-infected rat brain and cells were present in the Freon-extracted preparations. When these preparations were treated with RNase A prior to RNA-extraction, only negative-stranded, genomic RNA was detected in Northern blot hybridizations using sense and antisense RNA probes [20]. No loss of infectivity was observed after RNase A digestion of Freon-extracted material. Therefore, all data available suggest that BDV appears to be a negative, single-stranded RNA virus.

The Freon-treated preparations have also been used in ultrastructural studies. The only virus-like structures which could be identified by transmission electron microscopy were particles of 60–82 nm diameter. The presence of a rim-like structure surrounding these particles suggests the presence of an envelope. Similar particles were described previously by Ludwig and Becht [15]. It cannot be assumed with certainty that these virion-like structures correspond to the original native virus particle, because few of these structures were found and the virus preparations had been treated with the lipid solvent Freon-113.

In conclusion, BDV-specific cDNA clones encoding the p24 and p38 BDV-proteins were isolated from persistently BDV-infected tissues. Four poly(A) mRNAs of 10.5, 3.6, 2.1, and 1.40–0.85 kb were recognized by negative-stranded RNA probes, whereas RNA transcripts of 10.0, 3.5 and 1.7 kb were recognized by positive-stranded RNA probes. No hybridization was detected in infected or uninfected MDCK or rat brain DNA, indicating that the p24 and p38 cDNAs were not host encoded and that BDV is unlikely to be a DNA virus or a retrovirus. The nature and polarity of the BDV genome was determined after isolation of BDV with Freon-113 from infected tissues. Treatment of Freon 113-treated virus preparations with RNase A prior to RNA-extraction revealed, that only negative-stranded, genomic RNA could be detected in Northern blots using sense and antisense RNA probes. Therefore, BDV appears to be a negative, single-stranded RNA virus.

Borna virus infection in man

The nature of the behavioral disturbances observed in rats [18] and tree shrews [26] appear related to certain mental disorders in human beings. Therefore, sera and cerebrospinal fluid (CSF) of patients with neuropsychiatric diseases were examined for BDV-specific antibodies. Antibodies were present in 4–7% of sera derived from more than 5,000 psychiatric and neurological patients from Germany, USA and Japan [23, 24; Ikeda, pers. comm.]. In some of the seropositive patients, specific antibodies were concomitantly detected in the CSF, but always lower than those found in the corresponding sera [23]. The highest percentage of seropositive patients came from a region in Southern Germany, where BD is known to be endemic among horses and sheep. The antibodies from human patients reacted either with the p24 or p38 BDV-proteins or with both BDV-antigens in immunoblots. The BDV-specificity of the human antibodies was recently reinforced by the finding that antibodies from seropositive patients recognized the p24 protein (Fig. 1) expressed by a BDV-specific cDNA clone [30]. There is no evidence for a major clinical manifestation of infection with BDV in man. Seropositive patients were found with a broad range of mental disorders, with a predominance of schizophrenia, affective psychoses and certain personality disorders [1]. About 1% of 1,000 randomly collected sera from hospital patients also showed antibodies specific for BDV-proteins [1, 23]. This seems to indicate that inapparent infection with BDV or a BDV-related agent might take place in human beings, but clinical manifestation is only observed in yet unknown circumstances similar to naturally infected horses [13]. Serological examination of patients infected with human immunodeficiency virus (HIV) by Bode et al. (1990) has shown an incidence of BDV-specific antibodies of ~8%. The same investigators have reported a high incidence of BDV-specific antibodies among patients with chronic inflammatory neurological disorders such as multiple sclerosis.

Furthermore, efforts have been made recently to isolate BDV from the CSF of three seropositive patients. The CSFs were either applied to fetal rabbit brain cells or inoculated i.c. into rabbits, which are highly susceptible to BDV isolated from naturally or experimentally infected animals. In the cell cultures, small numbers of immunoreactive foci were found with BDV-specific antibodies 10–12 days after inoculation [23]. The cells, however, lost their antigen during subsequent passages. On the other hand, the i.c. inoculated rabbits developed no clinical signs of BD, no histological lesions in the CNS and no BDV-specific antigens could be found in the nervous system (NS) after an observation period of 5 months [23]. Nevertheless, these animals developed BDV-specific antibodies in their sera with titers ranging from 1:20 to 1:640. The brain homogenate from one rabbit was infectious for fetal rabbit brain cells as demonstrated by positive cell foci in immunofluorescence assay. However, again, the antigen disappeared during attempts to propagate the agent by subcultivation of the cells. These findings can be interpreted as typical of an abortive infection, and might indicate that infection in humans is caused by a BDV-related agent for which no suitable isolation method is available so far.

In conclusion, the presence of virus-specific antibodies and the results obtained during attempts to isolate the agent from CSF of seropositive patients support the concept, that humans can be infected with BDV or a related virus. From clinical studies it appears that this virus could induce acute and chronic meningoencephalitis with neurological symptomatology and might contribute to or initiate certain psychiatric disorders [1, 3]. There is also evidence that BDV might be horizontally transmitted from domestic animals to man [2]. Whether this agent induces disease might be dependent on genetic preposition of the host or other endogenous or exogenous factors. Evidence that such factors play a role in the outcome of BD has been described in naturally as well as experimentally infected animals. The advent of BDV-specific cDNA clones encoding the p24 and p38 BDV-proteins allows further studies on characterization, epidemiology and pathogenesis of BDV infection in animals and man.

Acknowledgement

The work done by the authors S.H., J.A.R. and R.R. were supported by the Deutsche Forschungsgemeinschaft (Ro 202/7-1; Ro 202/7-3; Ri 518/1-2), by the authors J.E.C., O.N. and J.P. by institutional grants from the Johns Hopkins University. We would like to thank Dr. Boschek for critically reading the manuscript and K. Haberzettl and E. Gottfried for excellent technical assistance.

References

1. Bechter K, Herzog S (1990) Über Beziehungen der Borna'schen Krankheit zu endogenen Psychosen. In: Kaschka WP, Aschauer HN (eds) Psychoimmunologie. Thieme, Stuttgart, pp 133–141
2. Bechter K, Schüttler R, Herzog S (1992a) Case of neurological and behavioral abnormalities: due to Borna disease virus encephalitis? (Letter). Psychiatr Res 42: 193–196
3. Bechter K, Schüttler R, Herzog S (1992b) Borna disease virus: Possible causal agent in psychiatric and neurological disorders in two families (Letter). Psychiatr Res 42: 291–294
4. Bode L, Riegel S, Ludwig H, Amsterdam JD, Lange W, Koprowski H (1988) Borna disease virus specific antibodies in patients with HIV infection and with mental disorders (Letter). Lancet ii: 689
5. Danner K, Mayr A (1979) In vitro studies on Borna virus. II. Properties of the virus. Arch Virol 61: 261–271
6. de la Torre JC, Carbone KM, Lipkin WI (1990) Molecular characterization of the Borna disease agent. Virology 179: 853–856
7. Duchala CS, Carbone KM, Narayan O (1989) Preliminary studies on the biology of Borna disease virus. J Gen Virol 70: 3507–3511
8. Haas B, Becht H, Rott R (1986) Purification and properties of an intranuclear virus-specific antigen from tissues infected with Borna disease virus. J Gen Virol 67: 235–241
9. Heinig A (1969) Die Bornasche Krankheit der Pferde und Schafe. In: Röhrer, H (ed) Handbuch der Virusinfektionen bei Tieren. VEB Fischer Jena 4: 83–148
10. Herzog S, Kompter C, Frese K, Rott R (1984) Replication of Borna disease virus in rats: age-dependent differences in tissue distribution. Med Microbiol Immunol 173: 171–177
11. Holmes KV (1990) Coronaviruses and their replication. In: Fields BN, Knipe DM (eds) Virology 2nd edn. Raven Press, New York, pp 841–856
12. Krey HF, Stitz L, Ludwig H (1982) Virus-induced pigment epithelitis in rhesus monkeys. Clinical and histological findings. Ophthalmologica 185: 205–213
13. Lange H, Herzog S, Herbst W, Schliesser T (1987) Seroepidemiologische Untersuchungen zur Bornaschen Krankheit (Ansteckende Gehirn-Rückenmarkentzündung) der Pferde. Tierärztl Umschau 12: 938–946
14. Lipkin WI, Travis GH, Carbone KM, Wilson MC (1990) Isolation and characterisation of Borna disease agent cDNA clones. Proc Natl Acad Sci USA 87: 4184–4188
15. Ludwig H, Becht B (1977) Borna disease: a summary of our present knowledge. In: ter Meulen V, Katz H (eds) Slow virus infections of the central nervous system. Springer, Berlin Heidelberg New York, pp 75–83
16. Ludwig H, Bode L, Gostonyi G (1988) Borna disease: a persistent virus infection of the central nervous system. Prog Med Virol 35: 107–151
17. Morales JA, Herzog S, Kompter C, Frese K, Rott R (1988) Axonal transport of Borna disease virus along olfactory pathways in spontaneously and experimentally rats. Med Microbiol Immunol 177: 51–68
18. Narayan O, Herzog S, Frese K, Scheefers H, Rott R (1983) Behavioral disease in rats caused by immunopathological responses to persistent Borna virus in the brain. Science 220: 1401–1403

19. Pyper JM, Richt JA, Brown L, Rott R, Narayan O, Clements JE (1993) Genomic organization of the structural proteins of Borna disease virus revealed by a cDNA clone encoding the 38 kd protein. Virology (in press)

20. Richt JA, Clements JE, Herzog S, Pyper J, Becht H, Wahn K, Narayan O, Rott R (1993) Analysis of virus-specific RNA species and proteins in Freon-113 preparations of the Borna disease virus (submitted)

21. Richt JA, VandeWoude S, Zink MC, Narayan O, Clements JE (1991) Analysis of Borna disease virus-specific RNAs in infected cells and tissues. J Gen Virol 72: 2251–2255

22. Richt JA, VandeWoude S, Zink MC, Clements JE, Herzog S, Stitz L, Rott R, Narayan O (1992a) Infection with Borna disease virus: Molecular and immunobiological characerization of the agent. Clin Infect Dis 14: 1240–1250

23. Rott R, Herzog S, Bechter K, Frese K (1991) Borna disease, a possible hazard for man? Arch Virol 118: 143–149

24. Rott R, Herzog S, Fleischer B, Winokur A, Amsterdam J, Dyson W (1985) Detection of serum antibodies to Borna disease virus in patients with psychiatric disorders. Science 228: 755–756

25. Schädler R, Diringer H, Ludwig H (1985) Isolation and characterization of a 14,500 molecular weight protein from brains and tissue cultures persistently infected with Borna disease virus. J Gen Virol 66: 2479–2484

26. Sprankel H, Richarz K, Ludwig H, Rott R (1978) Behavior abnormalities in tree shrews (tupaia glis, Diard, 1920) induced by Borna disease virus. Med Microbiol Immunol 165: 1–18

27. Stitz L, Schilken D, Frese K (1991) Atypical dissemination of the highly neurotropic Borna disease virus during persistent infection in cyclosporin A-treated, immunosuppressed rats. J Virol 65: 457–460

28. Thiedemann N, Presek P, Rott R, Stitz L (1992) Antigenic relationship and further characterization of two major Borna disease virus proteins. J Gen Virol 73: 1057–1064

29. Thierer J, Riehle H, Grebenstein O, Binz T, Herzog S, Thiedemann N, Stitz L, Rott R, Lottspeich F, Niemann H (1991) The 24 kd protein of Borna disease virus. J Gen Virol 73: 413–416

30. VandeWoude S, Richt JA, Zink MC, Rott R, Narayan O, Clements JE (1990) A Borna virus cDNA encoding a protein recognized by antibodies in humans with behavioral diseases. Science 250: 1278–1281

31. Zwick W, Seifried O, Witte J (1927) Experimentelle Untersuchungen über die seuchenhafte Gehirn- und Rückenmarksentzündung der Pferde (Bornasche Krankheit). Z Infektionskr Haustiere 30: 42–136

Authors' address: Dr. J.A. Richt, Institut für Virologie, Justus-Liebig-Universität Giessen, Frankfurterstrasse 107, D-35392 Giessen, Federal Republic of Germany.

Arch Virol (1993) [Suppl] 7: 111–133

Biology and neurobiology of Borna disease viruses (BDV), defined by antibodies, neutralizability and their pathogenic potential

H. Ludwig[1], **K. Furuya**[1,*], **L. Bode**[2], **N. Klein**[1,**], **R. Dürrwald**[3],
and **D.S. Lee**[1,***] (with technical assistance of **T. Leiskau**[1])

[1] Institute of Virology, Free University of Berlin, Berlin
[2] Department of Virology, Robert Koch-Institut, BGA, Berlin
[3] Institute of Microbiology, University of Leipzig, Leipzig,
Federal Republic of Germany

Summary. Borna disease viruses (BDV) isolated from more than 20 naturally infected horses, 2 sheep and a possible feline isolate were included in these studies. Most of these wild-type viruses were grown in rabbit cells. Specifically rabbit-adapted viruses establish persistent infection in immortalized cell lines of various animal species. Brain-, tissue culture-, and cell-free released viruses could all be neutralized with antibodies from naturally and experimentally infected animals (horse; hamster, rat, rabbit, mouse, and chicken), with highest titres in birds. Splenectomized rabbits, which were subsequently infected with BDV, efficiently produced high titres of neutralizing antibodies. All of the neutralizing sera and cerebrospinal fluids from infected animals inhibited tissue culture spread of BDV. Experimental infection and hyperimmunization induced antibodies directed against the major components of the soluble antigen (60, 40/38, 25 and 14.5 kD proteins). Analysis of the s-antigen complex with these sera and 6 stable monoclonal antibodies revealed that it consists of 40/38 and 25 kD proteins. Although each of these antibodies detected intracellular virus-specific structures they did not recognize outer plasmamembrane antigens, showed no cross-reactivity, and had no neutralizing capacity. Unifying pathogenetic concepts of this neurotropic virus and its structural elements are discussed.

Introduction

Neurological symptoms characteristic of Borna disease in horses have been recognized since the end of the 18th century, mainly in the endemic

Present addresses: * Hokkaido Institute of Public Health, Sapporo, 060, Japan
** Behringwerke, Marburg, Federal Republic of Germany
*** Department Veterinary Medicine, Cheju, National University 69C 756, Korea

areas of Sachsen, Thüringen and Baden-Württemberg in Germany. This neurological syndrome, which is characterized by a non-purulent encephalomyelitis, has been called Borna disease since 1900 and has been reviewed by Zwick [51], Hiepe [19] and Heinig [18]. Recently, we have summarized the present knowledge of experimental infections in laboratory animals and on the pathogenesis of this persistent virus infection of the central nervous system [33, 34]. Little information exists concerning the virus itself and controversial views have been presented regarding its morphology [3, 28, 41, 43]. The isolation of Borna disease virus (BDV) nucleic acid from brains [27] or from infected cells [49] has resulted in characterization of the virus as a negative-, single-stranded RNA virus, which is transcribed in the nucleus [3]. Virus-specific antigens, previously called s-antigen and its further defined major proteins, for example the 60 kD, 40/38 kD, 25 kD, and additionally 14.5 kD proteins have been analysed by ourselves [34] and other groups [16, 48]. Although virus neutralizability has been a controversial issue [17, 38, 39, 42], it has been postulated to occur by Danner et al. [8], and has been consistently demonstrated by our group [20, 21, 34].

This report contributes recent information on the virus, its antigens and neutralization with an emphasis on the reactivity of different poly-, oligo- and monoclonal antibodies. Due to the association between BDV infection of human patients [4, 5, 42] with neuropsychiatric diseases, the structural elements and pathogenic features of this virus warrant further investigation.

Materials and methods

Virus strains

The wild-type strains of BDV used in our laboratory originated from diseased horses or sheep. These strains comprised 6 equine strains, isolated over the previous 20 years [14], and 18 more recent equine isolates (Dürrwald, unpubl.). All of these equine strains were derived *post mortem* from animals which had shown clinical disease, and were isolated principally from pieces of the limbic system, cortex and retina. Currently one ovine strain from Germany (Dürrwald, unpubl.) and a further one from Italy (Caramelli, unpubl.) are under investigation. Preliminary evidence that a wild-type feline strain exists is reported elsewhere [35].

All other strains are laboratory animal- or tissue culture adapted. From these, the rat adapted [20, 39], the mouse adapted [21] and a variety of cell culture strains (there are BDV infected permanent cell lines of many animal species and of human origin) have been established [40]. All strains were derived from 10% tissue suspensions (w/v) of brain pieces or tissue culture cells in Eagle's medium (Dulbeccos modification). After sonication and centrifugation at 1,000 g for 10 min the supernatant was usually stored at −70°C until use for virus or antibody titrations, as well as animal inoculation.

Two BDV live-virus strains, both rabbit adapted, have been employed for active immunization of horses and sheep in Germany for half a century. One is strain V "Zwick-vaccine" [51], and this was used mainly in western countries. The other is strain "Dessau", and this is still in use in eastern countries of Germany. Both vaccines are derived from original equine isolates that have been passaged 50 times in rabbit brains. The vaccine preparation consists of lyophylized rabbit brain suspension. Before use 1,000 mg of dry substance is reconstituted to contain approximately 100 to 1,000 rabbit infectious units (IU) per ml. Both attenuated viruses are administered by sub-cutaneous injection (s.c.) at the rate of 10^3 to 10^4 IU per horse and 10^2–10^3 IU per sheep. Annual revaccination is recommended. This report includes information concerning serological responses of horses and sheep vaccinated with strain "Dessau".

Tissue culture systems and infectivity assay

All virus assays and antibody titrations were performed using our standard system [34] in tissue culture of young rabbit brain (JKG) cells at passages 2–4. BDV is not cytopathogenic in tissue culture. One infectious unit produces 20–60 antigen carrying cells in about 5 days, which are recognizable as a focus by the focus-immunoassay or a cell-ELISA. Thus infectivity is monitored as focus forming units per ml (ffu/ml). The cell-ELISA is based on the same principle as the focus-immunoassay, namely titration of BDV strains by a focus test which follows the rules of a plaque test. These assays are outlined in detail elsewhere [20, 21, 40a]. Cells from one day old (used in this laboratory) and from embryonic rabbits [17, 38] have proved to be most sensitive for these assays. Immortalized, persistently infected cell lines from a variety of animal species and of human origin have been established. All have undergone at least 5 passages after infection had been introduced either by co-cultivation with infected rabbit cells or by mere infection of the original cell line with infectious brain suspensions [29, 40]. BDV was finally characterized [3] following hypertonic medium induced release of virus [40] from persistently infected human oligodendroglia (Oligo/TL) and equine dermal (ED/TL) cell lines.

Sera and cerebrospinal fluids

They were collected from naturally infected horses and sheep, or from experimentally infected animals, mostly in the terminal stage of the disease. Hamster, rat [20], mouse [21], tree shrew [45] and chicken sera were all taken also from persistently infected animals [33, 34]. In addition 100 further serum samples were analysed from horses and almost 500 from sheep, all of which had been vaccinated with the live-virus vaccine strain "Dessau". It is known that naturally and experimentally infected animals produce oligoclonal immunoglobulins in the cerebrospinal fluid (CSF) [34].

Monoclonal antibodies

Two sets of monoclonal antibodies (moabs) were produced from persistently infected, asymptomatic Balb/c mice [21]. Briefly, newborns were infected with the mouse adapted BDV strain. Between one and two years post infection (p.i.) mice in which high ELISA or neutralizing antibody titres could be demonstrated were re-inoculated ("boosted") with 0.1 to 0.3 ml of the same total mouse brain suspension by intra-peritoneal (i.p.) inoculation. Three days after the i.p. injection their spleens were removed and used for the production of hybridomas. The procedure used was essentially that described by Köhler and Millstein [24] with some modifications [12]. Three stable cell lines were selected, following screening of a variety of fusions by an ELISA assay on formalin fixed cultures in microplates, and cloned at least twice by the limiting dilution method. The antibodies produced by these cell lines were named Kfu 1, 2 and 3 and their characterization is described under results.

Additional efforts were made to produce moabs by in vitro boosting. Partially purified soluble antigen obtained from infected mouse brain suspensions was employed as a booster antigen for stimulation of spleen B-lymphocytes from persistently infected mice in vitro. Three animals with neutralizing titres over 1:1,000 had been selected for these purposes. The stimulation of spleen lymphocytes followed the detailed method reported by Morisson et al. [37]. The in vitro stimulated cells were fused with NS-1 cells as outlined in the first set of experiments and hybridoma cultures were established. Three clones were produced which excreted detectable levels of antibody. Two of these clones subsequently lost their activity after one month. The third hybridoma cell line, which was again subcloned by limiting dilution and produced antibody designated Kfu 4, could not be established as a stable cell line (see Results).

In a second set of experiments moabs were produced from Balb/c mice which had been hyperimmunized with purified soluble antigen. This time FPLC purified s-antigen components were used for immunization. The resulting cell lines, again cloned twice by limiting dilution, produced 3 stable moab-producing hybridomas (see Results). These antibodies, which are now designated R13E2, W1H8, W10F11 had been partially characterized in a previous report where they were named BDA 1, 2, 3, respectively [1].

Neutralization test

The neutralization test is based on the principle of the plaque reduction test, and measures the reduction in BDV-induced foci by sera. For standard conditions rat brain suspension, passage 5 and titrated on JKG cells, was used. Experiments with sonicated, infected tissue culture suspensions or with tissue culture-released virus, were also performed. The neutralization experiments were done with decomplemented sera or CSFs, using 10^3 ffu/ml mixed with aliquots of the geometrically diluted antibody solution and incubated for 1 h 37°C. In all experiments a rabbit pool serum or its purified IgG, and a normal rabbit serum or its IgG, were employed as positive and negative controls. The number of foci read at a given serum dilution (V) divided by the number of foci in the controls (V_0) is plotted against the reciprocal of serum dilution. Optimal conditions are reached when V_0 is approximately 50 ffu per well. Under ideal conditions the V/V_0 quotients fall on a line (compare also Fig. 2 in [34]). All titres give the dilution where the number of foci was reduced by 50%. This focus neutralization test was used to evaluate the neutralizing capacity of 1. the sera and/or CSFs collected from a variety of experimentally and naturally infected animal species, 2. the hyper-immune sera of rabbits, and 3. the moabs.

In a further test the inhibitory activity of antibodies in serum and CSF samples, or moabs was investigated in a virus cell-spread inhibition assay. For this purpose 50 foci per well were seeded into JKG cells. After absorption for 4 h the inoculum was removed, and the cell culture overlaid with 1:10 or 1:100 antibody solution diluted in medium. Five days after infection the cells were fixed and the number of antigen carrying cells was counted per focus using an immunoflourescence test (IFT) or the cell-ELISA.

Antibody detection methods other than neutralization: Immunofluorescence test,
cell-ELISA, immunoelectrophoretic techniques, immunoprecipitation test,
SDS-PAGE and Western blot

All our assay systems are based on an indirect immunofluorescence test (IFT) or a cell-ELISA using infected JKG cells. For antibody detection and titration the IFT was made highly specific by including an internal control in the form of a monoclonal antibody (Kfu 3), which detects the 40/38 kD protein (see also Results). Details of this IFT have been published recently [5]. The cell-ELISA, which was mainly employed for screening large numbers of sera, and for monitoring virus isolates, is based on the POD-technique as described by Pauli et al. [40a].

Cell or brain lysates of infected or uninfected material were prepared for immuno-precipitation or Western blotting tests. Immunoprecipitation was performed by in-cubation of the lysates with human or animal sera. Antigen/antibody complexes were then precipitated on to species-specific IgG agarose beads. Further steps were carried out in the Lämmli system, and details of this test, SDS-PAGE, and immunoblotting (Western blot) have been provided [2, 5]. Standard immuno-electrophoretic techniques were employed in the rabbit system to produce BDV antigen-antibody complexes by crossed or line electrophoresis [30, 32]. It is of importance that in some cases, after the line precipitate had formed, the anode and cathode were reversed and the run was continued overnight. Thus the line precipitate was electrophoretically purified. The lines were cut out and used for hyperimmunization of rabbits (see also Animal experiments).

Moabs were characterized by crossed immuno-electrophoresis using s-antigen from rat brain separated by two-dimensional electrophoresis in agarose gels on glass slides [30]. In the second dimension electrophoresis run, antigen was precipitated by rabbit antibodies. The gels were then cautiously washed 3 times in PBS (pH 7.2) to remove all non-precipitated proteins and each gel was incubated with an optimized dilution of the moab in 50 ml of PBS + 0.05% Tween 20 for 2 h at 37°C. This protocol was adapted from a procedure reported by Skjodt et al. [44]. After incubation the gels were washed 3 times for 30 min in PBS/Tween (s.a.). For immuno-enzymatic staining the gels were incubated for 2 h at 37°C in 50 ml of alkaline phosphatase-conjugated goat anti-mouse IgG (Dianova, Hamburg), diluted 1:3,000 in 50 mMol Tris (pH 8.0)/Tween (s.a.). After one further wash in Tris-buffer the gels were incubated in freshly prepared naphtol phosphate/Fast Red-substrate solution for 15–30 min at RT according to the staining of Western blots. After a final washing step with distilled water (2 × 15 min) the gels were air-tried and could be stored for an unlimited period.

Animal experiments

All infectivity assays were done by intracerebral inoculation (i.c.) of 0.3 ml of brain- or tissue culture suspension into 5–8 week old rabbits [29] or 0.025 ml i.c. into 1 day old mice, rats, or chicken [20, 21]. Animals were kept under standard conditions.

During studies with BDV infected rabbits which were immunosuppressed [26] we learned that splenectomy might have an influence on the course of the disease. Ten 4–6 week old rabbits were pre-bled and then splenectomized. The rabbits were then inoculated by i.p. inoculation of 1.0 ml of rabbit brain suspension (strain V), containing approx. 10^3 ffu of BDV, when the laparotomy incision had healed some 3–4 weeks later. Two further splenectomized rabbits which were inoculated with uninfected brain suspension were employed as controls. The infected and control rabbits were bled serially, weighed daily and observed over long time periods (2 to 4 ys). Details of such rabbit experiments have been reported [26, 29, 34].

Hyperimmune sera against brain s-antigen precipitated in the homologous system were prepared as follows. The line cut out from the agarose gel (see immuno-electrophoresis) was mixed with an equal volume of Freund's complete adjuvant (A), or with incomplete adjuvant (B), and kept at −20°C until use. Rabbits were inoculated intradermally in 3–5 sites with suspension A, and 3 weeks later were boosted twice at intervals of 10 days with suspension B. They were bled at monthly intervals.

Monospecific antisera against the subcomponents of the s-antigen, the purified 24 kD protein and the 38/40 kD protein [23] were produced in rabbits by inoculation of the immuno-affinity chromatography purified proteins in complete adjuvant, and then in incomplete adjuvant subcutaneously (s.c.) at day 0 and day 14.

Results

Borna viruses

BDV strains are poorly characterized. The only *constant* finding until now is that defined antibodies, for example those found in the CSF of infected horses or those produced against strain V, specifically "stain" tissue culture foci. Wild-type virus brain suspension usually produces fatal infection in rabbits following intracerebral inoculation. Our tissue culture studies using JKG cells show that almost half of such rabbit-virulent suspensions can also infect rabbit brain cells in the first passage. Recent epidemiological studies of Borna disease in endemic areas of Eastern Germany have yielded virus after one passage in tissue culture from 18 out of 29 histologically diagnosed equine cases of BDV infection. As in our earlier studies [14] certain regions of the brain harbour 10^3 to 10^4 ffu per gram of brain. The same is true for the wild-type ovine virus and this also grows efficiently in JKG cells (Table 1).

Attempts to culture feline BDV [35] on rabbit cells have proved unsuccessful. There is, however, evidence that this virus may have a different cell tropism from other BDV isolates.

It has been reported for a variety of animal species that passage of BDV in cell lines originating from the same species may lead to adaptation of the BDV isolate to that species [39, 20]. Thus the titre of BDV achieved in JKG cells usually increases with increasing passage level. Rat- and mouse-passaged virus can reach 10^4 ffu per gram of brain in JKG cells. A general phenomen is that the rabbit adapted strains can

Table 1. Borna disease virus (BDV) "strains"

Code	Number	Wild-type	Experimentally adapted	Virus titre (ffu/ml) in JKG cells	Collected by (reference)
	6	horse		10^4	H. Ludwig [14]
	18	horse		10^3	R. Dürrwald (unpubl.)
He/80	1	horse	rabbit, rat		R. Rott [17]
Dü/92	1	sheep		10^3	R. Dürrwald (unpubl.)
Ca/92	1	sheep		10^3	M. Caramelli (unpubl.)
Lun/92	1	cat		none[a]	A. Lundgren [35]
vac. str. "Zwick"	1		rabbit, rat, mouse, chicken	10^3	W. Zwick [51]
vac. str. "Dessau"			rabbit	10^3	R. Dürrwald (unpubl.)

[a] Based on BDV-specific serum antibodies

infect rats, hamsters, tree shrews and chicken. In rats BDV infection results in either an acute and fatal disease, or a chronic persistent infection. All other animals can be persistently infected with the rabbit virus. It is not known whether the wild-type strains have a similarly broad spectrum. In a small number of cases we could recover horse viruses by infection of one day old rats.

There is no apparent correlation between the vaccination status of horses and sheep, vaccinated with the two live virus vaccines ("Zwick-vaccine" based on strain V, JKG titre: 10^3 ffu/ml; "Dessau vaccine", JKG titre: $1 - 5 \times 10^3$ ffu/ml), and the incidence of outbreaks of disease.

Cell culture adapted viruses like that from G-26 [29], Oligo/TL-, ED/TL- and Vero/TL-cells do not reproducibly kill rabbits. For example infection with the Oligo/TL virus leads to survival of rabbits and induces high antibody titres and no disease. Since no BDV "strain" has ever been cloned, we have no information on the genetic make up of various BDV virus populations.

The pathogenic potential of these viruses (expressed by their ability to fatally, chronically, persistently infect or cause an obesity syndrome [34]) is linked both with the species and age of the animal, and in addition, depends on the virulence of these "strains". The latter parameter can only be studied by experimental infection of test animals and is therefore poorly defined. Some examples for BDV "strains" with their host-range are given in Table 1.

Our present knowledge still indicates that destruction or dysfunction of neurons is the cause of disease and death in animals infected with BDV [15, 47]. Recent studies of the neurobiological potential of Borna viruses have shown that the virus may target cells other than the neuron [6, 7, 15], and may influence neurotransmitter balance [4, 11, 13]. The passaging of BD viruses by experimental infection may change cell and animal tropism [20, 21, 39]. Nevertheless, in the natural hosts BDV infection produces a central nervous system disease of horses, sheep and most probably cats [35].

Antigens and antibodies

In the Borna-system, antibodies have characterized BDV-specific antigens. The soluble antigen (s-antigen), which is prepared by ultracentrifugation (100,000 g) of supernatant from infected brain or tissue culture preparations, had earlier been reported to contain major components with relative molecular weights of 22 kD and 40 kD [29]. The 40 kD protein is the 40/38 kD double band protein. These proteins are precipitated by horse-, rabbit- [30] and even by sera of persistently infected tree shrews [45]. Later, Haas et al. [16] purified the 40/38 kD protein. Our group found that s-antigen preparations contain the 60, 40/38, 25 and 14.5 kD proteins as major immunogenic components [34]. Recently Klein [23] purified the s-antigen complex from ammonium sulfate concentrated s-antigen solutions combined with moab-immunaffinity chromatography. The eluates consisted of high amounts of 40/38 kD and 25 kD proteins with some high molecular weight proteins (Fig. 1a, b). Further analysis showed that the s-antigen complex contains the 40/38 and the 25 kD proteins (Fig. 1c). Mono-specific polyvalent rabbit antisera raised against the 40/38 kD or the 25 kD proteins each recognize only their homologous proteins (Fig. 1d). No crossreactivity could be observed and neither sera detected antigens on the plasmamembrane of BDV infected cells using the IFT. These findings are divergent from those of Thiedemann et al. [48], who determined some features of the above proteins using infected tissue lysates and cells and found antigenic

---

→

Fig. 1. BDV s-antigen proteins: SDS-PAGE with silver stain in (**a**), (**c**), and Western blot in (**b**), (**d**). **a** Proteins from infected rat brain suspension were ammonium sulfate precipitated and submitted to immune affinity chromatography on a moab W1 coupled sepharose column; eluted proteins are stained; **b** proteins of the eluate in (**a**) were identified as the major immunogenic components (40/38 and 25 kD) by Western blots using moab W1 in (*A*) and R13 in (*B*); **c** purified 40/38 kD (*A*) and 25 kD (*B*) proteins; **d** Western blot of rabbit monospecific antisera prepared against each of the purified proteins in (**c**): total s-antigen from infected (*A* and *C*) and uninfected (*B* and *D*) rat brain was separated by SDS-PAGE and treated with anti-25 kD serum (*A* and *B*) and anti-40/38 kD serum (*C* and *D*). No crossreaction can be detected

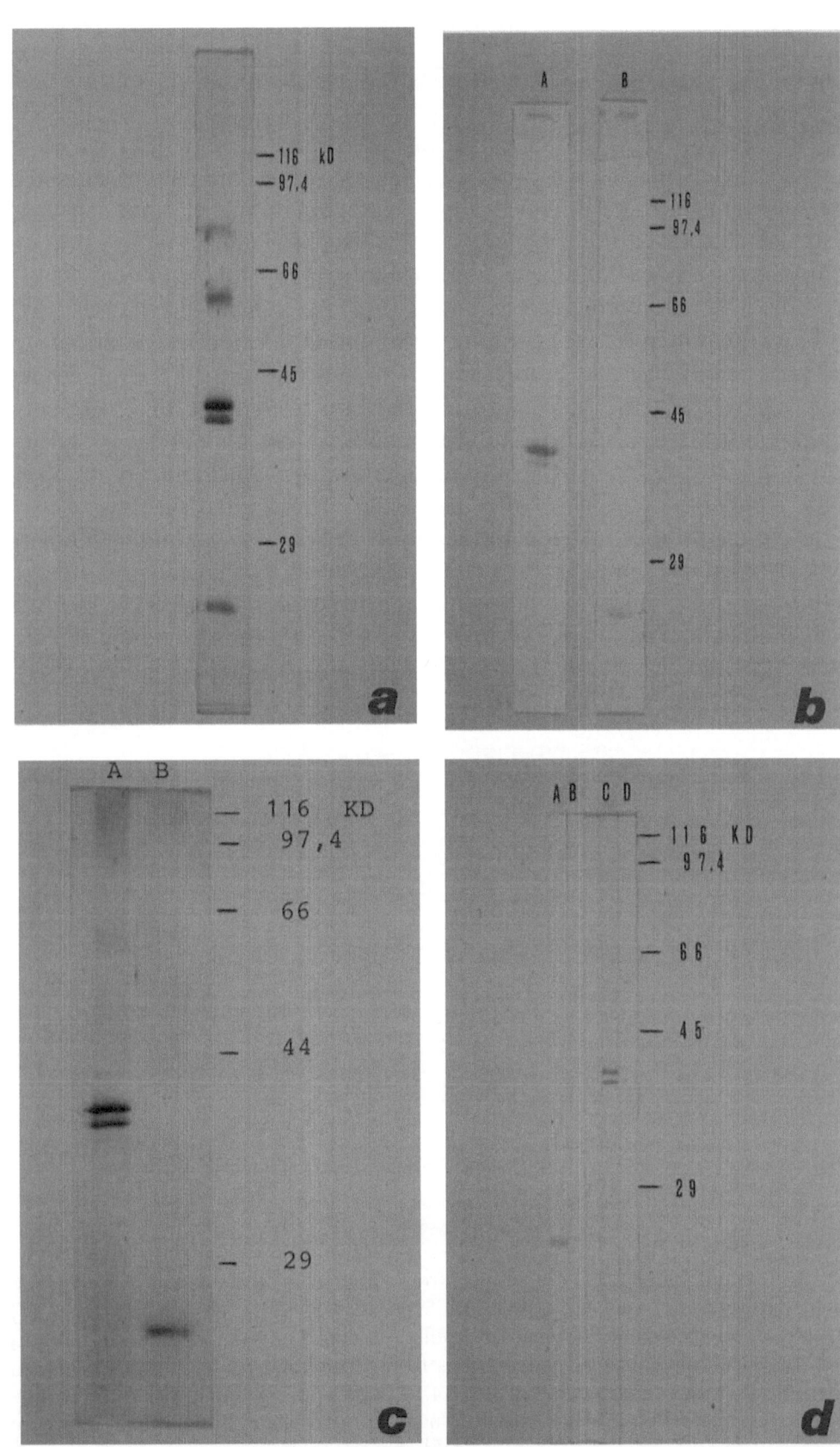

relationship together with a localization of the proteins on the cell surface.

No detailed information exists on the 60 kD protein [34] which is always present in infected brain suspensions. Our studies with infected horse brain and ED/TL (a persistently infected equine dermal cell line) suspensions showed that polyvalent sera, which recognize all the protein components of the s-antigen, detect the 60 kD, 40/38 kD and 25 kD proteins in the homologous system. In contrast, principally the 40/38 kD and low amounts of 25 kD protein were immunoblotted from extracts of the persistently infected human Oligo/TL cell line (Fig. 2). Preliminary data suggest that the 60 kD protein may be a complex of 25 kD sub-components (Briese, Bode, Ludwig unpubl.). This hypothesis is further supported by common epitope recognition by different moabs (see below).

Since antibodies of infected animals reflect the immuno-inducing capacity of major antigens, we have compared sera of terminally infected rabbits, rats and mice and anti-s-antigen hyperimmune sera. It became clear that some, but not all, terminally infected rabbit sera detect the 60, 40/38, 25 and occasionally the 14.5 kD proteins (compare Fig. 5 in [34]),

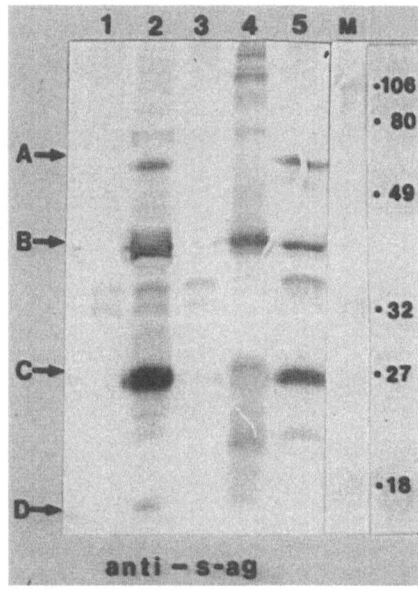

Fig. 2. Western blot of BDV s-antigen from different sources detected with BDV rabbit serum (EV1; see also Table 2). Separation of 10% suspensions from (2) persistently infected equine dermal (ED) cells, (4) human oligodendroglia cells, and (5) brain of a BD horse; control preparations from (1) normal ED cells and (3) human cells. A 60 kD; B 40/38 kD; C 25 kD, D 14.5 kD proteins. Immunostain with alkaline phosphatase-anti-rabbit IgG and Fast Red/naphthol-phosphate, as described elsewhere

[5]

Table 2. Borna disease virus (BDV)-specific rabbit hyperimmune sera against s-antigen components

No.	Code	IFT	Antibody titre neutralization	Recognition of viral antigens[a] in protein gels (non-reducing conditions)			
				60	40/38	25	14.5 (Kd)
1	EV 1	20,000	0	+	+	+	+
2	EV 3	20,000	0	+	+	+	
3	HB 1	1,000	0				
4	LL 2	2,000	0				
5	RNK 2	500	0	+		+	
6	RNK 3/3	800	0		+		

1–4: Rabbits were immunized with immunoprecipitated line-antigen (HB 1, LL 2) or electrophorectically purified line-antigen (EV 1, EV 3); for details see Materials and methods

5,6: Rabbits were immunized against the purified 25 Kd (RNK 2) or 38/40 Kd protein (RNK 3/3) of s-antigen from rat brain

[a] Western blot with s-antigen from BD horse brain

whereas hyperimmune sera against s-antigen constantly detect those major components in s-antigen preparations (Table 2). By using CSFs from different animal species for Western blots, we have already detected significant antibody titres against the 60 and 40/38 kD protein, which vary from individual to individual (compare Fig. 9 in [34]).

Table 3. Borna disease virus (BDV)-specific monoclonal antibodies (moabs)

Code[a]	IgG subclass	Mean titre (cell ELISA)	Recognition of viral antigens in				Neutralization[d]
			cells[b]	protein gels[c]			
				60*	38/40	25 (KD)	
Kfu 1	1	1:500	n/c	+		+	0
Kfu 2	1	1:1,000	n/c	+		+	0
Kfu 3	2b	1:1,500	n		+[(*)]		0
Kfu 4	2b	1:500	n/c	+		+	0
R13	2a	1:100	n/c	+		+	0
W1	2a	1:200	n/c		+		0
W10	2a	1:200	n/c		+		0

[a] Kfu moabs were established by K. Furuya, the other moabs by G. Pauli; the latter were partially characterized earlier [1] and named BDA 1, 2, and 3

[b] Monitored by IFT; n nuclear, c cytoplasmic fluorescence

[c] Determined by Western blot (WB) and immunoprecipitation (IP); * best under non-reducing conditions (Lämmli-system); (*) only IP positive, WB negative

[d] Assay with as well as without addition of complement

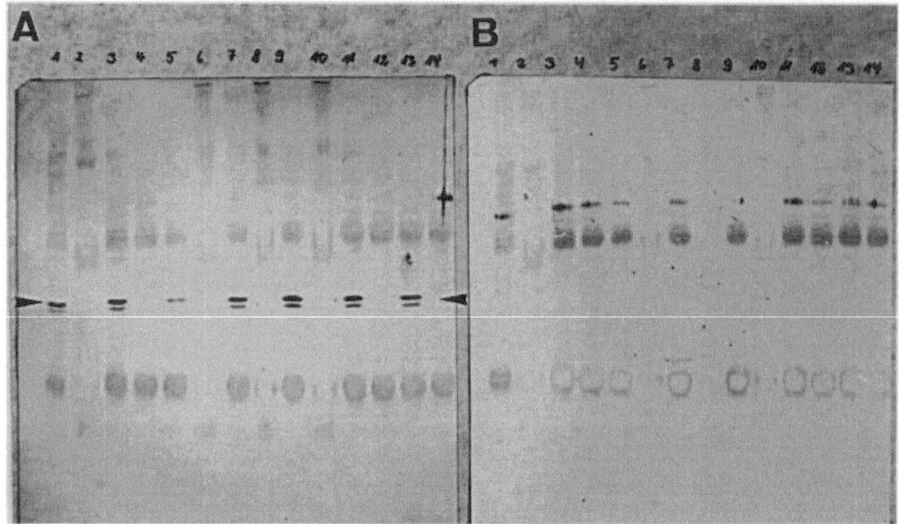

Fig. 3. Immunoprecipitation of lysates from BDV infected (numbers *1,3,5,...*) and uninfected (numbers *2,4,6,...*) Vero cells with undiluted hybridoma supernatants from subclones of moab Kfu 3. Immunostain of the blotted proteins with (**A**) rabbit antibodies against BDV, serum EV3, and (**B**) normal rabbit serum, following a protocol previously described [2, 5]; arrows mark the 40/38 kD protein on which Kfu 3 recognizes a discontinuous epitope

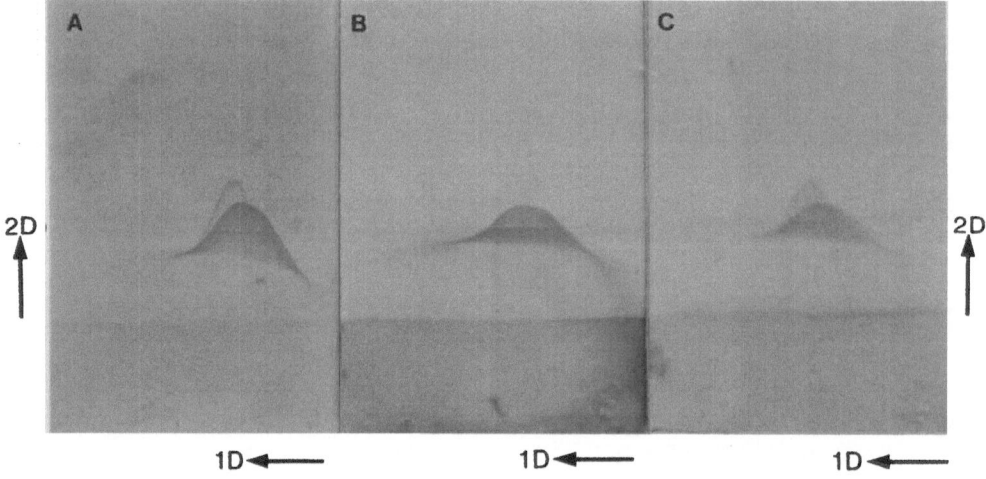

Fig. 4. BDV s-antigen from rat brain stained with moabs after precipitation by two-dimensional (2D) immunoelectrophoresis with BDV-specific rabbit antiserum. **A** Protein staining with Coomassie G 250. **B** Immunostain with moab Kfu 3 and anti-mouse alkaline phophatase conjugate/water-insoluble substrate as described in this paper. **C** Immunostain with moab Kfu 2; conjugate and substrate as in **B**. 1D is the first, 2D the second dimension of the electrophoresis

Fig. 5. Indirect immunofluorescent test with BDV infected, acetone fixed JKG (**A,B**) and ED/TL cells (**C**), using moab Kfu 1 (**A**), Kfu 3 (**B**), and horse CSF (**C**); human and cat antibodies result in similar antigen patterns (nuclear fluorescence) like shown in **B** and **C**

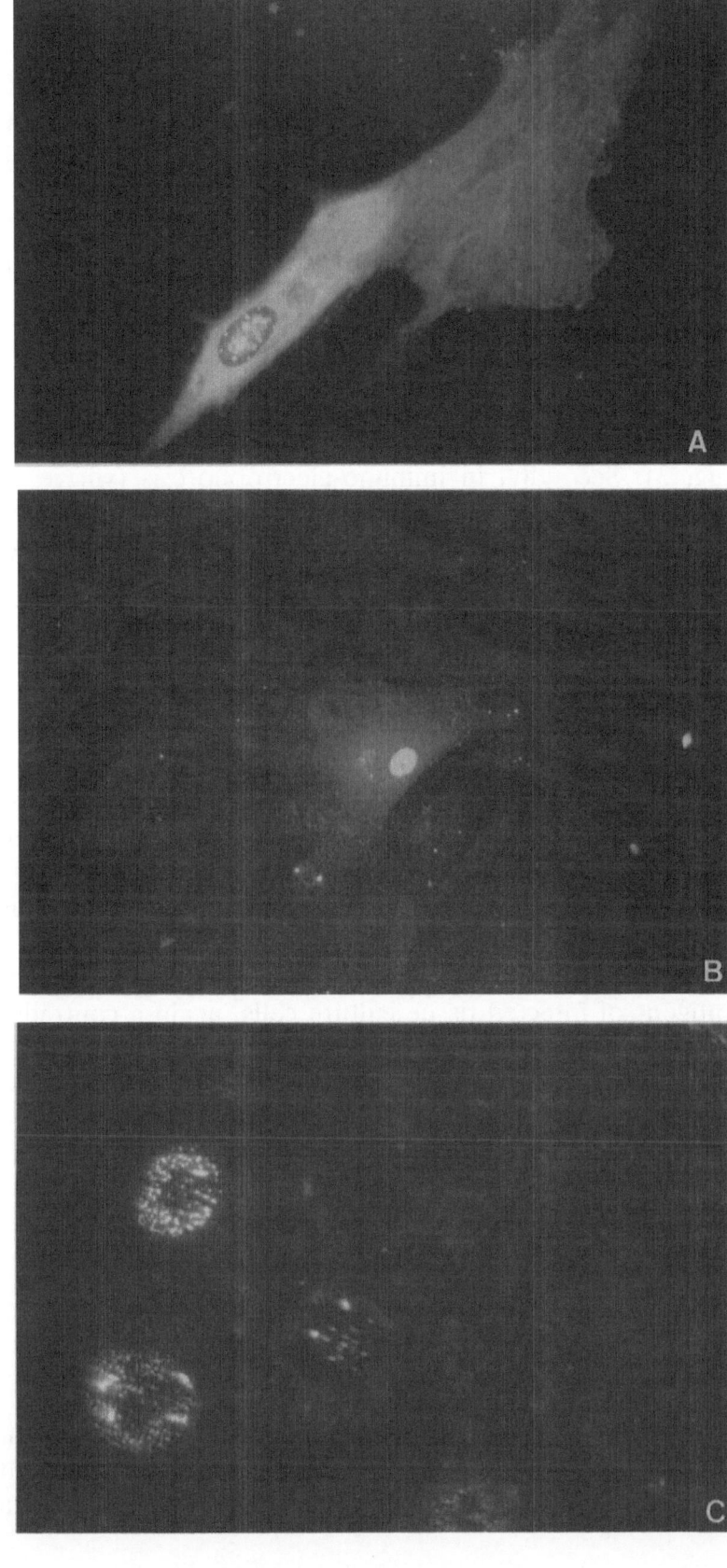

The BDV antigen-specific moabs are summarized in Table 3. It is apparent that they fall into two categories: 1. those recognizing the 60 kD and 25 kD protein (Kfu 1, 2, R 13) and 2. others detecting the 40/38 kD double band (Kfu 3, Kfu 4, W1, W10). In immunoblotting reactions (Western blots) most often the moabs bind with SDS-PAGE denatured proteins. A typical picture is shown in Fig. 1b. The data indicate that all except Kfu 3 recognize continuous epitopes. There are two reasons why we believe that Kfu 3 binds to a discontinuous antigenic site on the 40/38 kD proteins. Firstly, even at high moab concentrations no reaction is seen with denatured protein in Western blots, whereas in immunoprecipitation with s-antigen the 40/38 kD band is selectively bound. This immunoreactivity has been tested with a variety of Kfu 3 clones (Fig. 3). Secondly, in immuno-electrophoresis (where native s-antigen was immunoprecipitated by rabbit serum and usually 2 precipitation bands could be recognized), an additional incubation with Kfu 1 or 3 showed that Kfu 3 detected only epitopes in one band, namely the one containing the 40/38 kD protein (Fig. 4). Further confirmation comes from indirect immunofluorescence tests (see below).

In infected cells the antigen recognition by the different monoclonal antibodies was uniform. They all detected epitopes in the nucleus as well as the cytoplasm, illustrated in Fig. 5A, whereas Kfu 3 mainly detects nuclear antigen in acetone fixed infected cells (Fig. 5B). For this reason Kfu 3 was used for specificity tests in which antibodies in animal and human sera were investigated and titred [5, 35]. The antigen recognition picture of CSFs from horses and sera of humans and cats are similar to that of Kfu 3 (Fig. 5C).

In separate, reproducible experiments non of the moabs detected surface antigens of infected tissue culture cells, again a contradiction to the report of Thiedemann et al. [48]. With different kinds of sera and CSFs it had, however, occasionally been demonstrated that infected cell antigens exist on the plasma membrane [31]. Membrane fluorescence could be specifically observed with CSF when high neutralizing titres were present (see below). It is of considerable importance that all sera and moabs directed against the major antigenic components never showed neutralizing capacity (see below).

Neutralization

In the Borna-system, a complement independent IgG-specific neutralization of virus can be measured [34]. This has first been detected in the rabbit system. From 192 rabbits infected by different routes, 19 animals

Table 4. Neutralizing antibodies (n-abs) in various animal species after infection with Borna disease virus (BDV)

Species	Number of animals	Infected as	Number of n-ab positive animals	Neutralization titre[a] at weeks post infection		
				6	10	20
Mouse	75	newborn	75	10	100	160
Rat	20	adult	20	100	40	40
Hamster	5	newborn	5	ND	10	20
	4	adult	4	ND	10	20
Chicken	25	newborn	25	80	320	1,000
Rabbit	192	adult	19	between 10 and 80		
Treeshrew	3	adult	2	mean 20		

[a] Mean of five animals; *ND* not done

were positive with significant titres of between 1:10 to 1:80. In all other experimentally infected animal species, neutralizing antibodies were detectable as summarized in Table 4.

Highest titres in these series, by which kinetics of neutralization were measured, were present in the chicken. Independently from these experiments in 4 long-term survivors (four chicken infected at day 1), a rooster (see also Fig. 14a in [33]) had maximal titres of 1:2,000 three years after infection.

In another series, we found that infection of newborn rats (as had been the case in older rats) induced low titres up to 1:100, not exceeding 1:300, during the observation period up to 5 months p.i.. Newborn mice observed up till 7 months p.i. regularly reached titres of 1:500 to 1:1,000. Newborn hamsters (which respond in a similar way as 4 month old hamsters) observed until 6 months p.i. only had titres of 1:20 to 1:80. One day old chicken, observed until 4 months p.i., reached maximum titres of 1:1,000 to 1:5,000. Besides these series of experiments numerous serum samples (not listed here) of different animals species infected with different strains reproducibly had neutralizing antibodies. In a pilot experiment 5 quails, infected at day 1 also had neutralizing antibodies 2 months later. This clearly shows that under experimental conditions each animal species produces species-specific amounts of neutralizing antibodies, with a tendency of increasing titres in the following species: *hamster < rat < mouse < rabbit (splenectomized) < chicken.*

A reproducible experimental model for the production of neutralizing antibodies was found in the splenectomized rabbit. The increase of antibody titres over several years in 4 animals which survived i.p. infection are summarized in Table 5. It was of great interest that in animal BP-11 two months after a single i.p. inoculation, a titre of 1:1,000, increasing up to 1:4,000, was reached. This animal lived for 4 years without significant decline in antibody titre.

Successful experimental BDV infection has been demonstrated by ophthalmoscopy [25, 26]. Four weeks p.i. the typical signs of a multifocal retinopathy (Fig. 6) were seen. As illustrated by rabbit BP-11, these lesions healed when neutralizing antibodies appeared and increased with time. As in the other splenectomized animals, no clinical signs of blindness persisted. Virus could neither be isolated nor was antigen demonstrable by immuno-histology from the brain of BP-11 [14].

When the sera from naturally infected horses, collected from 1970 to 1988, were analysed it was found that nine out of fifty five animals had titres between 1:100 and 1:2,000. From samples from 30 animals originating from endemic areas of Sachsen and Thüringen (Germany) collected in 1991/92, three CSFs were positive (maximum titre 1:1,280). In

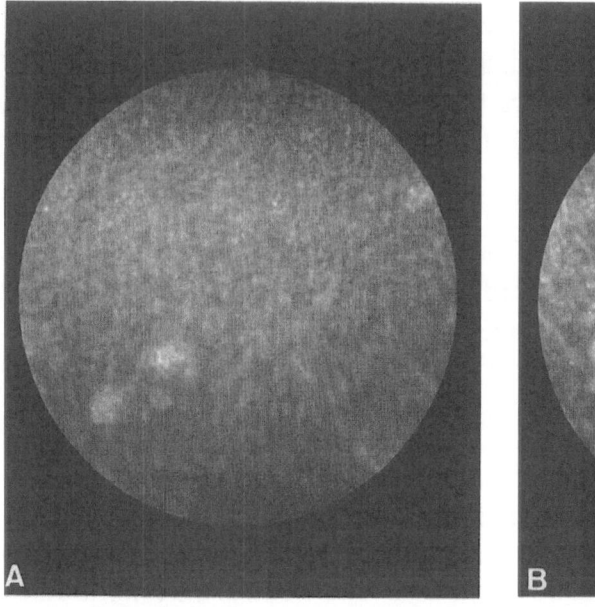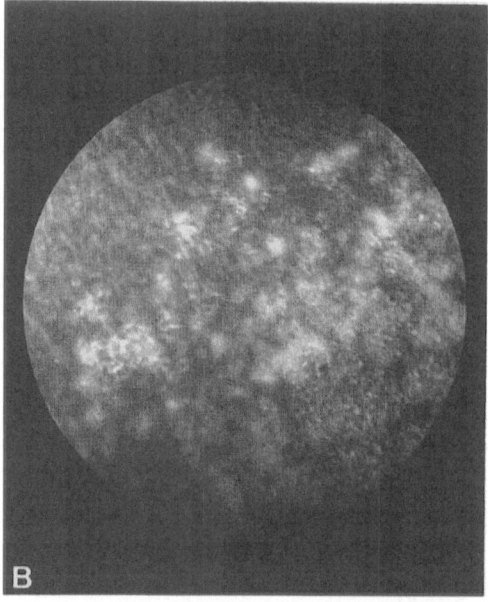

Fig. 6. Appearance (**A**) and progress (**B**) of fundus lesions during the BDV-induced multifocal retinopathy as shown with fluorescein angiorgraphy [25, 26] in rabbit BP-11 (splenectomized). During the healing process, high neutralizing antibody titres could be measured. We are grateful to Hauke Krey to lend us these pictures

each case the neutralization titres considerably exceeded the IFT titres. In some cases a comparison of CSF and serum titres was done and the results supported the earlier hypothesis of a local CNS immune response in BDV infection [30, 32].

Table 5. Kinetic of neutralizing antibodies in splenectomized rabbits intraperitoneally (i.p.) infected with Borna disease virus (BDV)

Rabbit code	Neutralizing antibody titre[a] at months post infection							
	1	1,5	2	3	6	10	13	18
BP-11	100	1,000	1,000		4,000	4,000	3,000	4,000
BP-12		100	500	1,000	500	500	400	500
BP-8	10		500	500	80	50	50	50
BP-18				60	80	40	40	40

[a] Mean of 4 assays

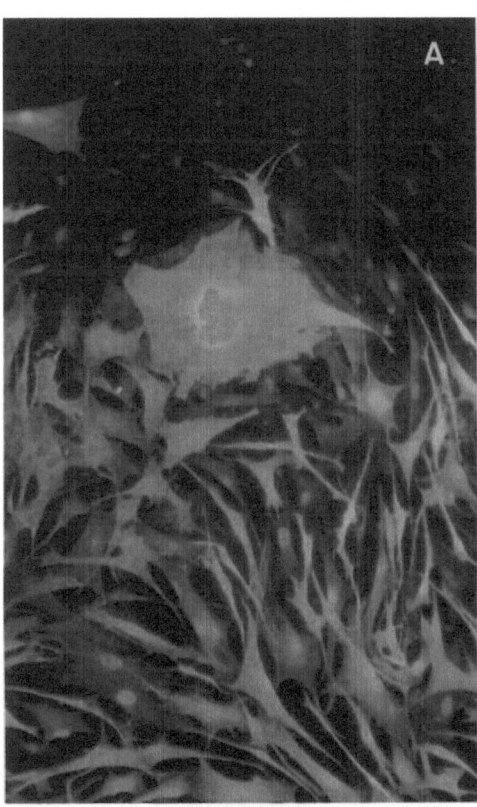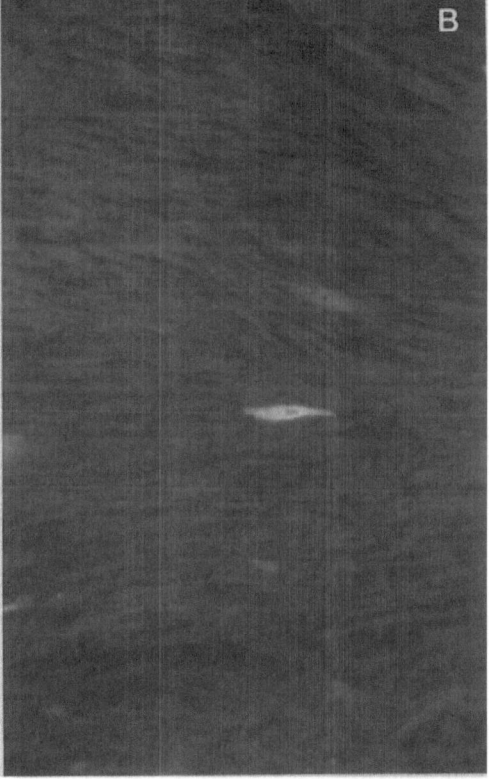

Fig. 7. BDV cell spread inhibition test (JKG cells). **A** No inhibition, if anti-s-antigen antibodies (poly- or monoclonal) were added to the overlay medium; **B** inhibition of virus spread, if neutralizing antibodies of any animal species (here BP-11; rabbit serum) were added

Since sheep farming has drastically decreased in endemic BD areas, relevant investigations could not be done, because diseased animals were not available.

The mortality in horses and sheep following BDV infection has been the reason for routine vaccination in endemic areas. When sera from immunized animals were analysed it was found that 28 out of 109 horses had neutralizing titres of 1:8 to 1:240 (average 1:32). 42 out of 472 vaccinated sheep had titres of 1:8 to 1:40. Since veterinarians in these areas have observed a decrease in BD cases after vaccination, this vaccine might be protective. It cannot be excluded that neutralizing antibodies with titres of 1:4 and higher may represent a reliable in vivo marker for protection against BD.

There is no information concerning the mechanism and functional importance of BDV neutralization. In a small number of in vitro experiments we have tested whether neutralizing rabbit sera (for example BP-11), horse CSFs (for example No. 149), other anti-s-antigen hyperimmune sera or moabs had any effect on the spread of BDV in JKG tissue culture cells. It was of interest that when neutralizing antibodies were added to the overlay medium shortly after adsorption (see Materials and methods), virus spread was consistently prevented, and was restricted to the first infected cell (Fig. 7). Sera containing only s-antigen antibodies or moabs did not reduce the number of antigen carrying cells in a focus (measured 5 d.p.i.). This was a reproducible finding with all the neutralizing sera obtained from rats, mice, hamsters, chicken and the CSFs of horses. In experiments where neutralizing antibodies were

Table 6. Comparison of neutralization of Borna disease virus (BDV) from brain and tissue culture

Neutralizing antibody	Wild-type (horse) brain virus	Adapted (rat p8) brain virus		Tissue culture (Oligo/TL)			
				cell virus		released virus	
	exp 1	exp 1	exp 2	exp 1	exp 2	exp 1	exp 2
Horse CSF 149		1,280	1,280	160	640	160	
Horse serum 149		320		10		10	
Horse CSF 4A	160	160		10		10	
Rabbit BP-11	1,280	1,280	320	160	160	320	80
Chicken H4			1,280		1280		320

In the neutralization assays, 50 ffu of virus were incubated with serial antibody dilutions; the titre is defined as highest dilution with 50% focus reduction; *exp 1* and *2* two different experiments

added to the medium at days 1, 2, 3, etc. a correlation with reduction of antigen carrying cells per focus was measured. The test was always read at 5 days p.i. (data not shown).

After the phenomenon of neutralization had been discovered, we used this test to determine whether serological variants in different strains could be observed. All efforts show that the wild-type and adapted strains (of brain or tissue culture origin) used as tissue suspensions were neutralized to the same extent by rabbit serum (BP-11) and horse CSFs. This indicates that no serological differences exist (Table 6).

When BDV released by hypertonic medium was compared with brain virus in similar experiments, a significantly lower neutralization effect was measured (Table 6).

Discussion

Refined infection tests for BDV released from tissue culture [40], its neutralizability [34], and the establishment of BDV-specific cDNA clones [27] have led to its characterization as a 8.5 kb negative-strand RNA virus [3]. It is generally accepted that the infectious virus is enveloped. It is likely from our data that none of the major immunogenic components (60, 40/38, 25 kD proteins), which are abundantly present in the s-antigen of infected tissues, are accessible on the virus envelope and the membrane of infected cells. These proteins may represent a virus structural element, although this remains speculative [36, 49]. Analysis of the BD system has been performed using BDV-specific cDNA clones. However, in view of the method of production of these clones [27], it is not possible to exclude the slight possibility that the major immunogenic proteins are of cellular origin, expressed under BDV infection.

Our findings suggest, in consensus with others [6, 16], that the 40/38 kD protein represents a typical nuclear antigen. This would be consistent with the accumulation of this antigen together with virus nucleic acid (demonstrable by in situ hybridization; Gosztonyi et al., unpubl.) of this intranuclearly transcribed virus [3] in intranuclear (Joest-Degen) inclusion bodies. Virus-like particles are also found at this same site [43].

It remains uncertain whether the BDV envelope with its proteins is of cellular or viral origin, although the phenomenon of virus neutralization implies the existence of a viral envelope containing virus-specific proteins. Preliminary data suggest that certain carbohydrate chains are involved in the neutralizing process (Stoyloff, unpubl.). The finding that neutralizing antibodies occur only after multiple infections of rabbits [8], contradicts our observation that one infection is sufficient. We believe that virus persistence and reactivation stimulates the production of neutralizing antibodies.

Neutralizing titres and recognition of cell membrane components by such antibodies [31; and unpubl. data] are, as yet, an unexplained phenomenon. Neutralizing antibodies have been consistently demonstrated in our experiments when various animal species (horses, sheep and laboratory animals) were infected with identified "strains" of BDV, although other groups have failed to demonstrate their existence [38, 42]. Why splenectomy enhances the generation of these antibodies remains unclear. Information on the role of the spleen in infectious processes is still poor [50]. Neutralizing activity has not been demonstrated in humans [4, 5] and cats [35]. These species appear to have their own, different, strains. The phenomenon of virus neutralization, which is known to represent a poorly understood mechanism [10, 22], awaits further elucidation.

The inhibition of BDV spread from cell to cell by neutralizing antibodies in tissue culture is reminiscent of the rabies system [9]. However, extensive in vivo experiments in the Borna system have not yet been done. Since both viruses are neuron-specific under natural conditions [14, 15], and seem to have many other properties in common, including evolutionary old sequences in RNAs [36], a search for further unifying concepts of these neurotropic viruses is indicated.

Acknowledgements

We are grateful to Josh Slater, Department of Clinical Veterinary Medicine, University of Cambridge, U.K., for reading the manuscript. These investigations have been supported by a BMFT-grant No. 07017660 and a DFG-grant (LU-142/5-1) to HL and in part by the BMG (LB).

References

1. Bause-Niedrig I, Pauli G, Ludwig H (1991) Borna disease virus-specific antigens: two different proteins identified by monoclonal antibodies. Vet Immunol Immunopathol 27: 293–301
2. Beutin L, Bode L, Özel M, Stephan R (1990) Enterohemolysin production is associated with a temperate bacteriophage in E. coli serogroup 026 strains. J Bacteriol 172: 6469–6475
3. Briese T, de la Torre JC, Lewis A, Ludwig H, Lipkin WI (1992) Borna disease virus, a negative-strand RNA virus, transcribes in the nucleus of infected cells. Proc Natl Acad Sci USA 89: 11486–11489
4. Bode L, Ferszt R, Czech G (1993) Borna disease virus infection and affective disorders in man. In: Kaaden OR, Eichhorn W, Czerny CP (eds) Unconventional

agents and unclassified viruses. Springer, Wien New York, pp 159–167 (Arch Virol [Suppl] 7)

5. Bode L, Riegel S, Lange W, Ludwig H (1992) Human infections with Borna disease virus: seroprevalence in patients with chronic disease and healthy individuals. J Med Virol 36: 309–315

6. Carbone K, Moench T, Lipkin WI (1991) Borna disease virus replicates in astrocytes, Schwann cells and ependymal cells in persistently infected rats; location of viral genomic and messenger RNAs by in situ hybridization. J Neuropathol Exp Neurol 50: 204–214

7. Carbone KM, Duchala CS, Griffin JW, Kincaid AL, Narayan O (1987) Pathogenesis of Borna disease in rats: Evidence that intra-axonal spread is the major route for virus dissemination and the determinant for disease incubation. J Virol 61: 3431–3440

8. Danner K, Lüthgen K, Herlyn M, Mayr A (1978) Vergleichende Untersuchungen über Nachweis und Bildung von Serumantikörpern gegen das Borna-Virus. Zbl Vet Med B 25: 345–355

9. Dietzschold B, Kao M, Zheng Y, Chen Z, Maul G, Fu Z, Rupprecht CE, Koprowski H (1992) Delineation of putative mechanisms involved in antibody-mediated clearance of rabies virus from the central nervous system. Proc Natl Acad Sci USA 89: 7252–7256

10. Dimmock N (1984) Mechanisms of neutralzation of animal viruses. J Gen Virol 65: 1015–1022

11. Dittrich W, Bode L, Kao M, Schneider K (1989) Learning deficiencies in Borna disease virus-infected but clinically healthy rats. Biol Psychiatry 26: 818–828

12. Furuya K, Noro S, Sakurada N (1984) Production of monoclonal antibodies to influenza virus A/USSR/92/77 (H1N1) and their application to epidemiological study. Rep Hokkaido Inst Publ Health 34: 1–10

13. Gosztonyi G, Ludwig H (1984) Neurotransmitter receptors and viral neurotropism. Neuropsychiatr Clinica 3: 107–114

14. Gosztonyi G, Ludwig H (1984) Borna disease of horses: An immunohistological and virological study of naturally infected animals. Acta Neuropathol 64: 213–221

15. Gosztonyi G, Dietzschold B, Kao M, Rupprecht CE, Ludwig H, Koprowski H (1993) Rabies and Borna disease: A comparative pathogenetic study of two neurovirulent agents. Lab Invest 68: 285–295

16. Haas B, Becht H, Rott R (1986) Purification and properties of an intra-nuclear virus-specific antigen from tissue infected with Borna disease virus. J Gen Virol 67: 235–241

17. Herzog S, Kompter C, Frese K, Rott R (1984) Replication of Borna disease virus in rats: age-dependent differences in tissue distribution. Med Microbiol Immunol 173: 171–177

18. Heinig A (1969) Die Bornasche Krankheit der Pferde und Schafe. In: Röhrer (ed) Handbuch der Virusinfektionen bei Tieren, Vol 4. Gustav-Fischer, Jena, pp 83–148

19. Hiepe T (1958) Die Bornasche Krankheit. Habilitationsschrift, University of Leipzig

20. Hirano N, Kao M, Ludwig H (1983) Persistent, tolerant or subacute infection in Borna disease virus-infected rats. J Gen Virol 64: 1521–1530

21. Kao M, Ludwig H, Gosztonyi G (1984) Adaptation of Borna disease virus to the mouse. J Gen Virol 65: 1845–1849

22. Kennedy-Stoskopf S, Narayan O (1986) Neutralizing antibodies to Visna lentivirus: Mechanism of action and possible role in virus persistence. J Virol 59: 37–44

23. Klein N (1990) Reinigung und Charakterisierung Bornavirus-spezifischer s-Antigenkomponenten. Vet Med Diss, FU Berlin

24. Köhler G, Milstein C (1975) Continuous cultures of fused cells secreting antibody of predefined specificity. Nature 256: 495–497

25. Krey HF, Ludwig H, Boschek CB (1979a) Multifocal retinopathy in Borna disease virus infected rabbits. Am J Ophthalmol 87: 157–164

26. Krey HF, Ludwig H, Gierend M (1981) Borna disease virus-induced retino-uveitis treated with immunosuppressive drugs. Graefes Arch Clin Exp Ophthalmol 216: 111–119

27. Lipkin WI, Travin GH, Carbone KM, Wilson MC (1990) Isolation and characterization of Borna disease agent cDNA clones. Proc Natl Acad Sci USA 87: 4184–4188

28. Ludwig H, Becht H (1977) Borna disease- a summary of our present knowledge. In: ter Meulen, Katz (eds) Slow virus infections of the central nervous system. Springer, Berlin Heidelberg New York, pp 75–83

29. Ludwig H, Becht H, Groh L (1973) Borna disease (BD), a slow virus infection. Biological properties of the virus. Med Microbiol Immunol 158: 275–289

30. Ludwig H, Koester V, Pauli G, Rott R (1977) The cerebrospinal fluid of rabbits infected with Borna disease virus. Arch Virol 55: 209–223

31. Ludwig H, Pauli G, Gierend M (1984) Nachweis von Oberflächen Antigenen auf Borna Virus infizierten Zellen. Berl Münch Tierärztl Wochenschr 97: 47–51

32. Ludwig H, Thein P (1977) Demonstration of specific antibodies in the central nervous system of horses naturally infected with Borna disease virus. Med Microbiol Immunol 163: 215–226

33. Ludwig H, Kraft W, Kao M, Gosztonyi G, Dahme E, Krey HF (1985) Die Borna-Krankheit bei natürlich und experimentell infizierten Tieren: Ihre Bedeutung für Forschung und Praxis. Tierärztl Praxis 13: 421–453

34. Ludwig H, Bode L, Gosztonyi G (1988) Borna disease: A persistent virus infection of the central nervous system. Prog Med Virol 35: 107–151

35. Lundgren A-L, Czech G, Bode L, Ludwig H (1993) Natural Borna disease in domestic animals others than horses and sheep. J Vet Med B 40: 298–303

36. McClure MA, Thibault KJ, Hatalski CG, Lipkin WI (1992) Sequence similarity between Borna disease virus p40 and a duplicated domain within the Paramyxovirus and Rhabdovirus polymerase proteins. J Virol 66: 6572–6577

37. Morrison DK, Carter JK, Moyer KW (1985) Isolation and characterization of monoclonal antibodies directed against two subunits of rabbit poxvirus-associated, DNA-directed RNA polymerase. J Virol 9: 670–680

38. Narayan O, Herzog S, Frese K, Scheefers H, Rott R (1983) Pathogenesis of Borna disease in rats: Immune-mediated viral ophthalmoencephalopathy causing blindness and behavorial abnormalities. J Infect Dis 148: 305–315

39. Nitzschke E (1963) Untersuchungen über die experimentelle Bornavirus-Infektion bei der Ratte. Zentralbl Veterinarmed [B] 10: 470–527

40. Pauli G, Ludwig H (1985) Increase of virus yields and release of Borna disease virus from persistently infected cells. Virus Res 2: 29–33

40a. Pauli G, Grunmach J, Ludwig H (1984) Focus-immunoassay for Borna disease virus-specific antigens. Zentralbl Veterinarmed [B] 31: 552–557

41. Richt JA, Herzog S, Binz T, Niemann H, Clements JE, Narayan O, Rott R (1993) Borna disease virus: nature of the agent and significance of infection in man. In:

Kaaden OR, Eichhorn W, Czerny CP (eds) Unconventional agents and unclassified viruses. Springer, Wien New York, pp 101–109 (Arch Virol [Suppl] 7)

42. Rott R, Herzog S, Fleischer B, Winokur H, Amsterdam JD, Dyson W, Koprowski H (1985) Detection of serum antibodies to Borna disease virus in patients with psychiatric disorders. Science 228: 755–756

43. Sasaki S, Ludwig H (1993) In Borna disease virus infected rabbit neurons 100 nm particle structures accumulate at areas of Joest-Degen inclusion bodies. J Vet Med B40: 291–297

44. Skjodt K, Schou C, Koch C (1984) Assay for the specificity of monoclonal antibodies in crossed immunoelectrophoresis. J Immunol Methods 72: 243–249

45. Sprankel H, Richarz K, Ludwig H, Rott R (1978) Behavior alterations in tree shrews (Tupaia glis, Diard 1820) induced by Borna disease virus. Med Microbiol Immunol 165: 1–18

46. Stitz L, Sobbe M, Bilzer T (1992) Preventive effects of early anti-CD4 or anti-CD8 treatment on Borna disease in rats. J Virol 66: 3316–3323

47. Stitz L, Bilze T, Richt JA, Rott R (1993) Pathogenesis of Borna disease. In: Kaaden OR, Eichhorn W, Czerny CP (eds) Unconventional agents and unclassified viruses. Springer, Wien New York, pp 135–151 (Arch Virol [Suppl] 7)

48. Thiedemann W, Presek P, Rott R, Stitz L (1992) Antigenic relationship and further characterization of two major Borna disease virus-specific proteins. J Gen Virol 73: 1057–1064

49. Van de Woude S, Richt JA, Zink MC, Rott R, Narayan O, Clements JE (1990) A Borna virus cDNA encoding a protein recognized by antibodies in humans with behavorial diseases. Science 250: 1278–1281

50. Wara DW (1981) Host defense against streptococcus pneumoniae: The role of the spleen. Rev Infect Dis 3: 299–309

51. Zwick W (1939) Bornasche Krankheit und Encephalomyelitis der Tiere. In: Gildenmeister, Haagen, Waldmann (eds) Handbuch der Viruskrankheiten II. Fischer, Jena, pp 254–354

Authors' address: Dr. H. Ludwig, Institute of Virology, Free University of Berlin, Nordufer 20, D-13353 Berlin, Federal Republic of Germany.

Arch Virol (1993) [Suppl] 7: 135–151

Pathogenesis of Borna disease

L. Stitz[1], **T. Bilzer**[2], **J.A. Richt**[1], and **R. Rott**[1]

[1] Institut für Virologie, Justus-Liebig-Universität, Gießen
[2] Abteilung für Neuropathologie, Heinrich-Heine-Universität, Düsseldorf,
Federal Republic of Germany

Summary. Borna disease represents a unique model of a virus-induced immunological disease of the brain. Naturally occurring in horses and sheep, the mechanisms of pathogenesis have been studied in experimental animals, namely in the rat. Many investigations have revealed that the infection of the natural hosts principally follows the same pathogenic pathways as observed in rats, leading to a severe encephalomyelitis. This affliction of the central nervous system results in severe neurological disorders that again, are fully comparable in laboratory animals to those in the natural and the different experimental hosts. In addition, alterations have been reported which are also based on the infection of the brain and do not result in the classical encephalitic clinical picture but rather in alterations of behavior. However, to all of our knowledge, the various clinical pictures of Borna disease are not caused by the infecting virus itself but rather by the hosts immune response towards it, i.e. by a virus-induced cell-mediated immunopathological reaction. The importance of virus-specific CD4+ T cells as exemplified by a cultured T cell line and of CD8+ T cells as shown by immunomodulatory substances and specific antibody treatment in vivo for the pathogenesis of acute Borna disease will be elucidated here. In addition, evidence will be provided that virus-specific CD8+ T cells are also responsible for the dramatic brain atrophy in the chronic phase of the disease in rats. Therefore, Borna disease not only lends itself exquisitely well to the study of the pathogenesis of an immunopathological disease of the brain but also represents one of the few models for immune-mediated tissue destruction that eventually leads to brain atrophy and clinically to dementia.

Introduction

Afflictions of the central nervous system have not only been in the focus of scientific research since most recently, when pathological alteration caused by viruses or virus-like agents have attained public interest. Most prominently, the presence of HIV-1 antigen in the brains of human patients suffering from dementia [26] and the most recent appearance of "mad-cow disease" caused by the yet unidentified "Scrapie agent" might be mentioned. Many autoimmune and virus-induced immuno-pathological alterations of the brain have been studied in experimental animals and serve as models for animal and human diseases. In the classical example, infection of mice with the lymphocytic choriomen-ingitis virus and recently also in Theiler virus infection in mice, the presence and the pathogenic importance of a virus-specific immune response, especially mediated by T cells, has been demonstrated in the brain [6, 10]. Other models studied in rodents include corona virus and measles virus infections in rats where an autoimmune reaction is of crucial importance in the pathogenesis of virus-induced diseases of the brain [19]. In addition to HIV in AIDS encephalopathy, several obser-vations in human patients support this view, namely the detection of measles virus in cases of subacute sclerosing panencephalopathy [33], and of papovavirus in progressive multifocal leukoencephalopathy [22, 41]. Furthermore, several viruses have been suggested to be involved in the development of multiple sclerosis [18].

In general, the outcome of a viral infection depends on characteristics of the virus and the efficiency and speed of the immune system including natural defense mechanisms to react to the invading agent. Viruses that cause cytopathogenicity are a much greater threat to the host than viruses that are non-cytopathic. Consequently, the evolution of the im-mune system has been significantly influenced by the demand for an efficient and rapid elimination of cytolytic viruses to avoid widespread infection and reduce tissue destruction. However, this strategy of a highly efficient immune reaction bears a considerable disadvantage for the host in the case of infection with persistent viruses lacking cyto-pathogenicity, since the immune system is apparently not capable of distinguishing between cytolytic and non-cytolytic viruses. The result may be damage to the host by an immune reaction although the agent that induced the immune response is otherwise perfectly innocuous.

Borna disease and Borna disease virus

An example of such a non-cytolytic virus is Borna disease virus. The virus and, respectively, the disease has been named after the town of Borna in Saxony, where in 1895 an endemic among horses of a cavalry regiment resulted in the loss of a high number of animals. By that time, the disease had already been recognized for about 100 years but was known under various synonyms reflecting central nervous system disorders. A particular characteristic of the disease is the variably long incubation period that was the reason why BD has been grouped with "slow" virus infections. The disease is associated with disturbances of motility and in sensory functions and usually results in paralysis and death in affected natural hosts [reviewed in 16]. Pathohistologically, BD is classified as a progressive polioencephalomyelitis with pronounced inflammatory reactions in the basal cortex, the nucleus caudatus and the entire hippocampal area.

BD virus has been only recently characterized as an RNA virus [7, 15, 17, 30, 39] with strong evidence for negative-sense polarity [3, 7]. The virus which is tightly cell associated and apparently lacks cytopathogenicity in vitro and in vivo replicates preferentially in cells derived from the neural crest [12]. However, after cocultivation [12] or after repeated infection with virus-containing supernatants from the brain of infected rats, other cell types such as cultured astrocytes [27, 28] and skin cells [25] also can be directly infected in vitro.

After experimental infection, a wide variety of animals can be infected including species phylogenetically distant, such as birds and non-human primates. Recently, BDV-specific antibodies have also been detected in humans, indicating a possible role of BDV as a human pathogen [reviewed in 31].

The species that has been studied most intensively is the rat and most if not all progress in understanding the pathogenesis of this virus-induced disease of the central nervous systems was achieved by using this experimental animal.

After i.c. or i.n. infection of adult rats infectious virus and virus-specific antigen can be detected in high concentrations (Table 1) in the

Table 1. Consequence of BDV infection in adult rats

Disease	Acute	← →	Chronic
Symptoms	hyperactivity, aggressiveness ataxia, pareses (paralyses)		ataxia, somnolence, chron. debility, dementia
Pathology	↓ +++ inflammation ———————— ++ —— + —— ± —— ¯ —		
	→ brain cell degeneration ———→ cortical brain atrophy ———→		
Virus (brain)	± positive ————————————————————————→		
(retina)	± positive ———— negative ————————————————→		
	⇔		
↑	⊥	⊥	⊥
BDV	2 wk	4 wk	8 wk natural lifespan

brain, the retina, the cerebrospinal fluid, in peripheral nerves and in the adrenal gland [1, 4, 5, 8, 20, 21, 36]. At later stages of the infection the virus disappears from the retina, but a persistent productive infection is maintained in the other tissues mentioned above. In parallel to the loss of virus from the eye, the animals go blind. In immunocompetent infected rats, disease symptoms such as lack of grooming, ataxia, hyper-activity and aggressiveness can be seen at about day 14 (Table 1). This acute phase of the disease lasts for about 3 weeks and later results in apathy, somnolence, progressive ataxia and paresis and sometimes results in paralysis and death. The clinical symptoms are paralleled by the development of an inflammatory reaction in the brain that is localized mainly perivascularly (Fig. 1). However, encephalitic lesions are also found in the brain parenchyma. In general, the inflammation is initially centered in the limbic system but spreads to other areas of the brain during infection [8, 20]. The chronic phase of BD is clinically governed by increasing apathy and the rats remain in a severe somnolence, show signs of dementia and behavioral abnormalities [9, 21]. Some rats, however, develop an impressive obesity with body weights of up to 500–600 g without clinical disease as compared to 200 g of uninfected adult rats. The histopathological picture of the chronic disease, which can be diagnosed after day 60, is characterized by a significant decrease of the inflammatory reaction and the development of a severe hydrocephalus internus.

Immunopathogenesis of Borna disease

Significantly different pictures are seen in rats that show natural or drug-induced immunoincompetence. Newborn, athymic, cyclophosphamide or cyclosporine A treated rats do not show Borna disease or the acute inflammatory reaction (Table 2). However, virus-specific nucleic acid and infectious virus, in addition to virus-specific antigen, is found in these animals in amounts comparable to fully immunocompetent rats [4, 13, 20, 25, 36, 37]. Most strikingly, immunocompromised rats show no destruction of the retina, although the virus persists in the eye, i.e. the rats do not become blind despite the presence of virus in retinal layers (Table 2) [20]. This fact shows perfectly well that BDV has very low or even no direct cytopathogenicity in vivo. The importance of the immune response for the pathogenesis was further stressed by showing that adoptive transfer of lymphocytes from BDV-immune rats into immuno-incompetent animals resulted in full-blown BD [20, 37]. As a whole, these facts together demonstrate that BD is based on a virus-induced immunopathological reaction rather than on a direct virus-cell inter-

Table 2. Consequence of BDV infection in newborn, athymic or immunosuppressed rats

Disease		No symptoms			
Pathology		No inflammation			
Virus (brain)	±	positive ————————————————————→			
(retina)	±	positive ——————————————————→			
				⇔ ———————	
↑	⊥	⊥	⊥		
BDV	2 wk	4 wk	8 wk	natural lifespan	

Table 3. Inhibition of BD after long treatment with CSA

Wk after CSA treatment		Duration of CSA treatment			
		–	2	3	4
5	disease	+	+	+	–
	encephalitic lesions	+++	+++	+/++	–
	antibodies	2,560	<40	<40	<40
20	disease	+	+	+	–
	encephalitic lesions	+	+	+	–
	antibodies	5,120	5,120	<40	<40
60	disease	+			–
	encephalitic lesions	+/–	ND	ND	–
	antibodies	5120			<40

Rats were treated with CSA doses of 25 mg/kg/day given s.c. starting one day before infection for various time intervals; antibody titers are expressed as the reciprocal of end-point dilution

action. The possibility that antiviral antibodies play a significant role in the pathogenesis of BD can be excluded according to a variety of experimental evidence [13, 20]. The most decisive argument against the involvement of antibodies in the pathogenesis of BD comes from experiments with the immunosuppressive drug cyclosporine A. Rats treated with CSA at appropriate doses and for a sufficiently long time can be protected from BD and do not mount an antibody response (Table 3) [37]. However, suboptimal treatment of BDV infected rats with CSA results in the development of an encephalitis in the absence of an anti-BDV antibody response (Table 3). Although these findings clearly indicated that the pathogenesis of BD is closely related to the cellular immune response, they did not reveal the cellular basis of the immunopathological process resulting in inflammation of the brain. By characterizing the cells present in the inflammatory lesions, employing

140   L. Stitz et al.

Table 4. Adoptive transfer of a BDV-specific CD4+ T cell line into BDV-infected immunosuppressed rats

BDV infection	Immunosuppression	T cell transfer	Disease[a]	Encephalitis[a]
+	−	−	3/3	3/3
+	+	−	0/3	0/3
+	+	+	8/8	8/8
−	+	+	0/3	0/3

BDV-specific T cells were passively transferred i.v. immediately after in vitro restimulation at concentrations of 1×10^6 to 8×10^6. Recipients were immunosuppressed 1 day after i.c. infection by intraperitoneal injection of a single dose of 150 mg/kg cyclophosphamide.

[a] Number of rats per experimental group

immunohistochemical methods, a useful approach was found to solve this question. Immunohistological investigations into the quality of cells involved in the perivascular inflammatory reaction revealed the presence of CD4+ and CD8+ T cells in addition to numerous macrophages and B cells (Fig. 1) [8]. To elucidate the importance of T cell subsets in the pathogenesis of BD in a first approach, a homogeneous virus-specific T cell line was established. Lymphocytes obtained from regional lymph nodes after subcutaneous immunization with purified virus-specific antigen were cultured and restimulated in vitro employing a protocol for the cultivation of CD4+ T cells [29]. Analysis of this cell line revealed BDV-specificity, MHC class II restriction and the phenotypical markers of CD4+ helper/inflammatory cells. Adoptive transfer of this cell line into BDV-infected immunosuppressed healthy recipients resulted in severe disease and death as early as day 5 after the injection of effector cells (Table 4) [28, 29]. In contrast, passive transfer into uninfected rats did not result in encephalitis or disease, which shows that this BDV-specific T cell line by itself is not encephalitogenic. These results, together with the immunohistological characterization of inflammatory cells in the brain of BDV-infected rats, strongly suggest that BD is caused by a delayed type hypersensitivity reaction (DTH). The finding of the importance of MHC class II restricted T cells agrees with the presence of MHC class II antigen in the brain of BDV-infected rats. This self antigen is detected on various cell types upon immunohistological characterization, namely perivascularly but also on oligodendrocytes, microglial and ependymal cells (Fig. 2) [8, 29, 35].

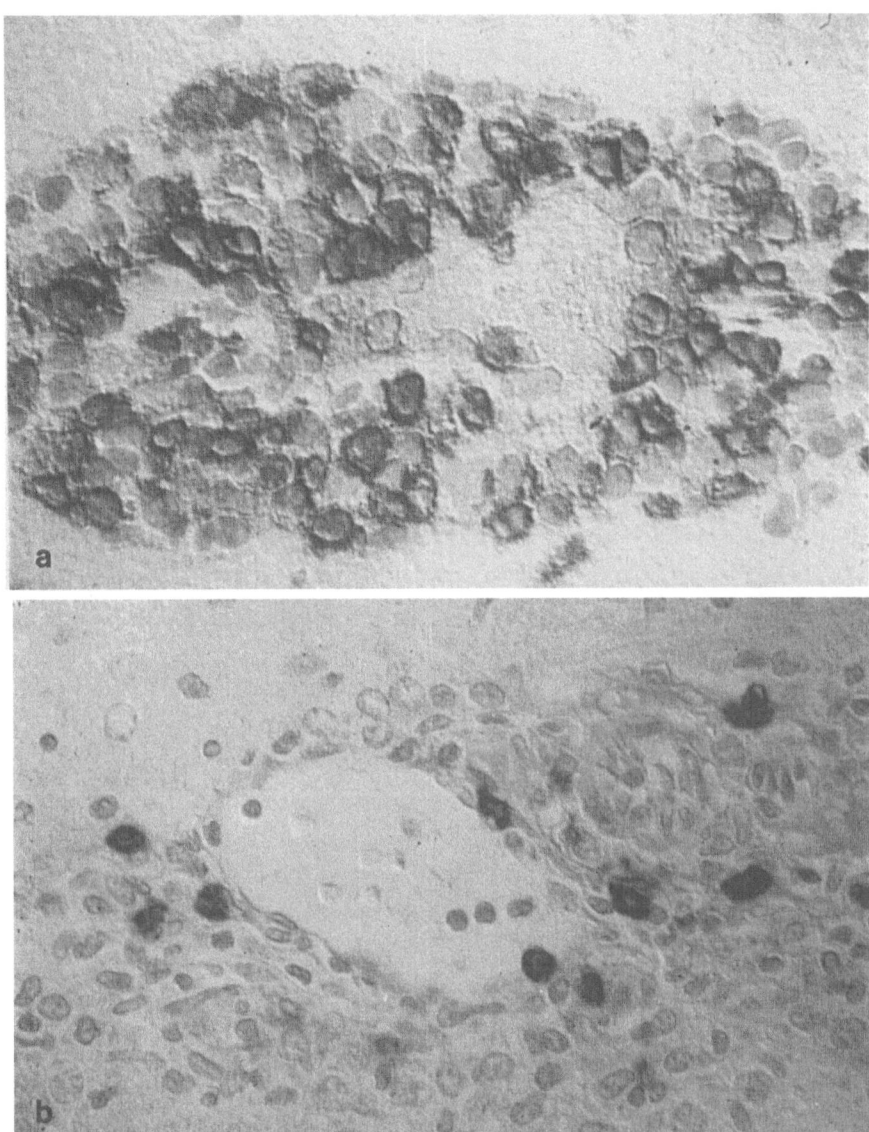

Fig. 1. Immunhistological characterization of cells in perivascular infiltrates of BDV-
infected rats. Frozen sections from brains were reacted in a peroxidase anti-peroxidase
reaction with a monoclonal antibody specific for CD4+ T cells (**a**) or specific for
CD8+ T cells (**b**)

The importance of these cells in the pathogenic mechanism has not
yet been determined, especially since it has not yet been possible to
demonstrate their role in antigen presentation of BDV-specific antigen in
the brain. However, some evidence has accumulated that provides better
insight into the mechanisms of pathogenicity and the T cell subsets
involved. With regard to MHC class II it was shown that the elevation of
the expression of this self antigen by IFN-γ increased the proliferative

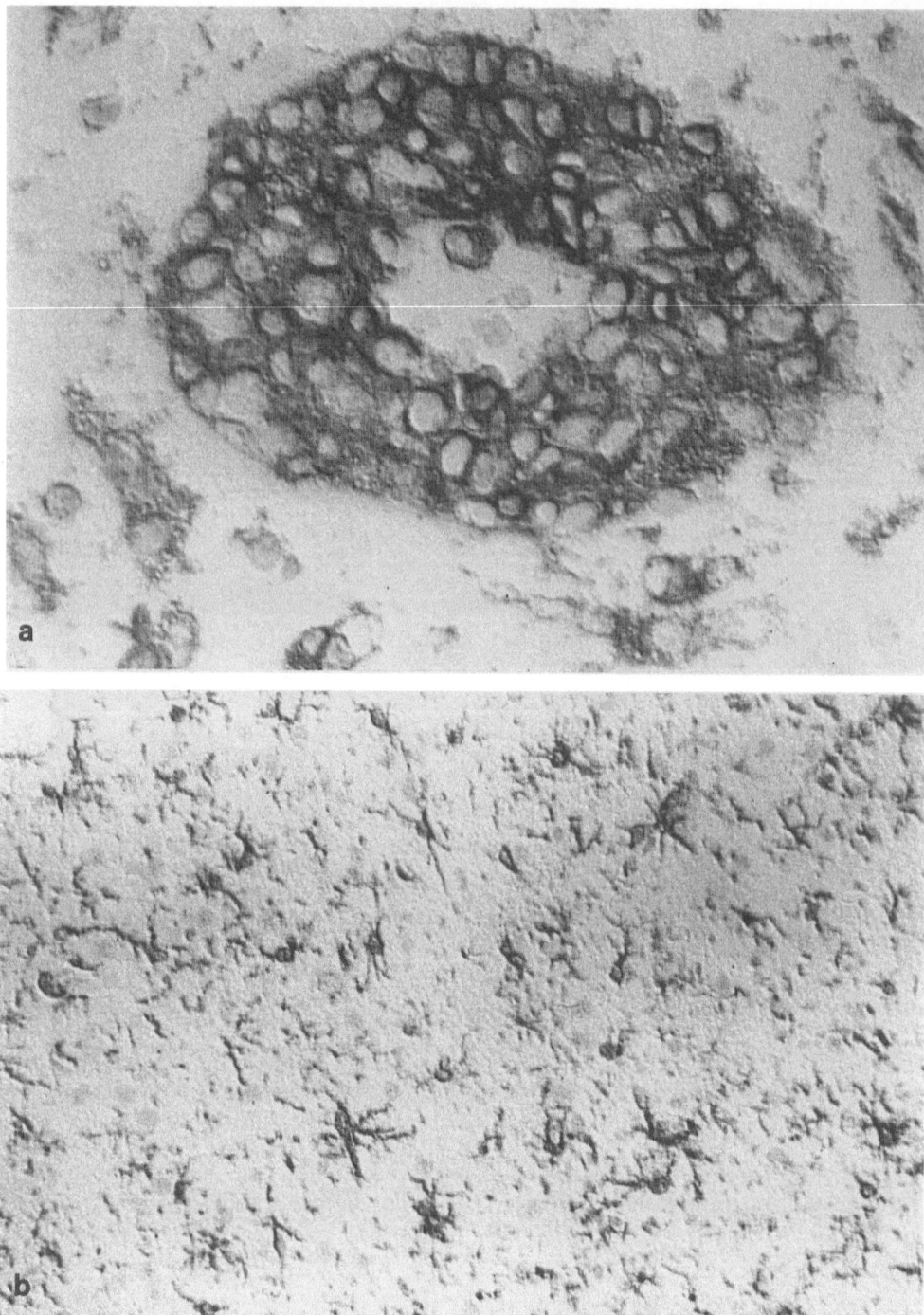

Fig. 2. Demonstration of MHC class II antigen expression in the brain of BDV-infected rats. Massive perivascular class II expression (**a**) and MHC class II-positive cells in the parenchyma, presumably activated microglia (**b**). Sections were incubated with an MHC class II-specific monoclonal antibody

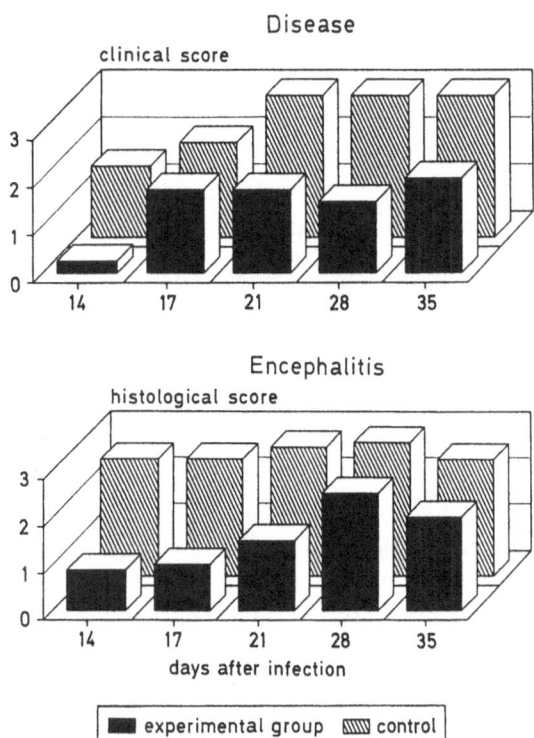

Fig. 3. Mean clinical and histological scores of BDV-infected rats treated with recombinant human transforming growth factor-β2 (TGF-β2; 1 μg/day i.p. 1 day before infection through day 7 after infection) and infected rats without further treatment

capacity of the BDV-specific CD4+ T cell line in vitro [29]. Likewise, the expression of MHC class II on BDV-infected astrocytes used as target cells proved to be a prerequisite for in vitro cytotoxicity by the same cell line [27].

These in vitro findings suggest that lymphokines and, obviously especially IFN-γ and, the cells capable of producing this type of interferon might be of crucial importance in the pathogenesis of BD [32]. This assumption could be substantiated in experiments employing the transforming growth factor β2 (TGF) in vivo [35]. TGF belongs to a class of polypeptides exhibiting diverse effects on cell growth and differentiation. These substances act as multifunctional cytokines with potent inhibitory activity on growth, differentiation and effector functions of activated T and B lymphocytes as well as macrophages [reviewed in 23, 40]. Experiments with TGF-β2 in BDV-infected rats revealed a reduction of the severity of clinical symptoms that was paralleled by a significant reduction of the inflammatory reaction in the brain (Fig. 3). However, the efficacy of the treatment was only transient. Immunhistological investigations revealed slightly reduced CD4+ T cell numbers

144 L. Stitz et al.

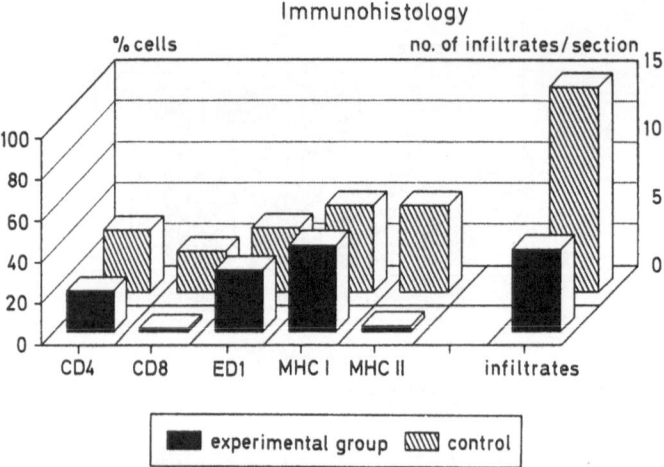

Fig. 4. Histological and immunohistological evaluation of encephalitic lesions in TGF-β2 treated and untreated BDV-infected rats. The columns on the right represent the mean number of infiltrates/section, the columns on the left show the percentage of positively immunostained cells in the infiltrates

and no changes in macrophage counts in encephalitic lesions of TGF treated rats. However, this study provided first evidence for the pathogenic importance of CD8+ T cells in BD (Fig. 4). Whereas other cell populations present in perivascular inflammatory lesions were not significantly altered, as mentioned above, CD8+ T cells were virtually absent from encephalitic lesions. Furthermore, the expression of MHC class II antigen was significantly reduced in the brain of TGF-treated Borna disease virus-infected rats, whereas MHC class I expression was not. We had shown previously that neither in the brain of infected rats [35] nor in astrocytic cultures in vitro [25, 27, 29] the virus by itself induced the expression of MHC class II, but IFN-γ was able to do so. Since CD8+ T cells are potent producers of immune interferon and IFN-γ on its turn regulates MHC class I and class II expression, the absence of CD8+ T cells in the brain of TGF-treated rats might result in the observed reduction of MHC class II antigen [34]. Our recent finding that BDV-infected astrocytes produce in vitro IFN of the α/β type that shows all characteristics of a previously described astrocyte-IFN [38] agrees with this interpretation [25]. α/β IFN and, in particular, IFN produced by astrocytes upregulates MHC class I but not class II expression [11, 38] which would explain the IFN-γ independent presence of MHC class I in TGF-treated rats. The experiments performed with TGF-β revealed an initial drastic reduction of the local immune response after BDV-infection. The relative absence of CD8+ T cells seemed to be decisive, since the production of soluble mediators that induce the expression of MHC antigen is hampered. Interestingly, and fully supporting this hypothesis, the increase of CD8+ T cells late after TGF treatment was directly correlated with an increase in the expression of MHC class II antigen in the brain and encephalitic lesions (Fig. 4). In all, the reduced expression of restriction elements for cell-mediated immune response leads to an initial inhibition of the encephalitic reaction and clinical symptoms due to a relative absence of restriction elements despite the presence of CD4+ T cells.

Additional evidence for the pathogenic importance of CD8+ T cells in BD comes from experiments in which BDV-infected rats were treated with monoclonal antibodies directed against various T cell markers [36]. Antibodies specific for CD4+ and CD8+ cells were able to decrease or even prevent the local inflammatory reaction if given early during the infection. However, CD8-specific monoclonal antibodies appeared to be much more effective and easily prevented the immunopathological disease whereas antibodies directed against CD4+ cells were significantly less effective (Table 5). These findings fully agree with the above mentioned result that CD8+ T cells play an important role in the pathogenesis of BD insofar as cells of this phenotype have been shown to be potent producers of cytokines. We therefore proposed a sequential role of an initial CD8+ T cell response that is decisive in triggering the local CD4+ T cell-mediated delayed type hypersensitivity reaction in the brain after BDV infection [34, 36].

From experiments employing monoclonal antibodies directed against CD8+ cells, an explanation for another characteristic feature of Borna disease, namely brain atrophy became obvious. As mentioned before, the chronic phase of BD is characterized by a prominent cortical atrophy (Fig. 5a) and chronic debility [20]. In a recent study we could demonstrate that, in addition to the inhibition of the immunopathological reaction, treatment of BDV-infected rats with anti-CD8 monoclonal antibodies resulted only in minimal brain-cell lesions and no obvious loss of brain substance could be seen even long after infection (Fig. 5b) [2]. In untreated rats necrobiotic changes of brain cells were found to be present from early stages of the disease and neuronal cell loss was one of the most prominent features of BD. By characterizing the brain cells that express MHC class I antigen it became evident that this self antigen could be demonstrated on neurons and astrocytes [2, 25]. The most intriguing finding upon histology was the coincidence of the occurrence of CD8+ cells and a dramatic increase of MHC class I on the one hand, and the presence of first neuronal degenerations on the other. In rats depleted of CD8+ T cells in vivo, no marked amounts of MHC class I antigen was detected in the brain parenchyma, whereas MHC class II expression was not different in any of the BDV-infected rats, regardless

Table 5. Effect of in vivo administration of monoclonal antibodies directed against CD4+ or CD8+ T cells on encephalitis and disease after BDV infection

Monoclonal antibody	Specificity	Administration on days	Weeks after infection					
			3		4		8	
			lesions	disease	lesions	disease	lesions	disease
OX-52	CD4, CD8	−1, +1	0.2	0	0	0	0	0
W3/25	CD4	−1, +1	3.0	2.0	1.5	3.0	1.5	2.0
		−1, +1, +7, +14	1.0	0.5	2.0	2.0	1.5	1.0
OX-38	CD4	−1, +1	3.0	2.25	2.5	2.75	n.d.	n.d.
		−1, +1, +7, +14	0	0	0.5	0.25	0	0
OX-8	CD8	−1, +1	0.5	0	0.5	0	0.25	0.25
control	untreated	−	2.75	2.5	2.75	2.5	2.0	2.75

Brain tissue sections were scored on an arbitraty scale ranging from 0 to 3.0 based on the numbers of infiltrates per section and the number of cell layers present in each infiltrate. The disease symptoms were scored on an arbitraty scale from 0 to 3.0 based on the general state of health and the appearance of neurological symptoms. n.d. Not determined

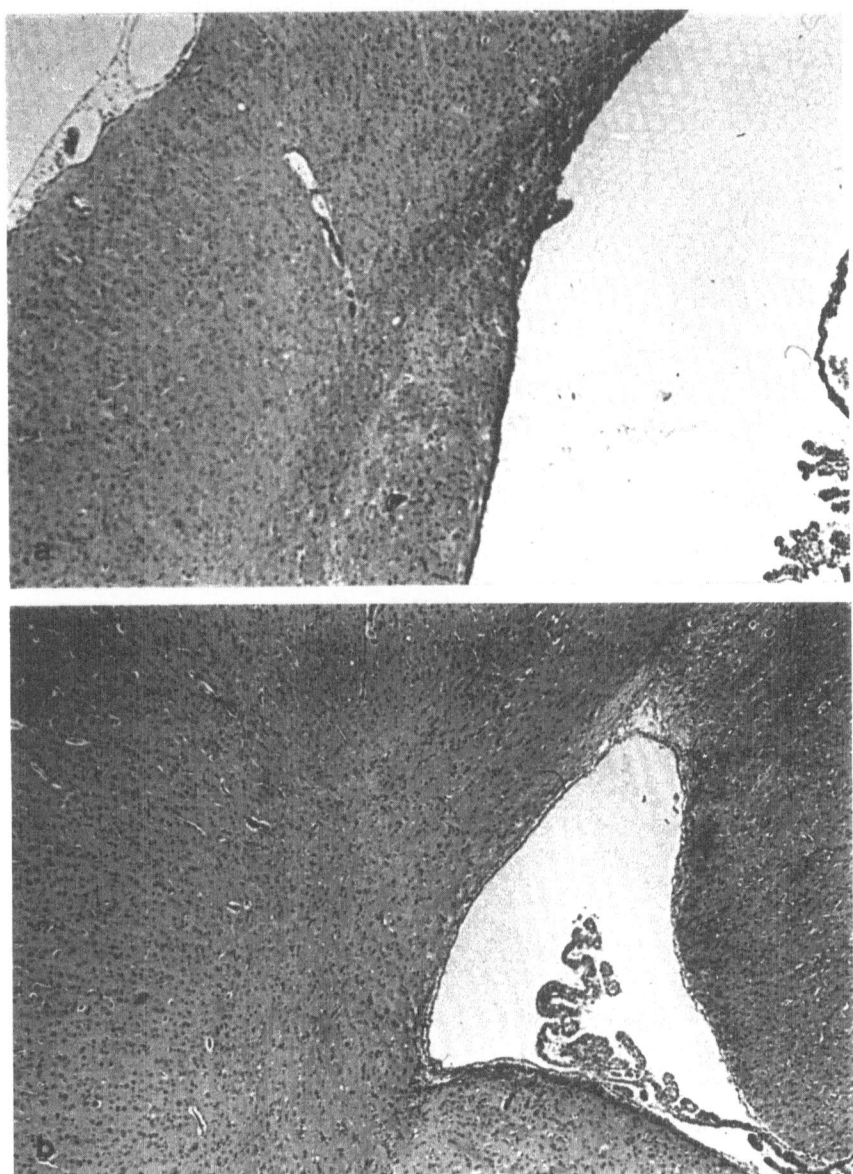

Fig. 5. Severe cortical brain atrophy in an adult untreated rat 8 weeks after BDV infection. Note the "burned-out" inflammation and the enormously dilatated ventricle (**a**). Brain section of a BDV-infected rat treated with a monoclonal directed against CD8+ T cells. Note the complete lack of atrophy or signs of inflammation (**b**)

of whether treated with anti-T cell antibodies or not. Since coexpression of MHC class I antigen in association with virus-specific proteins renders cells as targets for cytotoxic CD8+ T lymphocytes [reviewed in 42], we consequently looked for cytotoxic T cell activity in BDV-infected rats. Employing syngeneic and allogeneic BDV-infected target cells we could demonstrate virus-specific cytotoxic T cell activity (Table 6) [25].

Table 6. Cytotoxic activity in lymphocyte preparation isolated from the brain of BDV-infected rats

Type of target cell	% Lysis of target cell
Syngeneic BDV-infected	45
Syngeneic uninfected	3
Allogeneic BDV-infected	5
Syngeneic BDV-infected	40
Syngeneic uninfected	5
Syngeneic BDV-infected in the presence of Anti-MHC class I antibody	7

Lymphocytes from the brain were isolated on a modified RPMI/Ficoll gradient [25]. Specific lysis was determined after a 9 h coincubation of effector cells with ^{51}Cr labeled target cells. Effector: target ratio 30:1

Blocking experiments using antibodies directed against MHC class I antigen provided further evidence for the presence and activity of classical cytotoxic T lymphocytes in the brain of BDV-infected rats (Table 6).

Conclusion

In this short review we have summarized the present knowledge of the pathogenesis of Borna disease and presented data that help describe the cellular basis of this immunopathological disease of the brain. From all of our work it becomes apparent that interactions and the interplay among the components of the cellular immune response and between the cellular immune system and lymphokines are of crucial importance. As exemplified by immunosuppression and immunomodulation with various drugs, understanding of the pathogenetic pathways is the inevitable prerequisite to possible interference with disease processes. Here, we also show that Borna disease is not only a useful model of a virus-induced immunological disease but we have also provided evidence that the experimental disease in rats presently possibly represents the best model for studying *in vivo* cytotoxicity exerted by classical CD8+ cytotoxic T lymphocytes.

Acknowledgement

The studies mentioned here were performed partially in cooperation with H. Becht, S. Herzog, K. Frese, O. Narayan, O. Planz and M. Sobbe. We thank B. Boschek for critically reading the manuscript. L. Stitz is a recipient of a Hermann- and Lilly-Schilling Professorship for Theoretical and Clinical Medicine.

References

1. Bilzer T, Planz O, Stitz L, Lipkin WI (1993) Tropism of Borna disease virus. I. Distribution of virus specific nucleic acid and antigen in the nervous system of adult rats. Submitted for publication
2. Bilzer T, Stitz L (1993) Brain cell lesions in Borna disease are mediated by T cells. In: Kaaden OR, Eichhorn W, Czerny CP (eds) Unconventional agents and unclassified viruses. Springer, Wien New York, pp 153–158 (Arch Virol [Suppl] 7)
3. Briese T, De la Torre JC, Lewis A, Ludwig H, Lipkin WI (1992) Borna disease virus, a negative-strand RNA virus, transcribes in the nucleus of infected cells. Proc Natl Acad Sci USA 89: 11486–11489
4. Carbone KM, Duchala CS, Griffin JW, Kincaid AL, Narayan O (1987) Pathogenesis of Borna disease in rats: Evidence that intra-axonal spread is the major route for virus dissemination and the determinant for disease incubation. J Virol 61: 3431–3440
5. Carbone KM, Trapp BD, Griffin JW, Duchala CS, Narayan O (1989) Astrocytes and Schwann cells are virus-host cells in the nervous system of rats with Borna disease. J Neuropathol Exp Neurol 48: 631–644
6. Cole GA, Nathanson N, Prendergast RA (1972) Requirement for theta-bearing cells in lymphocytic choriomeningitis virus-induced central nervous system disease. Nature 238: 335–337
7. De la Torre JC, Carbone KM, Lipkin WI (1990) Molecular characterisation of Borna disease agent. Virology 179: 853–856
8. Deschl U, Stitz L, Herzog S, Frese K, Rott R (1990) Determination of immune cells and expression of major histocompatibility complex class II antigen in encephalitic lesions of experimental Borna disease. Acta Neuropathol 81: 41–50
9. Dittrich W, Bode L, Ludwig H, Kao M, Schneider K (1989) Learning deficiencies in Borna disease virus-infected but clinically healthy rats. Biol Psychiatry 26: 818–828
10. Doherty PC, Zinkernagel RM (1974) T cell-mediated immunopathology in viral infections. Transplant Rev 19: 89–120
11. Halloran PF, Urmson J, van der Meide PH, Autenried P (1989) Regulation of MHC expression in vivo: II. IFN-alpha/beta inducers and recombinant IFN-alpha modulate MHC antigen expression in mouse tissues. J Immunol 142: 4241–4247
12. Herzog S, Rott R (1980) Replication of Borna disease virus in cell culture. Med Microbiol Immunol 168: 153–158
13. Herzog S, Wonigeit K, Frese K, Hedrich HJ, Rott R (1985) Effect of Borna disease virus infection in athymic rats. J Gen Virol 66: 503–508
14. Hirano N, Kao M, Ludwig H (1983) Persistent, tolerant or subacute infection in Borna disease virus infected rats. J Gen Virol 64: 1521–1530
15. Lipkin WI, Travis KM, Carbone KM, Wilson CM (1990) Isolation and characterisation of Borna disease agent cDNA clones. Proc Natl Acad Sci USA 87: 4184–4188
16. Ludwig H, Bode L, Gosztonyi G (1988) Borna disease. A persistent virus infection of the central nervous system. Progr Med Virol 35: 107–151

17. McClure MA, Thibault KJ, Hatalski CG, Lipkin WI (1992) Sequence similarity between Borna disease virus p40 and a duplicated domain within the paramyxo- and rhabdovirus polymerase proteins. J Virol 66: 6572–6577
18. Meulen ter V, Stephenson JR (1983) The possible role of viral infections in MS and other related demyelinating diseases. In: Hallpike JF, Adams CWM, Tourtellotte WW (eds) Multiple Sclerosis. Chapmann and Hall, London, pp 241–274
19. Nagashima K, Wege H, Meyermann R, Ter Meulen V (1978) Corona virus induced subacute demyelinating encephalomyelitis in rats: a morphological analysis. Acta Neuropathol 44: 63–70
20. Narayan O, Herzog S, Frese K, Scheefers K, Rott R (1983) Behavioral disease in rats caused by immunopathological response to persistent Borna disease virus in the brain. Science 220: 1401–1403
21. Narayan O, Herzog S, Frese K, Scheefers K, Rott R (1983) Pathogenesis of Borna disease in rats: Immune-mediated viral ophthalmoencephalopathy causing blindness and behavioral abnormalities. J Infect Dis 148: 305–315
22. Padgett BL, Walter DL, zuRhein GM, Ederoode RJ, Dessel BH (1971) Cultivation of papova-like virus from human brain with progressive multifocal leukoenceph-alopathy. Lancet i: 1257–1260
23. Palladino MA, Morris RE, Starnes HF, Levinson DA (1990) The transforming growth factor-betas: A new family of immunoregulatory molecules. Ann NY Acad Sci 593: 181–185
24. Planz O, Bilzer T, Lipkin WI, Stitz L (1993) Tropism of Borna disease virus. II. Presence of Borna disease virus in non-neural tissue of immunosuppressed and newborn-infected rats. Submitted for publication
25. Planz O, Bilzer T, Sobbe M, Stitz L (1993) Lysis of MHC class I-bearing cells in Borna disease virus-induced degenerative encephalopathy. J Exp Med 178: 163–174
26. Price RW, Brew BJ (1988) The AIDS dementia complex. J Infect Dis 158: 1079–1083
27. Richt JA, Stitz L (1992) Borna disease virus infected astrocytes function in vitro as antigen- presenting and target cells for virus-specific CD4-bearing lymphocytes. Arch Virol 124: 95–109
28. Richt JA, Stitz L, Deschl U, Frese K, Rott R (1990) Borna disease virus-induced meningoencephalomyelitis caused by a virus-specific CD4+ T-cell mediated im-mune reaction. J Gen Virol 71: 2565–2573
29. Richt JA, Stitz L, Wekerle H, Rott R (1989) Borna disease, a progressive meningoencephalomyelitis as a model for CD4+ T cell-mediated immunopathology in the brain. J Exp Med 170: 1045–1050
30. Richt JA, VanDeWoude S, Zink MC, Narayan O, Clements JE (1991) Analysis of Borna disease virus-specific RNA's in infected cells and cultures. J Gen Virol 72: 2251–2255
31. Rott R, Herzog S, Bechter K, Frese K (1991) Borna disease, a possible hazard for man? Arch Virol 118: 143–149
32. Shankar V, Kao M, Hamir AN, Sheng H, Koprowski H, Dietzschold B (1992) Kinetics of virus spread and change in levels of several cytokine mRNAs in the brain after intranasal infection of rats with Borna disease virus. J Virol 66: 992–998
33. Stephenson JR, Ter Meulen V (1979) Subacute sclerosing panencephalitis: char-acterization of the etiological agent and its relationshio to morbilli virus. In: Tyrell DAJ (ed) Aspects of slow and persistent virus infection. Martinus Nijhoff, The Hague/Boston/London, pp 61–75

34. Stitz L, Planz O, Bilzer T, Frei K, Fontana A (1991) Transforming growth factor β modulates T cell-mediated encephalitis caused by Borna disease virus. Pathogenic importance of CD8+ cells and suppression of antibody formation. J Immunol 147: 3581–3586

35. Stitz L, Schilken D, Frese K (1991) Atypical dissemination of the highly neurotropic Borna disease virus during persistent infection in cyclosporine A-treated, immunosuppressed rats. J Virol 65: 457–460

36. Stitz L, Sobbe M, Bilzer T (1992) Preventive effects of early anti-CD4 or anti-CD8 treatment on Borna disease in rats. J Virol 66: 3316–3323

37. Stitz L, Soeder D, Deschl U, Frese K, Rott R (1989) Inhibition of immune-mediated meningoencephalitis in persistently Borna disease virus infected rats by Cyclosporine A. J Immunol 143: 4250–4256

38. Tedeschi B, Barrett JN, Keane RW (1986) Astrocytes produce interferon that enhances the expression of H-2 antigens on a subpopulation of brain cells. J Cell Biol 102: 2244–2253

39. VanDeWoude S, Richt JA, Zink MC, Rott R, Narayan O, Clements JE (1990) A Borna virus cDNA encoding a protein recognized by antibodies in humans with behavioral disease. Science 250: 1278–1281

40. Wahl SM, McCartney-Francis N, Mergenhagen SE (1989) Inflammatory and immunoregulatory role of TGF-β. Immunol Today 10: 258–262

41. Weimer LP, Herdon RM, Narayan O, Johnson RT (1972) Further studies of simian virus-40-like virus isolated from human brain. J Virol 10: 147–152

42. Zinkernagel RM, Doherty PC (1979) MHC-restricted cytotoxic T cells: Studies on the biological role of polymorphic major transplantation antigens determining T cell restriction-specificity, function and responsiveness. Adv Immunol 27: 52–142

Authors' address: Dr. L. Stitz, Institut für Virologie, Justus-Liebig-Universität, Frankfurterstrasse 107, D-35392 Gießen, Federal Republic of Germany.

Arch Virol (1993) [Suppl] 7: 153–158

Brain cell lesions in Borna disease are mediated by T cells

T. Bilzer[1] and **L. Stitz**[2]

[1] Department of Neuropathology, Heinrich-Heine-Universität Düsseldorf, Düsseldorf,
and [2] Institute of Virology, Justus-Liebig-Universität Giessen, Giessen,
Federal Republic of Germany

Summary. Experimental Borna Disease (BD) in rats is characterized by severe lymphocytic encephalitis and by massive brain cell lesions finally leading to brain atrophy. Treatment of BDV-infected rats with monoclonal antibodies directed against CD4$^+$ and CD8$^+$ T cells could almost completely inhibit the immunopathological reactions and revealed less BDV-infected neurons and astrocytes that expressed MHC class I antigen. Brain cell lesions were minimal, and no obvious brain atrophy could be observed even late after infection. Since BDV itself is not known to exert cytopathic effects and since brain cell damage was independent of antibody titers, brain cell destruction correlates well with the intracerebral presence of CD8$^+$ T cells and the expression of MHC class I antigens. Moreover, BDV-infected brain cells in vitro could be demonstrated to be lysed in a MHC class I-restricted manner. These findings provide evidence that virus-infected neurons can be destructed by T cell mediated cytotoxicity which results in organ atrophy and dementia.

Introduction

Viruses that persist in the CNS can induce severe immunopathological reactions and brain cell lesions. As shown in lymphocytic choriomeningitis in mice, CD8$^+$ T cells can either be protective and thus eliminate the virus from the host [2, 10, 20] or they contribute to immunopathology [1, 5]. Experimental Borna disease which is caused by the highly neurotropic non-cytopathic ssRNA Borna disease virus [BDV, 9, 19], is a unique model of a persistent virus infection of the central nervous system leading to T cell-mediated immunopathology [11, 14–16]. Intracerebral infection of adult Lewis rats with BDV leads to a severe meningoencephalitis and progressive nerve cell loss [12]. In the postencephalitic stage, marked brain atrophy is a regular finding. We have recently described that treatment of BDV-infected rats with mab against T cells can drastically reduce inflammatory reactions and cell damages in the brain [18]. These findings provide further evidence that brain atrophy in BD is immune-mediated.

Inflammation and brain cell lesions in experimental Borna disease

Clinical, virological, immunological and neuropathological details of the encephalis that develops after i.c. infection of Lewis rats with BDV have previously been described [3, 6, 17, 18]. Beyond day 14 after intracerebral infection of adult Lewis rats, infectious virus can be detected in the brain, and virus-specific serum antibody titers persist at 1:5120. Rats suffer from incoordination and ataxia, pareses and finally apathy and paralysis. Virus-specific antigens can be demonstrated by western blotting and visualized in brain cells, especially in neurons and astrocytes. BDV infection is preferentially located in the cerebral cortex and parts of the limbic system including the gyrus cinguli and the hippocampus formation. Massive lymphocellular reactions and progressive brain cell necroses regularly develop between 14 and 21 days after infection (Fig. 1). Most of the perivascular inflammatory cells were characterized to be $CD4^+$ cells (more than 40%), about 10 to 20% to be $CD8^+$ cells and about 30% $ED1^+$ cells. $OX-8^+$ cells and $ED1^+$ cells can also be found within the brain parenchyma. Interestingly, MHC class I and II antigen is expressed in the gray matter and obviously localized not only on inflammatory cells but also on brain cells. Simultaneous to the first signs of inflammation (day 14 p.i.), astrocytic edema and necrobiotic changes of neurons occur, especially in the pyramidal layers of the neocortex and the hippocampus. Neuronophagia and reactive gliosis including microglial proliferation can be observed. In the postencephalitic stage, i.e. later than 60 days after BDV-infection, animals often reveal an extreme reduction of brain substance (Fig. 3), and an extraordinary degree of hydrocephalus internus. Clinically, they are in a state of chronic debility.

Influence of mab directed against T cells on BDV-encephalopathy

Virustiters in the brain of BDV-infected rats treated in vivo with monoclonal antibodies directed against T cells were comparable to those of untreated BDV-infected rats [18]. In contrast to untreated rats, most of the experimentally treated rats revealed only little or even no clinical symptoms and only individual small perivascular lymphocytic reactions, mainly restricted to the meninx (Fig. 2). Surprisingly, also brain cell damages were only rarely to be observed. Whereas untreated virus-infected rats revealed an extreme reduction of brain substance, brain atrophy was not obvious after mab-treatment (Fig. 4). Immunohistochemistry of the brains of mab-treated rats revealed markedly fewer lymphocytic cells. CD8$^+$ cells were almost completely absent, whereas other inflammatory cells, including CD4$^+$ cells were still present, although reduced. In contrast to the rats that developed severe encephalitis, MHC class I antigen was primarily expressed on infiltrating cells, whereas MHC class II antigen showed essentially the same distribution. To support the evidence that brain cell damage is mediated by lymphocytes, cytotoxicity of lymphocytes isolated from the brain of BDV-infected rats was determined. Syngeneic BDV-infected target cells were lysed in an in vitro assay, whereas uninfected cells were not. Cytotoxicity was completely inhibited in the presence of an anti-MHC class I antibody [13].

Discussion

Dramatic brain atrophy in which the cortical hemispheres can be reduced to less than 1 mm in diameter and the resulting hydrocephalus internus e vacuo in the post-encephalitic stage have often been observed in BDV-infected animals [3, 11, 12]. We found neuronal cell lesions from early stages of the disease throughout the whole experiment until 12 weeks after BDV-infection. The following factors contributing to brain atrophy after BDV infection can be ruled out: (i) there is no evidence for brain cell lesions resulting secondarily from hydrocephalus occlusus [8], or damages of blood vessels (hypoxia), since we found no vascular pathological alterations; (ii) there is only circumstantial evidence for in-vivo-cytopathogenicity of BDV [4], whereas BDV has never been shown to be cytopathic in vitro [7]. Thus, the enormous loss of brain cells, especially neurons, has been postulated to be the result of an immune-mediated cell lysis [3, 11]. However, the factors contributing to a BDV-specific T cell mediated brain cell lysis, i.e. the presentation of MHC I antigens on BDV-infected cells and the role of cytotoxic T cells have not yet been investigated, although the presence of MHC class I antigen has recently been demonstrated in the brain of BDV-infected rats [4]. We found a very low level of MHC I expression at day 6 p.i., but a dramatic increase of MHC I antigen expression on BDV-infected neurons from day 10 to 14 p.i. Since first neuronal degenerations coincide with MHC class I expression, and CD8$^+$ T cells appear to play an important role for the development of BD [17, 18], it seemed reasonable to investigate the role of these cells in brain cell destruction. We have previously shown the protective effect of the mab-treatment against T cells with regard to the inflammatory reactions [18]. To our surprise, the protected animals not only were free of encephalitis, but also showed no signs of pro-

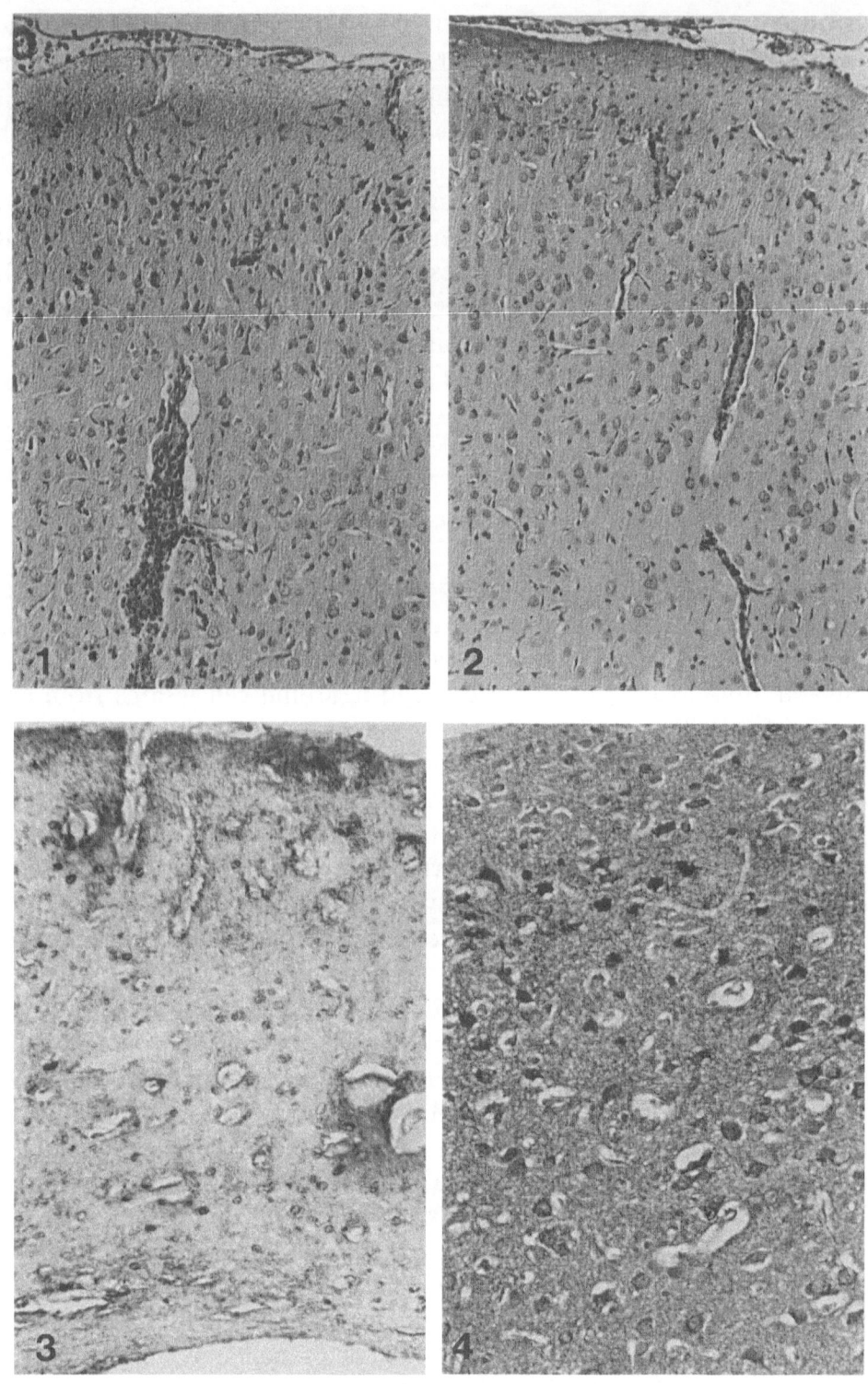

gressive brain cell loss resulting in brain atrophy. The remarkable differences between BDV-infected animals with encephalitis, and those protected from encephalitis, provide evidence for an immune-mediated brain cell death, in which cytotoxic T cells play a central role. This evidence could be strengthened by the finding that brain lymphocytes of BDV-infected rats are able to lyse effectively BDV-infected neurons in vitro [13]. Taken together, BD in rats provides a unique model of immunopathology in the brain in which the causative agent persists in the brain throughout the life despite a vigorous immune response and in which inflammatory encepalitic reactions are paralleled by a massive brain cell loss. There is now strong evidence that T cells play a central role in this process leading to dementia and finally chronic debility.

References

1. Baenziger J, Hengartner H, Zinkernagel RM, Cole GA (1986) Induction or prevention of immunopathological disease by cloned cytotoxic T-cell line specific for lymphocytic choriomeningitis virus. Eur J Immunol 16: 387–393
2. Byrne JA, Oldstone MBA (1984) Biology of cloned cytotoxic T lymphocytes specific for lymphocytic choriomeningitis virus: virus clearance in vivo. J Virol 51: 682–686
3. Carbone KM, Trapp BD, Griffin JW, Duchala CS, Narayan O (1989) Astrocytes and Schwann cells are virus host cells in the nervous system of rats with Borna disease. J Neuropathol Exp Neurol 48: 631–644
4. Carbone KM, Park SW, Rubin SA, Waltrip RW II, Vogelsang GB (1991) Borna disease: Association with a maturation defect in the cellular immune response. J Virol 65: 6154–6164
5. Cole GA, Nathanson N, Prendergast RA (1972) Requirement for Q-bearing cells in lymphocytic choriomeningitis virus-induced central nervous system disease. Nature 238: 335–337

Fig. 1. Inflammatory reactions in the brain of an adult Lewis rat 25 days after intracerebral infection with BDV. Perivascular lymphocytic cuffing, intraparenchymal lymphocytes as well as neuronal cell damages can be seen. HE staining, ×130. Fig. 2. Brain of an adult Lewis rat treated with anti-CD8 antibody OX-8 also 25 days after intracerebral infection with BDV. There is nearly no inflammation and also no obvious neuronal cell damage. HE staining, ×130. Fig. 3. Persistent BDV-infection in the brain of an adult rat 90 days after BDV-infection. In this stage encephalitis has "burned out", the brain is severely damaged and the cortical substance and especially neurons and dramatically reduced. Most of the remaining BDV-infected cells seem to be glia or endothelial cells. Fig. 4. Brain of an adult Lewis rat treated with anti-CD8 antibody OX-8 also 90 days after BDV-infection. Many neurons, astrocytes and vascular endothelia are infected, but no inflammatory reactions and no neuronal cell loss are obvious. There is however a moderate status spongiosus. Figs. 3 and 4. Immunohistochemical staining using the anti-38/39 kD BDV-antigen mab Bo-18, the ABC-method and diaminobenzidine as substrate, counterstained with hemalum, ×200

6. Deschl U, Stitz L, Herzog S, Frese K, Rott R (1990) Determination of immune cells and expression of major histocompatibility complex class II antigen in encephalitic lesions of experimental Borna disease. Acta Neuropathol 81: 41–50

7. Herzog S, Rott R (1980) Replication of Borna disease virus in cell cultures. Med Microbiol Immunol 168: 153–158

8. Irigoin C, Rodriguez EM, Heinrichs M, Frese K, Herzog S, Oksche A, Rott R (1990) Immunocytochemical study of the subcommissural organ of rats with induced postnatal hydrocephalus. Exp Brain Res 82: 384–392

9. Lipkin WI, Travis GH, Carbone KM, Wilson CM (1990) Isolation and characterization of Borna disease agent cDNA clones. Proc Natl Acad Sci USA 87: 4184–4188

10. Moskophidis D, Cobbold SP, Waldmann H, Lehmann-Grube F (1987) Mechanism of recovery from acute virus infection: Treatment of lymphocytic choriomeningitis virus-infected mice with monoclonal antibodies reveals that Lyt-2$^+$ T-lymphocytes mediate clearance and regulate the antiviral antibody response. J Virol 61: 1867–1874

11. Narayan O, Herzog S, Frese K, Scheefers H, Rott R (1983) Pathogenesis of Borna disease in rats: Immune-mediated viral ophtalmoencephalopathy causing blindness and behavioral abnormalities. J Infect Dis 148: 305–315

12. Narayan O, Herzog S, Frese K, Scheefers K, Rott R (1983) Behavioural disease in rats caused by immunopathological response to persistent Borna virus in the brain. Science 220: 1401–1403

13. Planz O, Bilzer T, Sobbe M, Stitz L (1993) Lysis of MHC class I bearing cells in Borna disease virus-induced degenerative encephalopathy. J Exp Med 178: 163–174

14. Richt JA, Stitz L, Wekerle K, Rott R (1989) Borna disease, a progressive meningoencephalitis as a model for CD4$^+$ T-cell mediated immunopathology in the brain. J Exp Med 170: 1045–1050

15. Rott R, Herzog S, Richt J, Stitz L (1989) Immune-mediated pathogenesis of Borna disease. Zentralbl Bakteriolog Mikrobiolog Hyg (A) 270: 295–301

16. Stitz L, Bilzer T, Richt J, Rott R (1993) Pathogenesis of Borna disease. In: Kaaden OR, Eichhorn W, Czerny CP (eds) Unconventional agents and unclassified viruses Arch Virol [Suppl] 7. Springer, Wien New York, pp 135–152

17. Stitz L, Planz O, Bilzer T, Frey K, Fontana A (1991) Transforming growth factor-beta modulates T-cell-mediated encephalitis caused by Borna disease virus. Pathogenetic importance of CD8$^+$ cells and suppression of antibody formation. J Immunol 147: 3581–3586

18. Stitz L, Sobbe M, Bilzer T (1992) Preventive effects of early anti-CD4 or anti-CD8 treatment on Borna disease in rats. J Virol 66: 3316–3323

19. VanDeWoude S, Richt JA, Zink MC, Rott R, Narayan O, Clements JE (1990) A Borna virus cDNA encoding a protein recognized by antibodies in humans with behavioral disease. Science 250: 1278–1281

20. Zinkernagel RM, Welsh RM (1976) H-2 compatibility requirement for virus specific T-cell-mediated effector functions in vivo. I. Specifity of T-cells conferring antiviral protection against lymphocytic choriomeningitis virus is associated with H-2 K and H-2 D. J Immunol 117: 1495–1502

Authors' address: Dr T. Bilzer, Department of Neuropathology, Heinrich-Heine-Universität Düsseldorf, Universitätsstrasse 1, D-40225 Düsseldorf, Federal Republic of Germany.

Arch Virol (1993) [Suppl] 7: 159–167

Borna Disease virus infection and affective disorders in man

L. Bode[1], R. Ferszt[2], and G. Czech[3]

[1] Robert Koch-Institute, Department of Virology, Free University Berlin, Berlin
[2] Klinikum Steglitz, Department of Psychiatry, Free University Berlin, Berlin
[3] Institute of Virology, Free University Berlin, Berlin, Federal Republic of Germany

Summary. Borna Disease virus (BDV) can persistently infect the central nervous system of a broad spectrum of animal species. The clinical course varies from slight behavioral disturbancies to a fatal neurological syndrome. In-vivo diagnosis is based on the strong humoral immune response to BDV antigens. Since also human infections could be confirmed by specific antibodies and increased seroprevalence was found in patients with chronic neurologic or immunologic disorders, the contribution of BDV or a BDV-like human variant to syndromes with yet unknown etiology became of great interest. We presented the first data of a current follow-up study on 70 psychiatric patients who were tested three times each after hospitalization. In contrast to previously found low prevalence of antibody carriers by screening (2–4%), we now found 20% positives by follow-up testing. Furthermore, of the randomly selected patients with different psychiatric diagnosis, the highest proportion of antibody carriers was detected among patients with major depression (more than 30%), compared to only 8% among patients with dysthymia (neurotic depression). This led us to hypothesize that Bornavirus infection might contribute somehow to the syndrome of major depressive illness by altering neuronal cells in the limbic system.

Introduction

Borna Disease virus (BDV) is as yet an unclassified infectious agent which causes a neurologic disease in horses and sheep [12]. Recently, molecular cloning experiments gave strong evidence that BDV is a single and negativ-strand RNA virus [7, 11, 15]. BDV is transmissible to several animal species provoking different clinical pictures, varying from fatal disease to almost asymptomatic life-long virus persistence [10, 13].

Humoral immune response is induced irrespective of disease. Antibodies recognize at least two of three proteins of the BDV soluble (s)-antigen complex [12] which is found to be not only virus-induced, but most probably a virus component [11]. Thus, s-antigen-specific antibodies are reliable infection markers for in vivo-diagnosis. Basic detection methods are immunofluorescence (IF) and enzymeimmunoassay (EIA), supplemented by Western-blot (WB) and immunoprecipitation (IP).

Seroepidemiological studies recently showed strong evidence of worldwide (latent) BDV infection also in man based on specific antibody detection [3, 6]. Patients with chronic diseases of the brain and/or the immune system showed an at least six times higher antibody prevalence than healthy subjects [6]. However, since the main purpose of those previous studies was to evaluate the existence and distribution of the new human infection by screening of a large number of individuals, the contribution of a BDV infection to human disease, whatsoever, remained unclear.

The purpose of this report was to investigate the prevalence of such an infection in psychiatric patients after onset of acute disease/ reactivated symptoms of chronic disease, in order to elucidate any specific relationship of BDV to certain mental disorders. We present here the first data on antibody development and frequencies among acute patients with ten different psychiatric syndromes.

Materials and methods

Subjects

In this report, 71 psychiatric patients from a current study started at July 1991 were included. A subcohort of 43 of these patients was already under previous investigations [4, 5]. All patients were hospitalized due to acute symptoms for on average one month in the psychiatric department of the Klinikum Steglitz (Free University, Berlin). Two third of the patients were women, one third men, but both sexes had a similar mean age of around 40 years. The majority of the patients (60%) were suffering from chronic psychiatric disorders (at least since one year) which means that they had experienced several disease episodes before. About 30% had a subacute disease (duration up to two months) and only 7% of the patients were for the first time mentally ill (< two weeks). From ten psychiatric diseases, the most frequent diagnoses in our cohort were major depression, neurotic depression and paranoid psychosis.

The sera were usually collected in weekly intervals with a mean of three samples per patient. All samples were coded before use and tested twice; storage was at −20°C.

Antibody detection method

As basic routine technique for BDV antibody detection in the sera of our patient cohort, a modified (double-stain) immunofluorescence test (IFT) was used according to the previously described protocol [6, 4]. Briefly: BDV infected young rabbit brain cells (7×10^{-3} focus forming units per cell), seeded on cover slips and after 5 days fixed with acetone, were incubated (1 h at 37°C) with equal volumes of human serum and a specific monoclonal antibody (mab), followed by washing and further incubation (1 h at 37°C) of adaequate dilutions of FITC-labeled goat anti-human IgG and TRITC-labeled rat anti-mouse IgG (Dianova, Hamburg). Human samples which showed the same staining pattern as the control mab were scored specific positive. Antibody titres were determined by endpoint dilution.

Results

Detection of BDV-specific antibodies

All previously tested human serum samples of various origins (including psychiatric patients) recognized nuclear antigen in BDV infected cells by IFT [3–6]. A representative result is given in Fig. 1 comparing three patients with different diseases (see also [5]). The specificity for BDV is defined by the identical staining pattern of human serum and mab, the latter known to bind to the nuclear antigen [6].

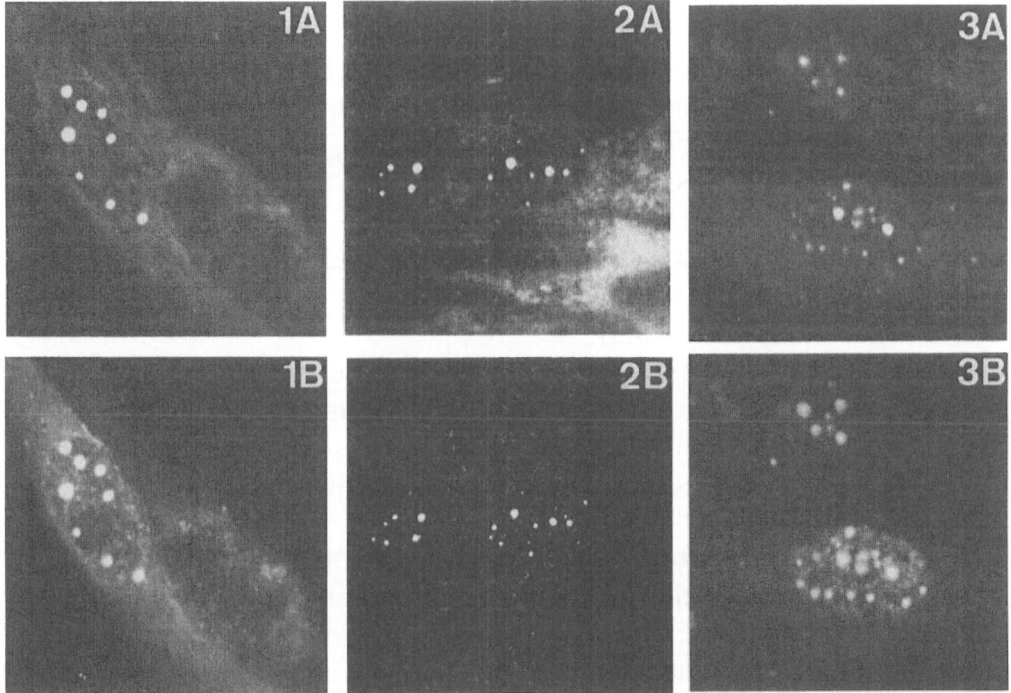

Fig. 1. Immunofluorescence (IFT) of BDV infected young rabbit brain cells with (**A**) human sera (1–3) and (**B**) corresponding BDV-specific monoclonal antibody (mab). Stained with **A** FITC anti-human IgG and **B** TRITC anti-mouse IgG, **1A** serum of a psychiatric patient, **2A** serum of a HIV infected patient, **3A** serum of a patient with multiple parasitic diseases (see also [5, 6])

In accordance with those previous findings, all positive sera of our psychiatric patients showed exactly the same nuclear fluorescence, with mean antibody titres of 1:80. This nuclear antigen had been identified as the 38/40 KD protein of the BDV s-antigen complex [6].

Antibody development (seroconversion)

From the majority of the patients, three to four consecutive serum samples were available to investigate a chronological antibody profile, starting 3–5 days after patients' admission in the hospital. Most of the positive patients tended to develope detectable antibody levels later than seven days after hospitalization which means that their first serum samples gave negative results. The mean period for seroconversion was 17.2 days (S.D. 11.7).

Antibody prevalence

We found a total frequency of antibody carriers of 20% (23.3% based on 43 patients; 19.7% based on 71 patients) as shown in Fig. 2 (see S1 and S2). Among male patients, a slightly higher percentage than in females was detected. Additionally, the seropositive men were found to be older than the whole male cohort (48 versus 40 years, with S.D. of 18 and 13 years), in contrast to seropositive women who were younger than the female cohort (33 versus 43 years, with S.D. of 12 and 16 years). However, age differences could not be confirmed, if all positive patients (mean age 40, S.D. 16 years) were compared with the whole cohort (mean age 42, S.D. 15 years).

In contrast, obvious differences could be observed among the various psychiatric disorders. It is remarkable that the majority of seropositives were depressives presenting with major depression (Fig. 3, column 1), while patients with this particular diagnosis constituted a proportion of only one third in the whole cohort.

The most striking result was the significant difference between these more severely incapacitated patients and other depressives, formerly categorized as neurotic, now as dysthymic. We found the exorbitant rate of 37% antibody carriers among the major (unipolar) depressed, in contrast to only 6–7% among the dysthymic patients (Fig. 2, columns a and b); from corresponding subcohorts of patients (Fig. 2, S1 and S2), nearly identical results were obtained. Furthermore, clinical observations revealed that the BDV antibody carriers tended to a more severe clinical course than BDV negatives and belonged to those patients who had

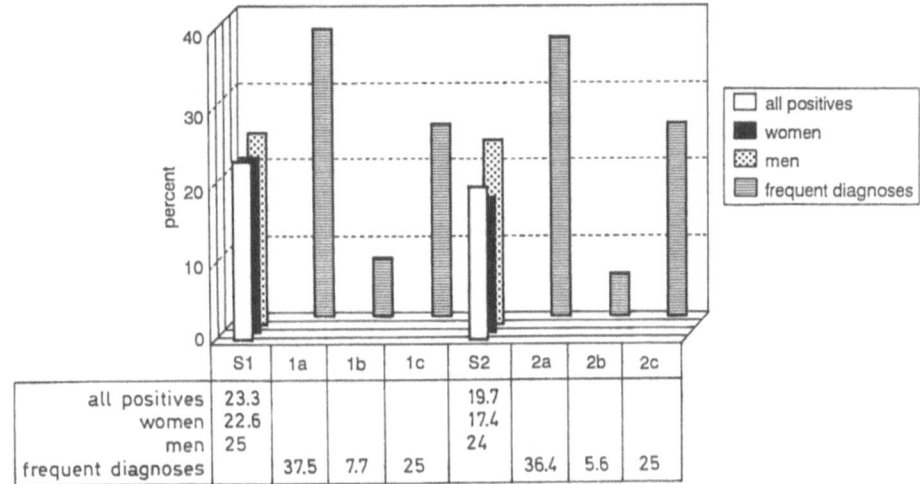

	S1	1a	1b	1c	S2	2a	2b	2c
all positives	23.3				19.7			
women	22.6				17.4			
men	25				24			
frequent diagnoses		37.5	7.7	25		36.4	5.6	25

Fig. 2. Prevalence of BDV antibodies in psychiatric patients. *S1* Antibody carriers (%) among a subcohort of 43 patients (31 women; 12 men) and *S2* among the total cohort of 71 patients (46 women; 25 men). *a–c* Antibody carriers (%) among the three most frequent diagnoses in S1 and S2: (*a*) unipolar (major) depression, (*b*) neurotic depression (dysthymia), (*c*) paranoid psychosis

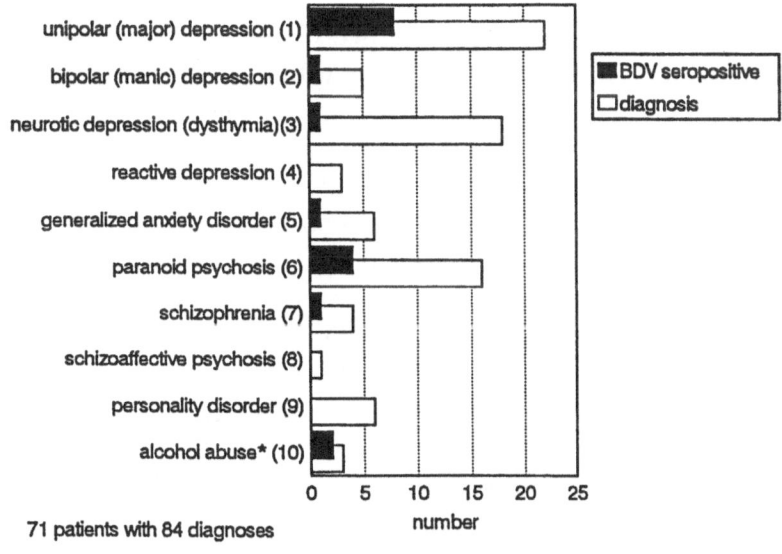

Fig. 3. Distribution of BDV antibody carriers (number) among different psychiatric diagnoses (*1–10*) based on the total cohort of 71 patients, but with 84 diagnoses because of some single patients with multiple syndromes. *Only patients with more than one diagnosis

undergone previous episodes of their diseases (chronic and subacute). With the exception of their psychic impairment, all the patients were in good general health with no immunosuppressive factor than depression as such.

Discussion

Since the first reports on BDV antibodies in man [3, 14] up to a recent comprehensive study [6], the investigation of psychiatric patients had gained certain interest, suggesting implications, whatsoever, between infection with a neurotropic virus and mental disorders. However, those aspects had necessarily been inferior for several years in favour of work on the specificity of the human antibodies and the establishment of reliable detection methods. Since we could demonstrate that human antibodies recognize the major BDV antigen of 38/40 KD which is present in the nucleus of infected cells (detectable by IFT) [6] and this particular protein was shown to be encoded by BD cDNA clones [11], seroprevalence of human BDV antibodies means the existence of human infection. In that previous study [6], we furthermore claimed a latent BDV infection, with none or rare reactivation events in people with low antibody prevalence (2%) and rather high reactivation events with antigen expression and subsequent humoral immune response in patients with significantly increased prevalence (12–14%); screening one sample per patient, the low rate had been found in healthy individuals, the high in patients with chronic brain/immune diseases. In contrast to the previous test design, this study was based on a temporal relationship between test samples and acute clinical signs; the importance for testing acute psychiatric patients had come also from a study, observing higher freqencies of multiple foci (cortex) in BDV seropositive acute patients compared to negatives [1].

By testing follow-up samples from the same patient after onset of acute disease, we detected ten times more BDV antibody carriers among psychiatric patients (>20%) than we would expect in the normal population. Furthermore, in most of the BDV positive patients, a seroconversion could be observed after a mean period of 17 days (S.D. 12 days), in other words, at least two consecutive samples were necessary to find detectable antibody levels. However, the most surprising result was that we picked up an exorbitant percentage (36–38%) of BDV positives among a particular subcohort of depressives (patients with major depression), while among a comparable number of other (less incapacitated) neurotic depressives (Fig. 3) only 6–8% could be detected (Fig. 2). Besides major depressives, only patients with another severe mental disorder (paranoid psychosis) reached similar antibody frequencies (25%). On the other hand, there was no evidence that age or the physical health status had any influence on the carrier status; none of them had anamnestic neurological diseases.

These results implicated that every third to every fourth patient with an affective psychosis should be considered as latently infected with Bornavirus expressing humoral response during acute disease episodes. Such high incidences suggest the likelihood of some kind of relationship between BDV infection and those particular disorders, but how specific could this relationship be? In case of major depression, the development of immunosuppression which may occur would at first sight argue against specificity. However, neuroendocrinological studies revealed that this immunosuppression during depressive episodes is not due to peripheral disturbancies, but caused by a remarkable decrease of certain neuro-transmitters (noradrenaline, serotonine) at synaptic sites of the limbic system [2]. Provided this down-regulation of neurotransmitters is the main "biological" factor during a depression episode, our present data pointed to a more specific association of BDV infection with major depressive illness rather than to unspecific effects. This conclusion can be supported by the striking affinity of Bornavirus to structures of the limbic system in animals [9], and by the previous finding of specific learning deficiencies and altered affection in persistently infected but clinically healthy rats [8]. From the present study, it seemed that not any disease with depression as such, but most probably a particular disorder like major depressive illness, could be related to BDV infection. Further support came lately from an investigation on chronic fatigue syndrome (CFS), CFS causes also severe mood disturbancies as well as memory impairments, but, in contrast, seems unlikely to be associated with this virus infection so far, since no BDV antibodies could have been found in the sera of 50 patients with CFS [6a].

Although future investigations are necessary, especially on tracing of the virus genome in autopsy samples by aid of the now available cDNA probes [7, 11, 15], our present study already implicates that Bornavirus infection should be considered as health risk for patients suffering from psychiatric diseases. In case of the severe affective disorders, even a causative effect of this neurotropic agent on the clinical picture cannot be excluded so far.

Acknowledgements

We are grateful to P. Reckwald for excellent technical assistance and to Dr. H. Querfurth and Dr. N. Rigas for collecting the blood samples and basic patients' data.

We also thank Prof. Dr. H. Ludwig, Institute of Virology, Free University Berlin, for kindly supporting this subproject on the Bornavirus research program by a DFG grant (Lu 142/5-1).

References

1. Bechter K, Herzog S, Fleischer B, Schüttler R, Rott R (1987) Kernspintomographische Befunde bei psychiatrischen Patienten mit und ohne Serum-Antikörper gegen das Virus der Bornaschen Krankheit. Nervenarzt 58: 617–624

2. Bleuler E (1983) Lehrbuch der Psychiatrie, 15th edn. Springer, Berlin Heidelberg New York Tokyo, pp 485–486

3. Bode L, Riegel S, Ludwig H, Amsterdam JD, Lange W, Koprowski H (1988) Borna Disease virus-specific antibodies in patients with HIV infection and with mental disorders. Lancet ii: 689

4. Bode L, Querfurth H, Ferszt R, Gosztonyi G, Rigas N, Czech G, Ludwig H (1991) Antibodies to Borna disease virus are four times as frequent in depressives as in controls. In: Wissenschaftswoche 1991, Forschungsprojekte am Klinikum Steglitz, Freie Universität Berlin, pp 269–271

5. Bode L, Czech G, Ferszt R, Ludwig H (1992) Bornavirus-Infektion beim Menschen: eine neue Zoonose? In: Bericht des 4. Hohenheimer Seminars "Aktuelle Zoonosen", September 16–17, 1992, Stuttgart. Deutsche Veterinärmedizinische Gesellschaft (DVG), Giessen, pp 138–147

6. Bode L, Riegel S, Lange W, Ludwig H (1992) Human infections with Borna Disease virus: seroprevalence in patients with chronic diseases and healthy individuals. J Med Virol 36: 309–315

6a. Bode L, Komaroff AL, Ludwig H (1992) No serologic evidence of Borna Disease virus in patients with Chronic Fatigue Syndrome (CSF). Clin Infect Dis 15: 1049

7. Briese T, de la Torre JC, Lewis A, Ludwig H, Lipkin WI (1992) Borna disease virus, a negative-strand RNA virus, transcribes in the nucleus of infected cells. Proc Natl Acad Sci Neurobiol USA 89: 11486–11489

8. Dittrich W, Bode L, Ludwig H, Kao M, Schneider K (1989) Learning deficiencies in Borna Disease virus-infected but clinically healthy rats. Biol Psychiatry 26: 818–828

9. Gosztonyi G, Ludwig H (1984) Neurotransmitter receptors and viral neurotropism. Neurophysiol Clin 3: 107–114

10. Hirano N, Kao M, Ludwig H (1983) Persistent, tolerant or subacute infection in Borna disease virus infected rats. J Gen Virol 64: 1521–1530

11. Lipkin WI, Travis GH, Carbone KM, Wilson MC (1990) Isolation and characterization of Borna disease agent cDNA clones. Proc Natl Acad Sci USA 87: 4184–4188

12. Ludwig H, Bode L, Gosztonyi G (1988) Borna Disease: a persistent virus infection of the central nervous system. Prog Med Virol 35: 107–151

13. Narayan O, Herzog S, Frese K, Scheefers H, Rott R (1983) Pathogenesis of Borna disease in rats: immune-mediated viral ophthalmoencephalopathy causing blindness and behavioral abnormalities. J Infect Dis 148: 305–315

14. Rott R, Herzog S, Fleischer B, Winokur H, Amsterdam JD, Dyson W, Koprowski H (1985) Detection of serum antibodies to Borna Disease virus in patients with psychiatric disorders. Science 228: 755–756

15. Vande Woude S, Richt JA, Zink MC, Rott R, Narayan O, Clements JE (1990) A Borna virus cDNA encoding a protein recognized by antibodies in humans with behavioral diseases. Science 250: 1278–1281

Authors' address: Dr. L. Bode, Robert Koch-Institut des Bundesgesundheitsamtes, Abteilung Virologie, Nordufer 20, D-13353 Berlin, Federal Republic of Germany.

Arch Virol (1993) [Suppl] 7: 169–183

Virus-host interactions in African swine fever: the attachment to cellular receptors

A. Angulo, A. Alcamí*, and **E. Viñuela**

Centro de Biología Molecular, Facultad de Ciencias, Universidad Autónoma, Cantoblanco, Madrid, Spain

Summary. Biochemical and morphological techniques have shown that African swine fever virus (ASFV) enters susceptible cells by a mechanism of receptor-mediated endocytosis. The virus binds to a specific, saturable site in the cell and this interaction is required for a productive infection. A structural ASFV protein of 12 kDa (p12) has been identified to be involved in the recognition of the cellular receptor, on the basis of the specific binding of the polypeptide to sensitive Vero cells. Protein p12 is externally located in the virus particle, forming disulfide-linked dimers with an apparent molecular mass of 17 kDa. The gene has been mapped within the central region of the BA71V strain genome. Sequencing analysis has shown the existence of an open reading frame encoding a polypeptide of 61 amino acids characterized by the presence of a putative transmembrane domain, and a cysteine rich region in the C-terminal part which may be responsible for the dimerization of the protein. Transcripts of the p12 gene were only synthesized during the late phase of the infectious cycle. No posttranslational modifications of the polypeptide, such as glycosylation, phosphorylation or fatty acid acylation, have been found. The comparison of the amino acid sequence of protein p12 from 11 different virus strains has revealed a high degree of conservation of the polypeptide.

Introduction

African swine fever virus (ASFV) is an enveloped icosahedral deoxyvirus responsible of a devastating disease of swine [16, 19, 44, 45]. The virus infects domestic pigs and wild boars and multiplies in soft ticks, which can act as a natural reservoir and a way of transmission, difficulting its control [47]. A peculiar aspect of the infection is the absence of

*Present address: Sir William Dunn School of Pathology. University of Oxford, Oxford, U.K.

neutralizing antibodies, even in recovered or chronically infected pigs or in ASFV-resistant animal species inoculated with the virus, a fact that has prevented the obtention of a conventional vaccine [17]. The virus particle contains a nucleoprotein core of 70–100 nm diameter surrounded by a lipid membrane probably associated with the morphological units of its icosahedral capsid. The extracellular virions have an additional external lipoprotein envelope [15, 34]. ASFV particle is composed by more than 50 proteins ranging from 10 to 150 kDa [12, 20]. The viral genome is a linear molecule of double stranded DNA of approximately 170 kbp, with covalently closed ends and terminal inverted repetitions [8, 24, 35]. The virus shows a strict host range and cell tropism in natural infections, replicating in mononuclear phagocytes and in a small fraction of polymorphonuclear leukocytes [32]. However, the virus has been adapted to grow in several established cell lines from pigs and other animal species [44], and most of the studies on the biochemistry and molecular biology of the virus have been performed using cell-adapted virus. ASFV gene expression is regulated in a temporal fashion, and proteins are classified as early or late depending on whether their synthesis requires viral DNA synthesis as well [20, 40, 41]. More than 100 virus-induced polypeptides have been described in ASFV-infected cells [20, 41]. For many years, ASFV has been classified as an iridovirus, mainly on the basis of its morphology [31]. Other properties, such as the absence of structural glycoproteins [18] and the existence of an initial nuclear stage in the virus DNA replication [21] are shared by both ASFV and members of the *Iridoviridae* family. However, the genome structure [24, 42], the presence in the virion of the enzymes required for early RNA synthesis and processing [39], and several aspects concerning its molecular biology makes ASFV close to poxviruses [16, 45]. Altogether, the characteristics of ASFV have led to the establishment of a yet unnamed new family, of which ASFV is the only representative [11].

The initial event in the viral infectious cycle is the attachment of the virion to the host cell membrane [30]. This interaction is mediated by the binding of a viral attachment protein to a component on the cell surface acting as a viral receptor. The adsorption step is a major determinant of the host range and tissue tropism of viruses and has proven to be an ideal target for agents that prevent infection by inhibiting the attachment of virus to the host cell receptor. In order to develop these agents, it is necessary to identify the binding domains of the viral attachment protein and of the virus receptor, and determine how they interact at a molecular level.

The present review will focus on the initial events in the ASFV infectious cycle, the adsorption to the plasma membrane and subsequent penetration into the host cell, and on the identification and characteriza-

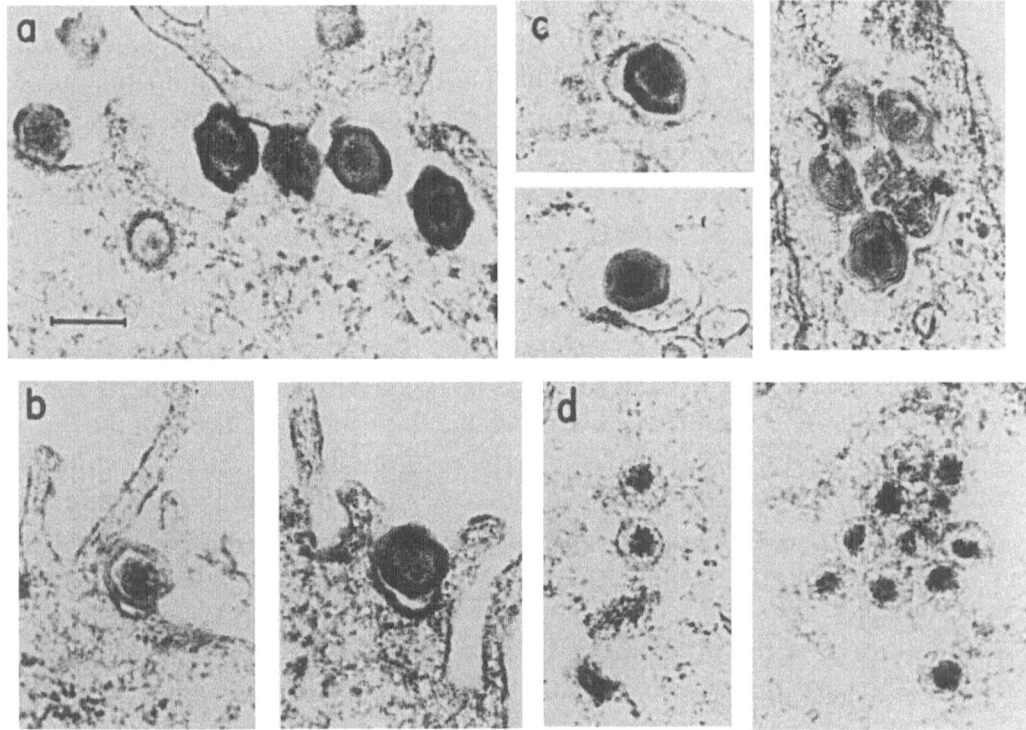

Fig. 1. Entry of ASFV into Vero cells. Vero cells infected with 3,000 PFU per cell at 4°C for 2 h were thin-sectioned and examined by electron microscopy at different times after shifting the temperature to 37°C. **a** Virus particles adsorbed to the plasma membrane, **b** in membrane invaginations, **c** in cytoplasmic vesicles, and **d** cores in the cytoplasm. Bar represents 200 nm

tion of the viral attachment protein. These studies have been performed with the Vero-adapted ASFV strain (BA71V) which maintains the capacity to infect swine macrophages, due to the possibility to obtain highly purified preparations of virus particles.

ASFV entry into susceptible cells

Morphological data obtained by electron microscopy have shown that ASFV enters susceptible Vero cells and swine macrophages by endocytosis [3, 4, 22, 43]. Virus particles have been frequently found adsorbed to cytoplasmic invaginations (Fig. 1) similar to the clathrin-coated pits involved in the entry of ligands through a receptor-mediated endocytosis mechanism [36]. ASFV production in Vero cells was inhibited by lysosomotropic agents, obtaining 50% of inhibition with concentrations of 5.0, 2.5, 0.2, 0.1 and 0.006 mM of methylamine, ammonium chloride, amantadine, dansylcadaverine and chloroquine, respectively [3]. In the presence of dansylcadaverine and chloroquine, ASFV particles were retained in cytoplasmic vacuoles and neither viral cores nor early viral RNA synthesis were detected, indicating that the low pH environment of the intracellular vacuoles is required for virus uncoating, as it has been described for other enveloped animal viruses.

Binding experiments of [3H]-labeled virus particles to cells have shown the presence of saturable binding sites for virus attachment on the plasma membrane of Vero cells and swine macrophages [2, 4]. The Scatchard analysis of the binding data at equilibrium has shown that there are about 10^4 cellular receptor sites per Vero cell with a dissociation constant of 70 pM, values similar to those reported for other animal viruses [29]. The fact that the saturable binding and the uptake into Vero cells of [3H]-labeled ASFV were competed by similar amounts of unlabeled virus indicates that virus entry into the cell is mediated by a saturable component in the plasma membrane. Similarly, the early viral protein synthesis and virus production in Vero cells and swine macrophages were inhibited by doses of UV-inactivated virus that were able to compete the attachment to saturable sites [2]. Altogether, these results indicate that specific virus receptors mediate the entry of virus particles that lead to a productive infection.

ASFV binding to virus-resistant L cells and rabbit macrophages was not mediated by saturable binding sites [2]. This fact, together with the failure of viral particles to enter L cells, suggests that the absence of specific receptors for ASFV is a factor that determines the resistance of L cells to the infection. Although the interaction with rabbit macrophages was not mediated by receptors, the virus penetrated into these cells, probably due to the high phagocytic activity of macrophages, and initiated the synthesis of some early viral proteins [4]; however, the viral replication cycle was aborted since viral DNA synthesis did not occur. A similar abortive infection was detected in macrophages from other virus-resistant animal species. The failure of high doses of UV-inactivated virus to inhibit the early viral protein synthesis in rabbit macrophages supports that ASFV attachment and entry into these cells is not mediated by a saturable component.

It has been described for other viruses that the virus receptor is not the only binding site for virus entry into macrophages, since receptors for the Fc portion of the immunoglobulins mediate the entry of virus-antibody complexes into macrophages, facilitating virus infection instead of protecting the host [37]. However, titration experiments in swine macrophages have shown that ASFV infectivity was not enhanced in the presence of antiviral antibodies [5]. The early viral protein synthesis and the viral DNA replication in swine macrophages infected with virus-antibody complexes were inhibited in the presence of high doses of UV-inactivated virus, which saturated specific receptors, but not when Fc receptors were saturated with antibodies. These results indicate that ASFV does not infect swine macrophages through Fc receptors and, therefore, that antibody-dependent virus entry into the cell is not a mechanism that facilitates the progression of ASFV infection.

Identification of the ASFV attachment protein

The presence of cellular receptors that mediate ASFV binding to the target cell involves the existence of a viral attachment protein in the virus particle. Treatment of purified virus particles with octyl-glucoside (OG) allowed a differential release of viral proteins [14]. As shown in Fig. 2, treatment of ASFV virions with 0.5% OG solubilizes proteins p12, p14, p17 and p35, while a reduction to about 2% of the initial virus

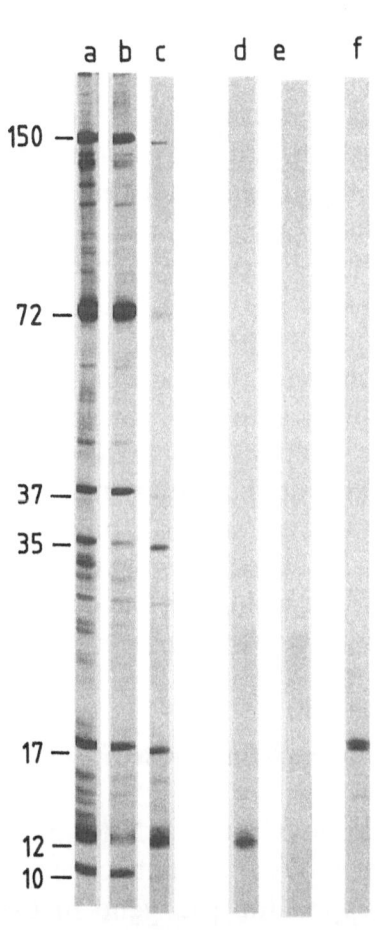

Fig. 2. Binding of OG-released ASFV proteins to sensitive and resistant cells. Proteins in the whole virus (*a*), and in the sediment (*b*), and supernatant (*c*) of the centrifugation after treatment of [^{35}S]-labeled ASFV with 0.5% OG for 1 h at 4°C are shown. Vero or L cell monolayers were incubated for 2 h at 37°C with [^{35}S]-labeled OG-released virus proteins, washed with phosphate buffered saline, and subjected to SDS-PAGE. (*d*) Proteins bound to Vero cells, (*e*) proteins bound to L cells, and (*f*) proteins bound to Vero cells analyzed in the absence of 2-mercaptoethanol. Molecular masses in kilodaltons are indicated

infectivity is observed. When different cell monolayers were incubated with this mixture of OG-released proteins, protein p12 bound specifically to virus-sensitive Vero cells and not to virus-resistant L cells. The subviral particles generated after the OG treatment had lost most of their capacity to bind to Vero cells. The biological relevance of the binding of protein p12 to sensitive cells was supported by the fact that it was competed by doses of unlabeled virus similar to those that abolish the binding of the labeled virus, indicating that both of them interact with the same saturable component in the plasma membrane of Vero cells.

A monoclonal antibody specific for protein p12 (MAb 24BB7) immunoprecipitated from purified virions a protein that, when analyzed in sodium dodecyl sulfate-polyacrylamide gel electrophoresis (SDS-PAGE) in the absence of the reducing agent 2-mercaptoethanol, showed a molecular mass of 17 kDa (p17) instead of 12 kDa as found under reducing conditions [14]. The relationship between these two proteins was confirmed by the total conversion of p17 to p12 when the former was isolated from polyacrylamide gels in the absence of 2-mercaptoethanol and subsequently treated with the reducing agent. The lack of the viral attachment protein in the supernatant obtained after immunoprecipitation with the MAb 24BB7 confirmed that the protein recognized by the monoclonal antibody was the same as that involved in the interaction of the virus with the cell. Thus, it seems that p12 is solubilized from the virion by OG treatment as a dimer of about 17 kDa and that this is the form of the protein that binds to sensitive cells.

To determine the location of protein p12 in ASFV particles, immuno-electron microscopy assays were performed using the MAb 24BB7 and protein A-gold complexes [13]. In intact particles, the labeling could be only detected in lateral protrusions that followed the external virus envelope. After a mild treatment of the virions with OG or ethanol, that partially disrupt the external membrane enabling the antibody to penetrate, the label was more uniformly distributed around the virus periphery. In addition, the absence of labeling in subviral particles, which lack the external proteins, and in virus capsids, indicates that protein p12 is situated in a layer above the virus capsid with the epitope recognized by the MAb 24BB7 not exposed in the virion surface.

Mapping and nucleotide sequence of the gene encoding protein p12

The hybridization of oligonucleotide probes derived from the N-terminal amino acid sequence of protein p12 to cloned DNA fragments covering the complete length of the viral genome mapped the p12 gene in the fragment EcoRI-O, located within the central region of the genome [1]

(Fig. 3A). The DNA sequence of an *Eco*RI-*Xba*I fragment showed an open reading frame (ORF), named ORF O61R, which is predicted to encode a polypeptide of 61 amino acids (Fig. 3B). The predicted molecular mass of the mature protein of 60 amino acids in length, since it was devoided of the initiator methionine, is 6.6 kDa, lower than that expected from its electrophoretic mobility in SDS-PAGE, 12 kDa. This disagreement might be explained by an unusual conformation of the protein in the presence of SDS, probably influenced by the existence of a long stretch of hydrophobic residues in its primary structure, as will be described later. To confirm that ORF O61R encoded protein p12, RNA selected from Vero cells at late times of infection by hybridization to fragment *Eco*RI-O, or transcripts synthesized in vitro by the T7 RNA polymerase using a recombinant plasmid that contains the p12 gene inserted immediately downstream the ϕ10 promoter, were translated in vitro [1]. The electrophoretic analysis of the cell free translation products showed a polypeptide of 12 kDa that was specifically recognized by the MAb 24BB7, indicating its identity with the protein p12 present in the virus particle. Additionally, the protein was expressed in *Escherichia coli* under the control of the T7 RNA polymerase, and a polypeptide of 12 kDa could be detected, that was also recognized by the MAb 24BB7. However, the protein expressed in these two systems did not form dimeric structures of 17 kDa when analyzed in the absence of 2-mercaptoethanol, and was not able to bind to sensitive Vero cells. The formation of the dimer of 17 kDa or, alternatively, the requirement of a posttranslational modification or a correct folding of the monomer, which do not take place in the two expression systems used, might be necessary for binding.

Properties of the polypeptide

As shown in Fig. 3C, the hydrophilicity profile indicates the absence of an N-terminal hydrophobic signal sequence that might initiate the translocation across the endoplasmic reticulum membrane. However, there is a stretch of 22 hydrophobic residues in the central region that may represent a transmembrane domain and could act as an anchor of the protein in the viral external envelope [23, 46]. A striking feature of the C-terminal part is the presence of a cysteine-rich region that may account for the dimerization of the protein through disculfide bonds. A potential site for asparagine-linked carbohydrate and a sequence that may function as a substrate for protein kinase C [28] are found in the predicted amino acid sequence. A search of the GenBank national data base revealed no significant homology with the entire sequence of

A

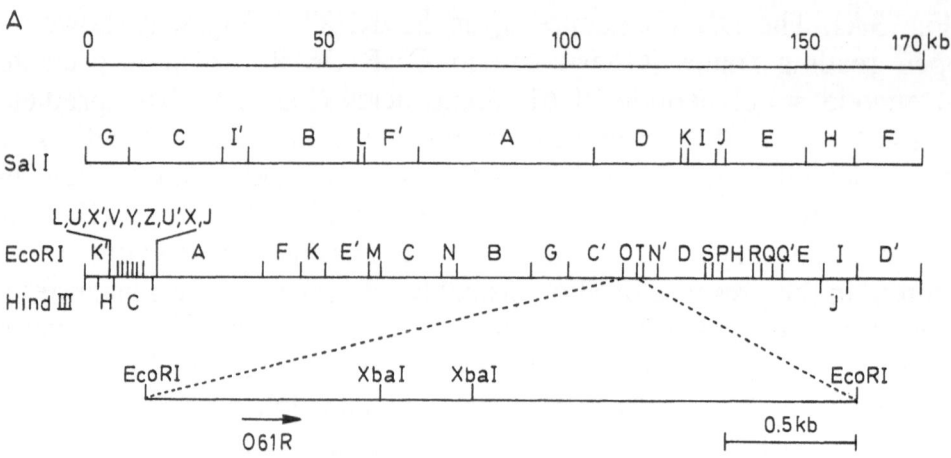

B

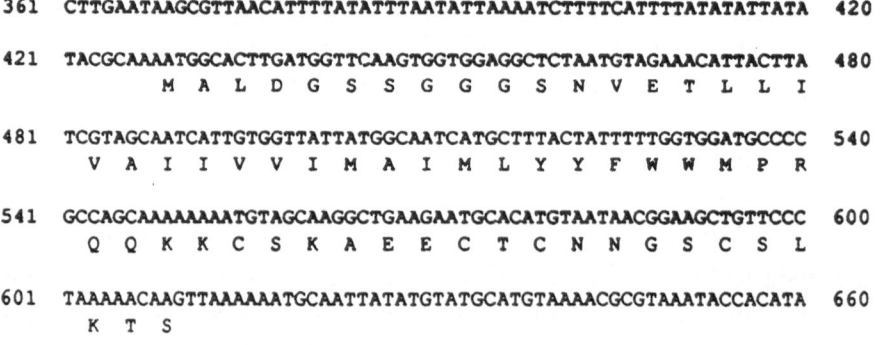

```
361  CTTGAATAAGCGTTAACATTTTATATTTAATATTAAAATCTTTTCATTTTATATATTATA  420

421  TACGCAAAATGGCACTTGATGGTTCAAGTGGTGGAGGCTCTAATGTAGAAACATTACTTA  480
          M  A  L  D  G  S  S  G  G  G  S  N  V  E  T  L  L  I

481  TCGTAGCAATCATTGTGGTTATTATGGCAATCATGCTTTACTATTTTTGGTGGATGCCCC  540
      V  A  I  I  V  V  I  M  A  I  M  L  Y  Y  F  W  W  M  P  R

541  GCCAGCAAAAAAAATGTAGCAAGGCTGAAGAATGCACATGTAATAACGGAAGCTGTTCCC  600
      Q  Q  K  K  C  S  K  A  E  E  C  T  C  N  N  G  S  C  S  L

601  TAAAAACAAGTTAAAAAAATGCAATTATATGTATGCATGTAAAACGCGTAAATACCACATA  660
      K  T  S

661  AAACTATAACATGTCAATCATGGAATCAACACTTTTATAATTTTCCGTAATATATTTTTC  720
```

C

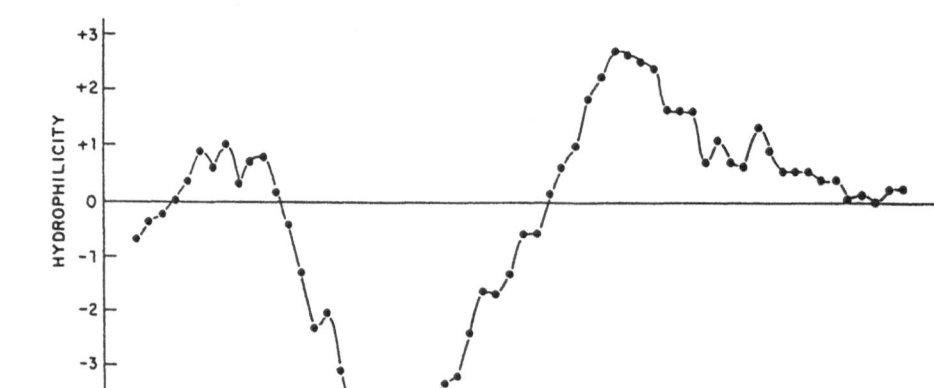

protein p12; however, it is interesting to note that the cysteine motif of protein p12 was also found in the β chain of integrins, which contain in their extracellular domain four tandem repeats of an eight-cysteine motif [26]. Four cysteines out of the eight that constitute each domain were found in the same positions in protein p12. It is tempting to speculate that both molecules, p12 and integrins, involved in cellular recognition, share a common structural motif.

Transcriptional mapping of protein p12

Primer extension and S1 nuclease analyses showed that the p12 gene was transcribed only during the late phase of the infectious cycle, giving rise to an mRNA of approximately 1.18 kb which lack poly(A) head at its 5′ end [6]. The p12 transcripts initiated 10 to 12 nucleotides upstream of the AUG codon of the ORF O61R and extended to approximately 980 nucleotides downstream of the termination codon of this ORF. It is interesting to note that the 3′ end of the p12 transcripts mapped within the first run of several consecutive thymidylate residues (9T) present in the downstream region of the ORF. This result, together with the fact that the 3′ ends of a number of ASFV early genes map within a conserved sequence motif formed by at least seven thymidylate residues (7T) [7], strongly suggests that the 7T motif might be involved in the 3′ end formation of both early and late ASFV mRNAs.

The characterization of the p12 gene transcription has provided the first clues about the organization of ASFV late mRNAs. In contrast to poxvirus [a virus that shares several genome structure peculiarities with ASFV] late mRNAs, which are heterogeneous in size and possess 5′ poly(A) heads [33], the p12 transcripts have a defined length and do not contain these peculiar 5′ end structures. Additionally, the TAAAT essential motif of most poxvirus late promoters [25, 38] was not present within the proximity of the transcriptional initiation site of the p12 gene.

Fig. 3. Mapping and sequence of the gene encoding protein p12. **A** Position of the ORF O61R in the ASFV genome and in the *Eco*RI-O fragment. The *Xba*I sites within the *Eco*RI-O fragment are indicated. **B** Nucleotide sequence and deduced amino acid sequence of ORF O61R and flanking regions. Numbers correspond to the position in the complete nucleotide sequence of the *Eco*RI-*Xba*I fragment of 913 bp. **C** Hydrophilicity profile of the predicted protein p12. It was obtained according to the method of Kyte and Doolittle [27]. The boxed region is the putative hydrophobic transmembrane segment. The phosphorylation site for protein kinase C is underlined. The potential N-linked glycosylation site is doubly underlined. Cysteines are indicated by asterisks

Synthesis of protein p12 in ASFV-infected cells

Labeling of infected cells with [^{35}S]methionine and subsequent immuno-precipitation with the MAb 24BB7 showed that the protein p12 was first detected after 15 hours of infection in Vero cells and 7 hours of infection in macrophages, as a dimer of 17 kDa when analyzed by SDS-PAGE in the absence of reducing agents [1]. The polypeptide was not synthesized in the presence of cytosine arabinoside, an inhibitor of DNA synthesis, confirming that the p12 gene belongs to the late class of viral genes. Results obtained by indirect immunofluorescence assays with the MAb 24BB7 performed in both Vero cells and swine macrophages (unpubl. results) are consistent with the time course of appearance of this protein by immunoprecipitation. The protein is initially localized in discrete perinuclear cytoplasmic areas, which correspond to viral factories where virus morphogenesis occurs. At later times, the protein is detected not only in viral factories, but also distributed throughout the cytoplasm in a speckled pattern, and accumulated in specific areas close to the plasma membrane that could correspond to specialized regions used for the egress of the virus from the cell, although no immunofluorescence was found in the periphery of infected cells under nonpermeable conditions.

The possibility that the protein might be posttranslationally modi-fied was investigated [1]. Experiments with infected cells labeled with [^{3}H]myristate and [^{3}H]palmitate followed by immunoprecipitation with the MAb 24BB7 indicate that the protein was not covalently acylated. Since a sequence that may function as a substrate for protein kinase C as well as a potential acceptor site for asparagine-linked carbohydrate are present in the predicted amino acid sequence, metabolic labeling of protein p12 in infected cells with [^{32}P]phosphate, or [^{14}C]glucosamine was carried out, followed by immunoprecipitation with the MAb 24BB7. However, none of these modifications were found in the protein. In order to ensure that protein p12, in contrast to most of the viral attach-ment proteins, was not glycosylated in the viral particle, protein p12 immunoprecipitated with the MAb 24BB7 from purified labeled virus was subjected to treatment with O and N-glycosidases. Additionally, a mixture of proteins released from the purified labeled virus particles with OG, including protein p12, was applied on a lectin affinity col-umn. No glycosylated component could be detected in the protein by either approach.

Comparison of the protein from different virus isolates

To determine if antigenic variation of the attachment protein could be a mechanism by which ASFV evades the host immune response, we

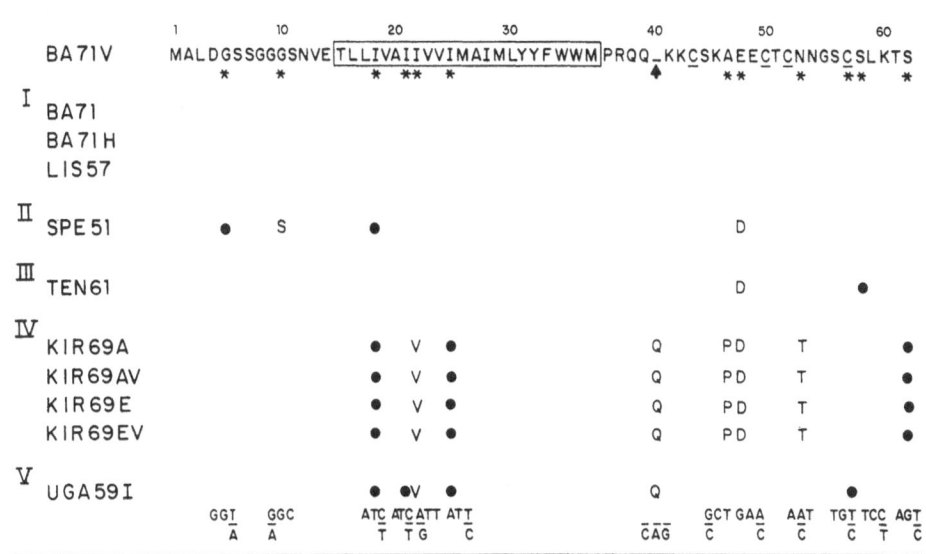

Fig. 4. Sequence of protein p12 deduced from the nucleotide sequence of the gene in different ASFV isolates. The complete amino acid sequence of the protein in the BA71V strain, with the hydrophobic segment boxed and the cysteine residues underlined, is shown for comparison. The asterisks indicate the positions of the polypeptide chain that undergo mutations in other virus strains and the arrow shows the site of addition of a glutamine residue. Silent point mutations (•) and the amino acids that change with respect to the sequence of BA71V are indicated. The triplets corresponding to these positions are shown in the lower part of the figure, where the nucleotide present in the BA71V strain (underlined) and the mutation (below) are indicated. The viruses are distributed in five groups according to the positions of the *Sal*I sites in the central region of the viral genome [10]

analyzed the variability of the gene encoding this protein in 11 different virus strains corresponding to 6 natural virus isolates, a virus passaged 100 times in porcine macrophages and 4 viruses adapted to grow in tissue culture [9]. The isolates selected for this study were geographically and genetically diverse and showed different degrees of pathogenicity in animals. The comparison of the predicted amino acid sequences revealed a high degree of conservation among different ASFV strains, since mainly conservative or silent point mutations were found in the different isolates (Fig. 4). When compared with the BA71V sequence previously established, there was an addition of a glutamine residue in some of the isolates, giving rise to a polypeptide of 62 amino acids; however, this insertion is located in a group of residues (QQKK) that are predicted to constitute the highest hydrophilic segment of the polypeptide, and therefore it would not greatly alter the properties of this region. The cysteine motif of the C-terminal region was conserved in all the viruses and the only variation found in the long stretch of hydrophobic residues was a conservative change of an isoleucine to a valine.

The sequence of the 51 nucleotides upstream of the initiator codon of the ORF O61R, where the 5′ end of the p12 specific mRNA has been mapped, was identical in all the viruses. The conservation of this sequence is in accordance with the existence of transcription signals required for the efficient transcription of this gene during infection. In contrast, the sequence of the 48 nucleotides downstream of the ORF O61R, which corresponds to an intergenic region, contained direct repeats in tandem that at the same time constituted inverted repeats and presented a high variation in length among the different virus isolates. However, the biological significance of these differences at the present time is unknown.

Labeling experiments with [^{35}S]methionine and immunoprecipitation with the MAb 24BB7 showed that the protein is translated during the infection of swine macrophages with all the viruses tested. This fact, together with the conservation of the polypeptide sequence, suggests an essential role of the protein p12 in the virus replication cycle and is in agreement with the attachment function that has been attributed to the polypeptide.

Concluding remarks

The ASFV protein involved in the virus attachment to the specific cellular receptor has been shown to be protein p12 and its sequence and main characteristics have been determined. The following steps will be the overexpression and purification of proteins modified by site-directed mutagenesis and their assay in binding experiments, in order to precisely identify the p12 determinants and residues responsible for the virus-cell recognition. Hopefully, these studies will ultimately lead in the future to the development of agents that could block the infection at this very early stage and contribute to the eradication of African swine fever.

Acknowledgements

This work was supported by grants from Comisión Interministerial de Ciencia y Tecnología, Junta de Extremadura and by an institutional grant of Fundación Ramón Areces.

References

1. Alcamí A, Angulo A, López-Otín C, Muñoz M, Freije JMP, Carrascosa AL, Viñuela E (1992) Amino acid sequence and structural properties of protein p12, an African swine fever virus attachment protein. J Virol 66: 3860–3868

2. Alcamí A, Carrascosa AL, Viñuela E (1989) Saturable binding sites mediate the entry of African swine fever virus into Vero cells. Virology 168: 393–398

3. Alcamí A, Carrascosa AL, Viñuela E (1989) The entry of African swine fever virus into Vero cells. Virology 171: 68–75

4. Alcamí A, Carrascosa AL, Viñuela E (1990) Interaction of African swine fever virus with macrophages. Virus Res 17: 93–104

5. Alcamí A, Viñuela E (1991) Fc receptors do not mediate African swine fever virus replication in macrophages. Virology 181: 756–759

6. Almazán F, Rodríguez JM, Angulo A, Viñuela E, Rodriguez JF (1993) Transcriptional mapping of a late gene coding for the p12 attachment protein of African swine fever virus. J Virol 67: 553–556

7. Almazán F, Rodríguez JM, Andrés G, Pérez R, Viñuela E, Rodriguez JF (1992) Transcriptional analysis of the multigene family 110 of African swine fever virus. J Virol 66: 6655–6667

8. Almendral JM, Blasco R, Ley V, Beloso A, Talavera A, Viñuela E (1984) Restriction site map of African swine fever virus DNA. Virology 133: 358–270

9. Angulo A, Viñuela E, Alcamí A (1992) Comparison of the sequence of the gene encoding African swine fever virus attachment protein p12 from field virus isolates and viruses passaged in tissue culture. J Virol 66: 3869–3872

10. Blasco R, Agüero M, Almendral JM, Viñuela E (1989) Variable and constant regions in African swine fever virus DNA. Virology 168: 330–338

11. Brown M, Faulkner P (1977) A plaque assay for nuclear polyhedrosis using a solid overlay. J Gen Virol 36: 361–364

12. Carrascosa AL, del Val M, Santarén JF, Viñuela E (1985) Purification and properties of African swine fever virus. J Virol 54: 337–344

13. Carrascosa AL, Sastre I, González P, Viñuela E (1993) Localization of the African swine fever virus attachment protein p12 in the virus particle by immunoelectron microscopy. Virology 193: 460–465

14. Carrascosa AL, Sastre I, Viñuela E (1991) African swine fever virus attachment protein. J Virol 65: 2283–2289

15. Carrascosa JL, Carazo JM, Carrascosa AL, García N, Santisteban A, Viñuela, E (1984) General morphology and capsid fine structure of African swine fever virus particles. Virology 132: 160–172

16. Costa J (1990) African swine fever virus. In: Darai G (ed) Molecular biology of iridoviruses. Kluwer Academic Publishers, Boston, pp 247–270

17. De Boer CJ (1967) Studies to determine neutralizing antibody in sera from animals recovered from African swine fever and laboratory animals inoculated with African virus with adjuvants. Arch Ges Virusforsch 20: 164–179

18. Del Val M, Carrascosa JL, Viñuela E (1986) Glycosylated components of African swine fever virus. Virology 152: 39–49

19. Dixon LK, Wilkinson PJ, Sumption KJ, Ekue F (1990) Diversity of the African swine fever virus genome. In: Darai G (ed) Molecular biology of iridoviruses. Kluwer Academic Publishers, Boston, pp 271–295

20. Esteves A, Marques MI, Costa JV (1986) Two-dimensional analysis of African swine fever proteins and proteins induced in infected cells. Virology 152: 181–191

21. García-Beato R, Salas ML, Viñuela E, Salas J (1992) Role of the host cell nucleus in the replication of African swine fever virus DNA. Virology 188: 637–649

22. Geraldes A, Valdeira ML (1985) Effect of chloroquine on African swine fever virus infection. J Gen Virol 66: 1145–1148

23. Gierasch LM (1989) Signal sequences. Biochemistry 28: 923–930

24. González A, Talavera A, Almendral JM, Viñuela E (1986) Hairpin loop structure of African swine fever virus DNA. Nucleic Acids Res 14: 6835–6844

25. Hanggy M, Bannwarth W, Stunnenberg HG (1986) Conserved TAAAT motif in vaccinia virus late promoters: overlapping TATA box and site of transcription initiation. EMBO J 5: 1071–1076

26. Kishimoto TK, O'Connor K, Lee A, Roberts TM, Springer TA (1987) Cloning of the β subunit of the leukocyte adhesion proteins: Homology to an extracellular matrix receptor defines a novel supergene family. Cell 48: 681–690

27. Kyte J, Doolittle RF (1982) A simple method for displaying the hydropathic character of a protein. J Mol Biol 157: 105–132

28. Leader DP, Katan M (1988) Viral aspects of protein phosphorylation. J Gen Virol 69: 1441–1464

29. Lonberg-Holm K (1981) Attachment of animal viruses to cells: an introduction. In: Lonberg-Holm K, Philipson L (eds) Virus receptors: Part 2, animal viruses. Chapman and Hall, London, pp 1–20

30. Marsh M, Helenius A (1989) Virus entry into animal cells. Adv Virus Res 36: 107–151

31. Matthews REF (1982) Classification and nomenclature of viruses. Karger, Basel

32. Maurer FD, Griesemer RA, Jones TC (1958) The pathology of African swine fever virus, a comparison with hog cholera. Am J Vet Res 19: 517–539

33. Moss B (1990) Regulation of vaccinia virus transcription. Annu Rev Biochem 59: 661–668

34. Moura Nunes JF, Vigario JD, Terrinha AM (1975) Ultrastructural study of African swine fever virus. Arch Virol 49: 59–66

35. Ortín J, Enjuanes L, Viñuela E (1979) Cross-links in African swine fever virus DNA. J Virol 31: 579–583

36. Pearse BMF (1987) Clathrin and coated vesicles. EMBO J 6: 2507–2512

37. Porterfield JS (1986) Antibody-dependent enhancement of viral infectivity. Adv Virus Res 31: 335–355

38. Rosel JL, Earl PL, Weir JP, Moss B (1986) Conserved TAAATG sequence at the transcriptional and translational initiation sites of vaccinia virus late genes deduced by structural and functional analysis of the HindIII H genome fragment. J Virol 60: 436–449

39. Salas ML, Kuznar J, Viñuela E (1981) Polyadenylation, methylation and capping of the RNA synthesized in vitro by African swine fever virus. Virology 113: 484–491

40. Salas ML, Rey-Campos J, Almendral JM, Talavera A, Viñuela E (1986) Transcription and translation maps of African swine fever virus. Virology 152: 228–240

41. Santarén JF, Viñuela E (1986) African swine fever virus-induced polypeptides in Vero cells. Virus Res 5: 391–405

42. Sogo JM, Almendral JM, Talavera A, Viñuela E (1984) Terminal and internal inverted repetitions in African swine fever virus DNA. Virology 133: 271–275

43. Valdeira ML, Geraldes A (1985) Morphological study on the entry of African swine fever virus into cells. Biol Cell 55: 35–40

44. Viñuela E (1985) African swine fever virus. Curr Top Microbiol Immunol 116: 151–170

45. Viñuela E (1987) Molecular biology of African swine fever virus. In: Becker Y (ed) African swine fever. Martinus Nijhoff Publishing, Boston, pp 31–49

46. von Heijne G (1988) Transcending the impenetrable: how proteins come to terms with membranes. Biochim Biophys Acta 947: 307–333

47. Wilkinson PJ (1989) African swine fever virus. In: Pensaert MB (ed) Virus infections of porcines. Elsevier Science Publishers BV, Amsterdam, pp 17–37
48. Zuker M, Stiegler P (1981) Optimal computer folding of large DNA sequences using thermodynamics and auxiliary information. Nucleic Acids Res 9: 133–144

Authors' address: Dr. E. Viñuela, Centro de Biología Molecular, Facultad de Ciencias, Universidad Autónoma, Cantoblanco, E-28049 Madrid, Spain.

Arch Virol (1993) [Suppl] 7: 185–199

African swine fever virus genome content and variability

L.K. Dixon[1], **S.A. Baylis**[2], **S. Vydelingum**[1], **S.R.F. Twigg**[2],
J.M. Hammond[1], **P.M. Hingamp**[1], **C. Bristow**[1], **P.J. Wilkinson**[1],
and **G.L. Smith**[2]

[1] AFRC Institute for Animal Health, Pirbright, Woking
[2] Sir William Dunn School of Pathology, University of Oxford, Oxford, U.K.

Summary. A 55 kilobase pair (kb) region from the right end of the virulent African swine fever virus isolate, Malawi LIL20/1, has been sequenced. The 68 major open reading frames (ORFs) encoded are generally closely spaced and read from both DNA strands across the complete sequence. Comparison of the amino acid sequences of predicted ORFs with sequence databases identified 15 ORFs which encode proteins that are similar to proteins of known function. Two ORFs are homologous to copies of multigene family 360 (MGF360) and one ORF is homologous to copies of multigene family 110 (MGF110). Both of these multigene families have been described previously [6, 31].

Introduction

African swine fever (ASF) was first reported in 1921 as a devastating disease of domestic pigs which caused mortality approaching 100% [48]. Since then the disease has been reported from most African countries south of the Sahara and some other African countries. In 1957 ASF spread to Portugal and although that outbreak was thought to be eradicated [42] the disease reappeared in 1960. ASF has been endemic in the Iberian peninsula since that time, but is currently confined to Portugal and south west Spain. Sporadic outbreaks of ASF which occurred elsewhere in Europe, the Caribbean and Brazil, were successfully eradicated, but ASF remained endemic in Sardinia after its introduction in 1978 [69, 71]. Some more recent isolates of ASFV from regions where disease is endemic have reduced virulence for domestic pigs [46, 47, 69].

ASF is caused by a large cytoplasmically replicating icosahedral virus (ASFV), which contains a long linear double stranded DNA genome that varies in length between 170 and 190 kilobase pairs (kb) depending on the virus isolate [19, 24, 66, 67]. In addition to domestic and wild swine (*Sus scrofa*), ASFV infects warthogs (*Phacochoerus aethiopicus*) and bush pigs (*Potamochoerus porcus*), although there are no apparent signs of disease in the latter two species. ASFV also infects soft ticks of the genus *Ornithodoros*. *O. moubata* acts as a virus vector in Southern Africa [11, 51].

ASFV is structurally similar to the *Iridoviridae* and this, combined with its cytoplasmic location and large DNA genome, led to its original classification with the *Iridoviridae* [33, 43]. Subsequent analysis has shown that the genome structure and replication strategy of ASFV exhibits similarities to the *Poxviridae* [19, 65, 67]. ASFV has therefore been removed from the *Iridoviridae* but still remains unclassified [13].

The similarities between ASFV and Poxviridae include genomic covalently closed terminal crosslinks and terminal inverted repeats, which consist of tandem repeat arrays [7, 29, 32, 67]. ASFV transcription, also in common with Poxviruses, is not dependent on host cell RNA polymerase II and virions contain many enzymes, including those required for early mRNA transcription and processing [19, 49]. The pattern of ASFV gene expression is temporally regulated in a manner similar to that of Poxviruses and late gene expression is dependent on virus DNA replication. [25, 55, 64]. There are, however, differences in replication strategy between ASFV and Poxviruses. One such difference may be that an early stage of ASFV DNA replication takes place in the nucleus [27].

A number of different ASFV isolates have been compared by restriction enzyme site mapping of virus genomes which has enabled isolates from different geographical locations to be distinguished by gain or loss of restriction enzyme sites from the virus genome [9, 23, 62, 63, 68].

Genomes of different isolates could also be distinguished by length variations which occurred within the left terminal 48 kb, the right terminal 22 kb and at one position in the centre of the genome [10, 62]. Large length variations that occur between 8 and 20 kb from the left DNA terminus resulted from gain or loss of copies of a multigene family which encodes proteins of about 110 amino acids [1, 32]. Small length variations that occurred at one position about 22 kb from the right DNA terminus resulted from variation in the number and sequence of repeat units within an intergenic tandem repeat array [24].

To date limited sequence analysis has been carried out on the ASFV genome. We have, however, recently determined the complete nucleotide sequence of a 55 kb region from the right end of the genome of a virulent ASFV isolate (Malawi LIL20/1) [8]. The results provide comparative data on the organisation and content of the ASFV genome and provide the basic information with which to begin functional analysis of ASFV encoded proteins.

Materials and methods

Sequence analysis

The complete genome of the ASFV Malawi LIL20/1 isolate has previously been cloned in bacteriophage lambda and plasmid vectors, apart from short sequences that included the genome terminal crosslinks and inverted terminal repeats [21]. ASFV DNA inserts from selected clones were isolated, randomly sheared by sonication and cloned into M13 vectors. Single-stranded DNA was sequenced using dideoxynucleoside triphosphates, [^{35}S] dATP and Klenow or T7 DNA polymerase. To join the sequence into one contiguous fragment, sequences between the inserts from individual clones were amplified by PCR from alternative clones containing overlapping ASFV DNA and the PCR products were then cloned into M13 and sequenced. In some cases sequence was obtained directly from bacteriophage lambda clones by cycle sequencing. The sequence was determined of 10,565 bp of *Sal*I fragment g, the entire *Sal*I fragments h, i, j, k, and 11,041 bp of *Sal*I fragment l (Fig. 1).

Computer analysis

Random DNA sequences were read using a sonic digitiser and assembled into a contiguous sequence using the computer programmes DBUTIL and DBAUTO or SAP [60, 61]. Open reading frames (ORFs) were identified using the programme ORFFILE and files for individual protein sequences were created using DELIB (both kindly provided by Mike Boursnll, Cantab Pharmaceuticals, Cambridge, U.K.). Protein sequences were compared against the GENEMBL database using the programme TFASTA [40] and against the SWISSPROT (version 21) protein sequence database using the programme PROSRCH [18]. Protein sequences were aligned and secondary structures [38] and potential signal peptides [65] were predicted using programmes of the University of Wisconson Genetics Computer Group [20].

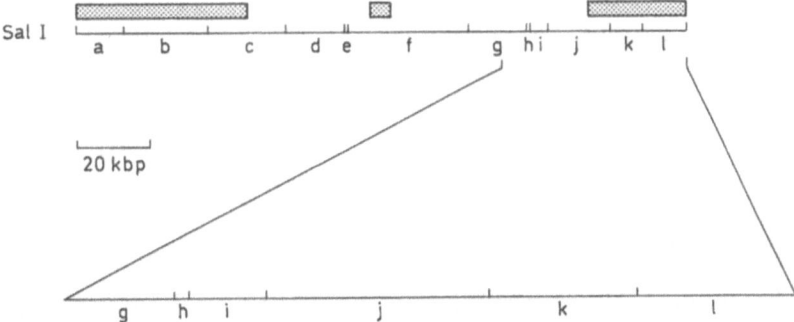

Fig. 1. *Sal*I Map of the ASFV Malawi LIL20/1 genome showing the 55 kb region sequenced. The *Sal*I map of the ASFV Malawi LIL20/1 genome is shown (Dixon 1988) at the top of the figure. The enlarged area underneath this map shows the 55 kb region sequenced in this study. Hatched boxes show regions of the virus genome which vary in length when different isolates are compared

Results and discussion

The region of the ASFV (Malawi LIL20/1 isolate) genome sequenced is shown in Fig. 1. Also marked on Fig. 1 are regions of the genome where length variations have been shown to occur when genomes of different virus isolates are compared. There is no evidence for the presence of introns within ASFV genes and major ORFs have been defined as those which start with methionine, are at least 60 amino acids long and do not substantially overlap other larger ORFs. ORFs are named according to the SalI fragment within which they are located and are numbered from left to right within the fragment. Those read towards the right of the genome are designated R and to the left L.

Transcriptional arrangement

The arrangement of ORFs is indicated on Fig. 2. There is no strong preference in coding strand and the 31 ORFs transcribed towards the right DNA terminus are interspersed with the 37 ORFs transcribed towards the left DNA terminus. This contrasts with the VV genome organisation, since most VV ORFs in the terminal regions of the genome are read towards the genome termini [30, 57]. ORFs are generally closely spaced and in some cases are overlapping. However, since transcription start sites have not been mapped, it is possible that some ORFs may begin at alternative downstream methionine codons to those predicted by computer analysis. A signal for termination of both early and late ASFV gene transcription consists of at least 7Ts [5] and this sequence occurs downstream of 19 ORFs. It remains to be determined whether transcription of the other ORFs terminates at specific sequences and if so what these are. Transcription regulatory sequences have not otherwise been defined and no consensus sequences are apparent from inspection of the sequence. At three positions longer sequences (507, 454 and 464 bp) are found between ORFs and within these regions and at two other shorter intergenic regions (164 and 167 bp) arrays of tandem repeats are found. These tandem repeat arrays are numbered I to V and are indicated on Fig. 2. The right end of tandem repeat array I overlaps with ORF i3L.

The sequence of repeat units within each of these arrays is different from those within any other array. This argues against a function for the sequences themselves as, for example, binding sites for DNA binding proteins. Possibly these tandem repeat arrays act as recombination "hot spots" or transcription regulatory signals. Frequent length variations, due to gain or loss of repeat units within arrays, occur within these loci

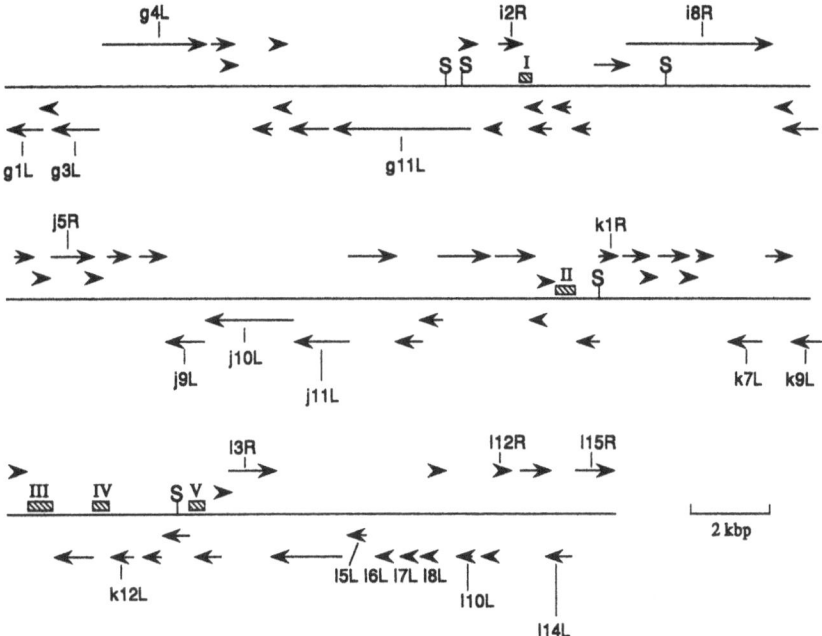

Fig. 2. The position of major open reading frames (ORFs) and principle homologies identified within the right terminal 55 kb of the ASFV Malawi LIL20/1 genome. The 55 kb region sequenced is indicated starting from the left at the top of the figure and finishing at the right at the bottom. The positions of major ORFs are indicated by arrows. The arrows point in the direction that ORFs are read. Homologies identified by comparison of amino acid sequences of predicted ORFs with sequence databases are indicated. ORFs described in Tables 1 to 3 are numbered. Hatched boxes number *I* to *V* show the position of inter ORF tandem repeat arrays described in the text. At the bottom right the scale is indicated. *Sal*I sites shown in Fig. 1 are marked with an *S*

and these length variations can be useful for distinguishing between virus isolates [22].

Amino acid homologies

The amino acid sequences of predicted ORFs were screened against protein sequence databases. Those ORFs encoding products with which homologies were found are shown in Table 1 as are ORFs containing protein motifs that are present in the PROSITE database. Some of these homologies are discussed below. As predicted from the cytoplasmic site of ASFV replication, a number of ORFs encode proteins which have homology with enzymes involved in DNA replication, repair or recombination and in mRNA transcription or processing. Of these a DNA topoisomerase type II (ORF i8R) [8, 26] and a DNA ligase (ORF g3L) [34] have been described. Three ORFs encode proteins with

Table 1. Open reading frames which encode proteins that are similar to sequences in the database or contain known protein motifs

ORF	From	To	Size (KDa)	Homology/Protein Motifs
g1L	4	864	32.4	RNA polymerase subunit
g3L	1,105	2,361	48.0	DNA ligase
g4R	2,427	5,030	99.8	mRNA capping enzyme large subunit
g7L	6,585	6,821	9.2	Calcium binding protein
g10L	8,160	11,453	125.0	Helicase motif
i2R	12,202	12,816	23.6	RNA polymerase subunit
i6R	14,616	15,434	31.4	RGD cell attachment site
i7R	15,472	19,044	134.9	DNA topoisomerase II
j3L	20,195	20,719	20.4	ATP binding site
j5R	21,151	22,167	39.8	Serine proteinase inhibitor
j9L	24,053	24,949	35.1	Protein kinase
j10L	24,927	27,044	80.4	Helicase motif
j11L	27,005	28,531	58.0	Helicase motif
k1R	34,711	35,196	17.8	dUTPase
k7L	37,955	38,755	30.9	mRNA capping enzyme small subunit
k9L	39,587	40,315	28.6	Transcription factor SII
k12L	42,517	43,155	24.4	Ubiquitin conjugating enzyme
l14L	53,370	53,921	21.3	Myeloid differentiation antigen

The designated name of ORFs is shown on the left. The nucleotide positions of the start and stop codons or ORFs is indicated and the molecular weight of the encoded protein. In the right column homologies with known sequences, or motifs in proteins are indicated

homology to subunits of RNA polymerase. One of these (g1L) is incomplete but encodes a predicted protein which is similar to the 147 kDa subunit of vaccinia virus (VV) RNA polymerase and to the corresponding subunits of both eukaryotic and prokaryotic RNA polymerases [15]. A second ORF (i2R) encodes a 23.6 kDa predicted protein which has 21% identical amino acids compared to the yeast 27 kDa RNA polymerase subunit and 27% identical amino acids compared to the human 23 kDa RNA polymerase subunit. A third ORF (k9L) encodes a predicted protein which has homology with the eukaryotic transcription elongation factor TFS II and with the *rpo* 30 subunit of VV RNA polymerase [2]. Two ORFs (g4R, k7L) encode proteins which have 21% and 23% identical amino acids respectively compared to the large and small subunits of VV mRNA capping enzyme.

Three ORFs (g10L, j10L, j11L) encode proteins which contain motifs characteristic of the helicase superfamily II [8]. One of these (j11L) is most similar to the VV A18R product, which is essential for VV replication in tissue culture. Temperature sensitive mutants in A18R have an abortive late phenotype which is characterized at the non-permissive temperature by abortion of late virus protein synthesis and degradation of ribosomal RNA [50]. The other two ORFs (g10L, j10L) which encode proteins containing helicase motifs have similar percentages of identical amino acids when compared both to each other and to the VV D6R and VV D11L products. The ASFV g10L predicted product has, however, a carboxy terminal extension not found in the other ORF products which contain helicase motifs. The VV D6R product is a component of the VV early transcription factor [14, 28] and VV D11L encodes an NTPase I that is an essential component of VV virions [15, 37, 54]. It remains to be determined whether the ASFV encoded putative helicases have similar functions to those encoded by VV.

One ORF (k1R) encodes a 17.8 kDa predicted protein which has between 20% and 23% identical amino acids compared to the *E. coli* and VV encoded deoxyuridine triphosphatases (dUTPase). The k1R product contains the five conserved motifs found in other dUTPases [44] and is similar to those dUTPases encoded by VV, some retroviruses and *E. coli* but different from those encoded by herpesviruses [44]. This dUTPase might function to prevent misincorporation of dUTP into DNA or in the production of thymidylate [44].

Two ORFs (j9L, k12L) encode predicted proteins which are homologous to enzymes (protein kinase, ubiquitin conjugating enzyme) involved in post-translational modification of proteins. The products of these ORFs have been expressed in *E. coli* and shown to have the predicted enzyme activity [35]. These enzymes might either be involved in regulating the virus replication cycle or in modulating host cell function.

Two ORFs (j5R, 114L) encode predicted proteins which have homology with known virulence factors encoded by other viruses. One of these (j5R) is similar to a family of serine proteinase inhibitors (serpins) [59, 72]. A cowpox virus encoded proteinase inhibitor, which shows homology to serpins was subsequently in vitro shown to inhibit the interleukin-1β converting enzyme, a cysteine proteinase [52]. Deletion of VV ORF K2L (another Poxvirus-encoded serpin) promotes virus induced cell fusion [39]. The putative ASFV j5R encoded serpin has between 16% and 22% identical amino acids compared with serpins encoded by VV B22R, VV K2L, *Manduca sexta* and *Xenopus laevis*. However, not all amino acids conserved in the other serpins are present in the j5R product and confirmation that j5R encodes a serpin awaits functional assay of the expressed product.

The 114L ORF encodes a 21.3 kDa predicted protein which has homology over a 56 amino acid domain at the carboxy terminus of the protein with both a myeloid differentiation primary response antigen [41] and the neurovirulence associated protein (ICP 34.5) of herpes simplex virus. (44% and 30% identical amino acids respectively) [17, 45].

The ASFV encoded proteins which have homology with known proteins include two enzymes, a ubiquitin conjugating enzyme and DNA topoisomerase type II, that have not been identified on other virus genomes. In general, the homology between ASFV proteins and other virus and cellular homologues is low and ranges between 20% and 25% identical amino acids. One exception is the ubiquitin conjugating (UBC) enzyme which belongs to a family of enzymes that is well conserved through evolution. The ASFV UBC enzyme has between 30% and 45% identical amino acids compared to other UBC enzymes [35, 54]. Other exceptions are the large and small subunits of ribonucleotide reductase which have up to 41% identical amino acids compared to other ribonucleotide reductase sequences [12].

The genome organization of these ASFV encoded enzymes is not colinear with that of the orthopoxviruses. This, combined with the sequence divergence observed between ASFV encoded proteins and other homologues, emphasises the distant relationship between ASFV and the Poxviruses which have been characterised so far.

Duplicated genes and genes containing repeated sequences

Six copies of a multigene family (MGF) encoding proteins of approximately 360 amino acids have been characterised on the ASFV Ba71V isolate genome [31]. Four of these are located at the left end of the genome and two at the right end. It was proposed that this gene family evolved by a process of gene duplication followed by sequence divergence of duplicated genes. These duplicated genes were then transposed from one end of the genome to the other and certain gene copies were deleted [31]. Similar transpositions of sequences from one end of the genome to the other occur on Poxvirus genomes. Two copies of MGF 360 are located near the right end of the Malawi LIL20/1 genome (Table 2, Fig. 2). The product of one of these ORFs (13R) has between 33% and 89% identical amino acids compared to proteins encoded by different copies of MGF 360 present on the Ba71V genome. The highest

Table 2. Open reading frames which encode duplicated genes

ORF	From	To	Size (KDa)	Duplicated ORFs and ORFs containing repeats
g8R	6,859	7,053	7.6	Internal repeats
i3L	12,975	13,681	7.4	Internal repeats
k14L	43,791	44,453	24.7	Internal repeats
13R	45,537	46,592	40.6	Multigene family 360
15L	48,430	48,852	16.8	Multigene family 100
16L	49,094	49,324	9.1	Multigene family 100
17L	49,647	49,952	12.3	Multigene family 100
18L	50,172	50,480	12.7	Multigene family 100
l10L	51,025	51,522	18.7	P22
l12R	52,196	52,507	12.4	Multigene family 110
l15R	54,168	55,094	36.2	Multigene family 360

The designated name of ORFs is shown on the left. The nucleotide positions of the start and stop codons is indicated and the molecular weight of encoded proteins. In the right column is indicated which duplicated genes the ORF is similar to. MGF360 is multigene family 360, MGF110 is multigene family 110, VP22 is virus protein P22

homology (89%) is observed between the 13R product and the Ba71V D'311 product and these two ORFs are located at similar positions on their respective genomes. In contrast, the 115R ORF, which is the final ORF sequenced at the right end of the LIL20/1 genome, encodes a MGF 360 product that is more similar to that of the MGF 360 copy furthest to the left of the Ba71V genome (74% identity with BA71V K'360 product), than to products of MGF 360 copies at the right end of the Ba71V genome. This analysis supports the model for evolution of MGF 360 by transposition of sequences from one end of the genome to the other but indicates that different copies of MGF 360 may have been deleted from the right end of the LIL20/1 genome compared to the Ba71V genome.

The LIL20/1 genome also encodes a copy of another multigene family, MGF 110, near the right end of the genome (ORF 112R) (Table 2, Fig. 2). Multiple copies of MGF 110 have been identified near the left end of the Ba71V and other ASFV isolate genomes [1, 6], but none have so far been identified at the right end of the genome. As discussed above, it is probable that the right ends of the Ba71V and LIL20/1 genomes have undergone separate DNA rearrangements that have resulted in deletion of different ORFs. These may include the MGF 110 copy from the right end of the Ba71V genome. Further evidence for sequence transpositions from one end of the genome to the other is provided by the presence at the right end of the LIL20/1 genome of an ORF (110L) which encodes a predicted protein with 41% identical amino acids compared to a protein, P22, that is encoded at the left end of the Ba71V genome. P22 is expressed on the membrane of infected cells and is thought to be a virus structural protein [16]. Four ORFs (15L, 167, 17L, 18L) (Table 2, Fig. 2) are similar to each other and have probably also evolved by a process of gene duplication followed by sequence divergence of duplicated copies. The products of these ORFs, which have been named multigene family 100, have between 20% and 53% identical amino acids compared to each other.

The functions of the duplicated genes are unknown, although virus clones which have lost all copies of MGF 110 from the left end of the genome are equally virulent in pigs and replicate in macrophages as well as virus clones which contain multiple copies of MGF 110 [1]. The sequence divergence between multiple gene copies is sufficiently great to permit variation in antigenicity and/or function of the encoded proteins. Transcriptional analysis has shown that different copies of MGF 110 are transcribed from the Ba71V genome during ASFV infection of Vero cells [5, 6]. This argues against a mechanism of switching transcriptionally active gene copies to provide an escape mechanism from immune surveillance.

In addition to the duplicated gene copies which are described above, three ORFs (g8R, i3L, k14L) encode predicted proteins which contain repeated blocks of amino acids. The g8R product contains three copies of a five amino acid repeat, the i3L product contains four copies of a different five amino acid repeat and the k14L product contains three copies of a twenty-one amino acid repeat.

Proteins containing putative transmembrane domains and signal peptides

The amino acid sequences of encoded ORFs were screened to predict putative transmembrane domains and signal peptides. Putative transmembrane domains were considered to consist of at least 16 uncharged hydrophobic amino acids. Signal peptides were predicted [65] using the programme SIGCLEAVE.

Three ORFs (j7R, j14L, k11L) encode proteins that contain predicted signal peptides (Table 3). Two of these (j7R, K11L) contain an additional putative transmembrane domain and may therefore be integral

Table 3. Open reading frames which encode putative transmembrane domains and signal peptides

ORF	From	To	Size (KDa)	Hydrophobic regions and signal peptides
i1L	11,795	12,145	13.1	central hydrophobic region
j6R	22,182	22,514	12.9	signal peptide
j7R	22,477	23,175	25.7	signal peptide and hydrophobic C terminal region
j14L	30,200	30,727	19.0	signal peptide
j19L	34,099	34,683	21.7	hydrophobic C terminal region
k11L	41,069	42,055	38.5	signal peptide and central hydrophobic region
k13L	43,313	43,795	18.4	signal peptide
l10L	51,025	51,522	18.7	signal peptide
l11L	51,617	51,850	9.1	signal peptide

The designated name of ORFs is shown on the left. The nucleotide positions of the start and stop codons is indicated and the molecular weight of encoded proteins. The presence of putative signal peptides and transmembrane domains is indicated in the right column

membrane proteins whereas the j14L product may be secreted from infected cells. Several Poxvirus encoded virulence factors are secreted from infected cells and have homology with the extracellular ligand binding domains of cytokine receptors [3, 36, 56, 58]. No homology was, however, detected between the j14L product and cytokine receptors, but this might reflect the generally low homology between ASFV encoded proteins and other virus and cellular homologues. Six additional ORFs (i1L, j6R, j19L, k13L, l10L, l11L) encode proteins that contain putative transmembrane domains. All of these contain a single putative transmembrane domain. The i1L, j6R, j19L, l10L, l11L products contain potential glycosylation sites. These putative membrane proteins may be expressed on the surface of infected cells and might also be incorporated into the virus structure, either in the external or internal lipid membrane. Six virus induced proteins have been detected on the surface of infected cells [4] and these may include some of the putative membrane proteins that we have identified.

A function has not been assigned to the proteins encoded by the majority of ORFs which we have sequenced. It has been suggested that ASFV was originally a virus infecting only ticks. Possibly some ASFV genes may function specifically during virus replication in the tick vector.

Acknowledgements

We would like to thank Mrs. Chris Chisholm and Mrs. Julia Duncan for typing the manuscript and AFRC and MAFF for support.

References

1. Aguero M, Blasco R, Wilkinson PJ, Vinuela E (1990) Analysis of naturally occurring deletion variants of African Swine fever virus: Multigene family 110 is not essential for infectivity or virulence in pigs. Virology 176: 195–204
2. Ahn BY, Jones EV, Moss B (1990) Identification of rpo 30, a vaccinia virus RNA polymerase gene with structural similarity to a eucaryotic transcription elongation factor. Mol Cell Biol 10: 5433–5441
3. Alcami A, Smith GL (1992) A soluble receptor for interleukin-1β encoded by vaccinia virus: A novel mechanism of virus modulation of the host response to infection. Cell 71: 153–167
4. Alcaraz C, Alvarez A, Escribano JM (1992) Flow cytometric analysis of African swine fever virus induced plasma membrane proteins and their humoral immune response in infected pigs. Virology 189: 266–273
5. Almazán F, Rodriguez JM, Andrés G, Pérez R, Vinuela E, Rodriguez JF (1992) Transcriptional analysis of multigene family 110 of African swine fever virus. J Virol 66: 6655–6667
6. Almendral JM, Blasco R, Vinuela E (1990) Multigene families in African swine fever virus DNA. Family 110. J Virol 64: 2064–2072
7. Baroudy BM, Venkatesan S, Moss B (1982) Incompletely base-paired flip-flop terminal loops link the two DNA strands of the vaccinia virus genome into one uninterrupted polynucleotide chain. Cell 28: 315–324
8. Baylis SA, Dixon LK, Vydelingum S, Smith GL (1992) African swine fever encodes a gene with extensive homology to type II topoisomerases. J Mol Biol 228: 1003–1010
9. Blasco R, Aguero M, Almendral JM, Vinuela E (1989a) Variable and constant regions in African swine fever virus DNA. Virology 168: 330–338
10. Blasco R, De La Vega I, Almazan F, Aguero M, Vinuela E (1989b) Genetic variation of African swine fever; variable regions near the ends of the viral DNA. Virology 173: 251–257
11. Botija CS (1963) Reservours of ASFV: a study of ASFV in arthropods by means of the haemadsorbtion test. Bull Off Int Epiz 60: 895–899
12. Boursnell M, Shaw K, Yanez RJ, Vinuela E, Dixon L (1991) The sequences of the ribonucleotide reductase genes from African swine fever virus show considerable homology with those of the orthopoxvirus, vaccinia virus. Virology 184: 411–416
13. Brown F (1986) The classification and nonenclature of viruses; Summary of results of meetings of the International Committee on taxonomy of viruses, Sendai, September 1984. Intervirology 25: 141–143
14. Broyles SS, Fesler BS (1990) Vaccinia virus gene encoding a component of the viral early transcription factor. J Virol 64: 1523–1529
15. Broyles SS, Moss B (1987) Identification of the vaccinia virus gene encoding nucleoside triphosphate phosphohydrolase I, a DNA-dependent ATPase. J Virol 61: 1738–1742

16. Camacho A, Vinuela E (1991) Protein P22 of African swine fever virus, an early structural protein that is incorporated into the membrane of infected cells. Virology 181: 251–257

17. Chou J, Roizman B (1990) The herpes simplex virus I gene for ICP 34.5, which maps in inverted repeats, is conserved in several limited-passage isolates but not in strain 17 syn +. J Virol 64: 1014–1020

18. Collins JF, Coulson AFW (1987) Molecular sequence comparison and alignment. In: Bishop M, Rawlings C (eds) Nucleic acid and protein sequence analysis: a practical approach. I.R.L. Press, Oxford, pp 323–358

19. Costa JV (1990) African swine fever virus. In: Darai G (ed) Molecular biology of iridoviruses. Kluwer Academic Publishers, Boston, pp 267–270

20. Devereux J, Haeberli P, Smithies (1984) A comprehensive set of sequence analysis programmes for the Vax. Anal Biochem 129: 216–223

21. Dixon LK (1988) Molecular cloning and restriction enzyme mapping of an African swine fever virus isolate from Malawi. J Gen Virol 69: 1683–1694

22. Dixon LK, Bristow C, Wilkinson PJ, Sumption KJ (1990b) Identification of a variable region of the African swine fever virus genome which has undergone separate DNA rearrangements leading to expansion of minisatellite like sequences. J Mol Biol 216: 677–688

23. Dixon LK, Wilkinson PJ (1988) Genetic diversity of African swine fever virus isolates from soft ticks (Ornithodoros moubata) inhabiting warthog burrows in Zambia. J Gen Virol 69: 2981–2993

24. Dixon LK, Wilkinson PJ, Sumption PJ, Ekue F (1990a) Diversity of the African swine fever virus genome. In: Darai G (ed) Molecular biology of iridoviruses. Kluwer Academic Publishers, Boston, pp 271–295

25. Esteves A, Marques ML, Costa JV (1986) Two dimensional analysis of African swine fever virus and proteins induced in infected cells. Virology 152: 192–206

26. Garcia-Beato R, Freye JMP, Lopez-Otin C, Blasco R, Vinuela E, Salas M (1992b) A gene homologous to topoisomerase II in African swine fever virus. Virology 188: 938–947

27. Garcia-Beato R, Salas ML, Vinuela E, Salas J (1992a) Role of the host cell nucleus in the replication of African swine fever virus DNA. Virology 188: 637–649

28. Gershon PD, Moss B (1990) Early transcription factor subunits are encoded by vaccinia virus late genes. Proc Natl Acad Sci USA 87: 4401–4405

29. Geshelin P, Berns KI (1974) Characterization and localization of the naturally occurring cross-links in vaccinia virus DNA. J Mol Biol 88: 785–796

30. Goebel SJ, Johnson GP, Perkins ME, David SW, Winslow JP, Paoletti E (1990) The complete DNA sequence of vaccinia virus. Virology 188: 637–649

31. Gonzalez A, Calvo V, Almazan F, Almendral JM, Ramirez JC, De La Vega I, Blasco R, Vinuela E (1990) Multigene families in African swine fever: family 360. J Virol 64: 2073–2081

32. Gonzalez A, Talavera A, Almendral JM, Vinuela E (1986) Hairpin loop structure of African swine fever virus DNA. Nucleic Acids Res 14: 6835–6844

33. Goorha R, Granoff A (1979) Icosahedral cytoplasmic deoxyviruses. Newly characterised vertebrate viruses. Comp Virol 14: 367–369

34. Hammond JM, Kerr SM, Smith GL, Dixon LK (1992) An African swine fever virus gene with homology to DNA ligases. Nucleic Acids Res 20: 2667–2671

35. Hingamp PM, Arnold JE, Mayer RJ, Dixon LK (1992) A ubiquitin conjugating enzyme encoded by African swine fever virus. EMBO J 11: 361–366

36. Howard ST, Chan YS, Smith GL (1991) Vaccinia virus homologues of the Shope fibroma virus inverted terminal repeat proteins and a discontinuous ORF related to the tumour necrosis factor receptor family. Virology 180: 633–647

37. Kahn JS, Esteban M (1990) Identification of the point mutations in two vaccinia virus nucleoside triphosphate phosphohydrolase 1 temperature sensitive mutants and role of this DNA-dependent ATPase enzyme in virus gene expression. Virology 174: 459–471

38. Kyte J, Doolittle RF (1982) A simple method for displaying the hydropathic character of a protein. J Mol Biol 157: 105–132

39. Law KM, Smith GL (1992) A vaccinia serine protease inhibitor which prevents virus induced cell fusion. J Gen Virol 73: 549–557

40. Lipman DJ, Pearson WR (1985) Rapid and sensitive protein similarity searches. Science 227: 1435–1441

41. Lord K, Hoffman-Lieberman B, Lieberman D (1990) Sequence of MyD116 cDNA: A novel myeloid differentiation primary response gene induced by 1L6. Nucleic Acids Res 18: 28232

42. Manso Ribeiro J, Rosa Azevedo JA, Texeira MJO, Braco Forte MC, Rodrigues Ribero AM, Oliveira E, Noronha F, Grave Pereira C, Dias Vigaro J (1958) Peste porcine provoquee par une souche different (Souche L) de la souche classique. Bull Off Int Epiz 50: 516–534

43. Matthews REF (1982) Classification and nomenclature of viruses. Intervirology 17: 1–99

44. McGeoch DJ (1990) Protein sequence comparisons show that the "pseudo-proteases" encoded by Poxviruses and certain retroviruses belong to the deoxyuridine triphosphatase family. Nucleic Acids Res 18: 4105–4110

45. McGeoch D, Barnett B (1991) Neurovirulence factor. Nature 353: 609

46. McVicar JW (1984) Quantitative aspects of the transmission of ASF. Am J Vet Res 45: 1535–1541

47. Mebus CA, Daidiri AH (1980) Western hemisphere isolates of ASFV-asymptomatic carriers and resistance to challenge inoculation. Am J Vet Res 41: 1867–1869

48. Montgomery RE (1921) On a form of swine fever occurring in British East Africa (Kenya colony). J Comp Pathol 34: 159–191, 243–262

49. Moss B (1990) Regulation of vaccinia virus transcription. Annu Rev Biochem 59: 661–688

50. Pacha RF, Meis R, Condit RC (1990) Structure and expression of the vaccinia virus gene which prevents virus induced breakdown of RNA. J Virol 64: 3853–3863

51. Plowright W, Parker J, Peirce MA (1969a) African swine fever in ticks (Ornithodoros moubata Murray) collected from animal burrows in Tanzania. Nature 221: 1071–1073

52. Ray CA, Black RA, Kronheim SR, Greenstreet TA, Sleath PR, Salvesen GS, Pickup DJ (1992) Viral inhibition of inflammation: Cowpox virus encodes an inhibitor of the interleukin-1 beta converting enzyme. Cell 69: 597–604

53. Rodriguez JF, Kahn JS, Esteban M (1986) Molecular cloning, encoding sequence and expression of vaccinia virus nucleic acid-dependent nucleotide triphosphatase gene. Proc Natl Acad Sci USA 83: 9566–9570

54. Rodriguez JM, Salas ML, Vinuela E (1992) Genes homologous to ubiquitin conjugating proteins and eukaryotic transcription factor SII in African swine fever. Virology 186: 40–52

55. Santaren JF, Vinuela E (1986) African swine fever virus induced polypeptides Virus Res 5: 391–405

56. Smith CA, Davis T, Anderson D, Solan L, Beckmann MP, Jerzy R, Dower SK, Cosman D, Goodwin RG (1990) A receptor for tumour necrosis factor defines an unusual family of cellular and viral proteins. Science 248: 1019–1023

57. Smith GL, Chan YS, Howard ST (1991) Nucleotide sequence of 42 kbp of vaccinia virus strain WR from near the right inverted terminal repeat. J Gen Virol 72: 1349–1374

58. Smith GL, Chan YS (1991) Two vaccinia virus proteins structurally related to the interleukin-1 receptor and the immunoglobulin superfamily. J Gen Virol 72: 511–518

59. Smith GL, Howard ST, Chan YS (1989) Vaccinia virus encodes a family of genes with homology to serine proteinase inhibitors. J Gen Virol 70: 2333–2343

60. Staden R (1982) Automation of computer handling of gel reading data produced by the shotgun method of DNA sequencing. Nucleic Acids Res 10: 4731–4751

61. Staden R (1990) An improved sequence handling package that runs on the Apple Macintosh computer. Appl Biosci 6: 387–393

62. Sumption KJ, Hutchings GH, Wilkinson PJ, Dixon LK (1990) Variable regions on the genome of Malawi isolates of African swine fever virus. J Gen Virol 71: 2331–2340

63. Thomson GR (1985) The epidermiology of African swine fever: the role of free-living hosts in Africa. Onderstepoort J Vet Res 52: 201–209

64. Urzainqui A, Tabares E, Carrascosa L (1987) Proteins synthesised in African swine fever infected cells analysed by two dimensional gel electrophoresis. Virology 160: 286–291

65. Von Heijne G (1986) A new method for predicting signal cleavage sites. Nucleic Acids Res 14: 4683–4690

66. Vinuela E (1985) African swine fever virus. Curr Top Microbiol Immunol 116: 151–170

67. Vinuela E (1987) African swine fever virus. In: Becker Y (ed) Developments in veterinary virology. Martinus Nijhoff, The Hague, pp 31–49

68. Wesley RD, Tuthill AE (1984) Genome relatedness among African swine fever virus field isolates by restriction endonuclease analysis. Prev Vet Med 2: 53–62

69. Wilkinson PJ (1981) African swine fever. In: Gibbs EPJ (ed) Virus diseases of food animals. A world geography of epidemiology and control, vol 2. Disease monographs. Academic Press, London, pp 767–786

70. Wilkinson PJ, Wardley RC, Williams SM (1981) In: Wilkinson PJ (ed) African swine fever. EEC Publication Eur 84: 66 En, pp 74–84

71. Wilkinson PJ (1989) African swine fever virus family. In: Pensaert MB (ed) Virus infections of porcines. Elsevier Publications BV, New York, pp 15–36

72. Zhou J, Sun XY, Fernando GJP, Frazer IH (1992) The vaccinia virus K2L gene encodes a serine protease inhibitor which inhibits cell fusion. Virology 189: 678–686

Authors' address: Dr. L.K. Dixon, AFRC Institute for Animal Health, Pirbright, Woking, Surrey, GU24 ONF, U.K.

Arch Virol (1993) [Suppl] 7: 201–214

The biology and molecular biology of scrapie-like diseases

J. Hope

AFRC & MRC Neuropathogenesis Unit, Edinburgh, U.K.

Summary. The transmissible spongiform encephalopathies (TSE's) are degenerative diseases of the central nervous system which naturally affect man (Creutzfeldt-Jakob disease [CJD], Gerstmann-Straussler syndrome [GSS], kuru), sheep and goats (scrapie), cattle (bovine spongiform encephalopathy [BSE]), mink (transmissible mink encephalopathy), mule deer, elk and antelope (chronic wasting disease). Spongiform encephalopathies have also been diagnosed in captive species of zoo antelope and in domestic cats. Much has been written about these maladies in the wake of the BSE outbreak, the tragic cases of CJD in recipients of cadaver-derived human growth hormone, sex hormones or dura mater and this has stimulated a continuing public health debate about the transmissibility, prevalence and clinical variability of scrapie, CJD and related ("prion") diseases. Prions (Weissmann, Liautard, this volume) and the human (Kretzschmar, this volume) and cattle (Wilesmith, Marsh, this volume) diseases are described in more detail elsewhere. This article presents a brief overview of the biology and molecular cell biology of scrapie and rodent models of these diseases.

Scrapie is a transmissible disease

Scrapie and related diseases can be transmitted by inoculation and the incubation period of the experimental disease is chiefly determined by dose, route of infection, host and donor species and genotype, and strain of pathogen [41]. Even so, the relative importance of host genetics and infection in the spread and epidemiology of the naturally-occuring diseases of man and animal remains controversial. Genetic loci have been identified in sheep (*Sip*) and mice (*Sinc*) which determine the incubation time of experimentally-induced disease [17], and the incidence of related diseases in man (familial CJD and GSS) is also genetically determined [3]. Transmission to susceptible animals may occur naturally by ingestion of diseased tissues or feed (as in BSE) (Wilesmith, this volume) but the transmissible factor or pathogen has yet to be fully

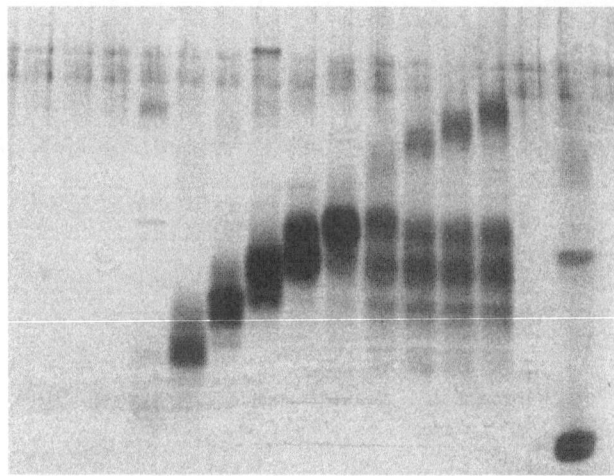

Fig. 1. Size fractionation of molecular weight variants of PrP in scrapie-associated fibrils. Fibrils were dissociated in 0.1%-SDS buffer and this extract was fractionated using a SDS-PAGE tube gel system on an Applied Biosystems HPEC A230. Protein complexes eluting from the tube gel were further disrupted in 2%-SDS buffer and analysed by slab gel SDS-PAGE and silver staining. The figure illustrates the range of monomeric protein and aggregates seen on a slab gel following analysis of SAF components in ME7-affected mouse brain. Further methods and data on the protein composition of SAF and BSE fibrils can be found in Hope et al. [24–26]

characterised. Progress has been hindered by the time and difficulties of quantitating infectivity (which requires animal bioassay) and the lack of an immune response to infection which might be used to monitor the development of disease.

Clinical symptoms are variable but can include mental deterioration and inco-ordination of limb movement in humans [4] while in domestic animals abnormal behaviour, impaired gait, itching and apprehension are frequently observed (Wilesmith, this volume). These neurological signs occur late in the pathogenesis of the disease and may take months or years to develop. Hence, one of the major problems in the control and eradication of these diseases is to identify the pre-clinical carrier of infection.

Membrane-bound cysts (vacuoles) in nerve tissue, an increase in the number and size of astro-glial cells and the accumulation of scrapie (or TSE) – associated fibrils (SAF) in brain are diagnostic features of these neurodegenerative diseases. SAF are an aggregated, protease-resistant form (PrPSc) of a neuronal membrane protein, PrP or PrPc, which co-purify with high titres of infectivity during tissue sub-cellular fraction-ation (Fig. 1) [6, 18, 36, 37; for review see 23].

Infectivity

In the 1950's, infectivity from extracts of scrapie-affected sheep was shown to pass through filters with pores small enough to retain all but viruses. The infectious factor does have some other attributes of a conventional virus [47], such as phenotypic variation (see below), but what sets scrapie apart from other diseases is the relative resistance of this viral-like activity to inactivation by normal virucidal procedures, such as boiling or exposure to ionising and ultra-violet radiation [1, 2]. This may indicate that, unlike a virus, a nucleic acid template is not required for replication of the scrapie phenotypes and that a protein may be the sole or an integral part of the infectious particle [44, 54, 55].

Apart from a virus [47], two novel structures for the infectious particle of the TSE's have been put forward to explain the biological and physico-chemical properties of infectivity: a prion or virino.

A PRION is a "proteinaceous infectious particle which resists inactivation by procedures which modify nucleic acids" and has come to be over-used as a general word for the scrapie-like agents. One version of a prion structure has some form of a protein as the sole component of the pathogen [20].

A VIRINO is a composite structure of host protein and a host-independent molecule (possibly a small nucleic acid) which determines the strain of scrapie [17].

Which, if either, of these model structures is correct remains to be seen. Much current research is focused on PrP either, in some form or other, as the prion protein, as the virino coat protein or as a cellular receptor for a more conventional pathogen.

PrP, prions and virinos

Prions

In the prion model, the infectious agent or prion is a modified form of PrPc. This protein is glycosylated at two asparagine residues, has a molecular mass of 33–35,000 Da and is anchored to the outside surface of the plasma membrane by a phosphatidyl-inositol glycolipid attached to its carboxy-terminal amino-acid [50, 51]. Most protein (and mRNA) is found in the brain but the gene is expressed at lower levels in non-neural adult and embryonic tissues [8, 31–33]. The normal isoform of the protein is completely degraded by proteases under conditions which leave a 27–30,000 Da, protease-resistant core (PrP27–30) of PrPSc.

PrP27–30 lacks the first 67 amino-acids of the mature PrPc protein [24–26].

There have been extensive studies of the co-valent structure of PrPc and PrPSc but no disease-specific, co-valent modification of PrP has been defined. While the accumulation of PrPSc in brain does appear to depend on the strain of agent (not genotype of host) [11, 22, 32], in situ hybridisation analysis has shown these phenotypes are not related to increased levels of PrP mRNA in specific areas of the brain during the development of disease [32]. This has implicated a post-transcriptional process in the conversion of PrPc to PrPSc, and it has been suggested that the only difference between PrPSc and PrPc might be conformational [24–26].

The protein is encoded in one exon by a single-copy host gene, so this gene structure precludes the modification of PrPc by an alternative splicing mechanism. In the prion model, the conversion of PrPc into the modified form (PrPSc) is proposed to occur by either direct interaction of PrPc with PrPSc (the "infectious agent"), or in the case of sporadic or familial disease (for example, GSS or CJD or natural scrapie) by a stochastic mechanism – a rare, chance event which sparks of the self-replicating, catalytic conversion of PrPc to PrPSc [54].

But what encodes strain variation?

A great many of the physico-chemical and epidemiological features of these diseases can be explained by this simple, one component model. However, as yet, the mechanism by which different strains of agent produce different phenotypes in the same strain of mouse (which encodes a single PrP amino-acid sequence) is more difficult to envisage. These phenotypes are defined by relative incubation periods and the anatomical distribution of lesions in the brains of mice of the three *Sinc* genotypes. One suggestion has been that strains of scrapie are encoded by multiple, metastable conformations of PrP [54], and that their replication is mediated by molecules designed to catalyse the folding/re-folding of nascent polypeptides, the chaperones (Liautard, this volume). Attempts to mimic this putative, protein-folding mechanism in vitro (without chaperones) have had little success [46, Hope and Lang, in prep.].

Virinos

The structure of the virino provides a more conventional explanation for strain variation. In this two component model, a second molecule binds

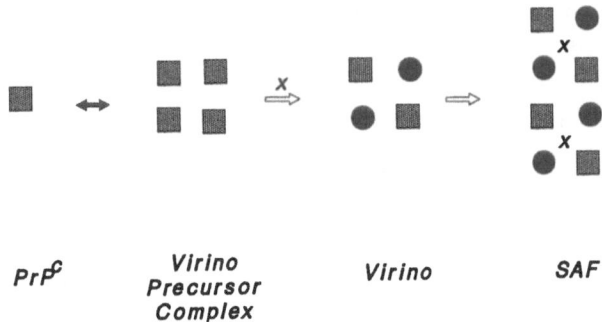

PrP^c Virino   Virino   SAF
             Precursor
             Complex

Fig. 2. Ideas about the binding of PrP and the infectious agent. In this model, a PrP (*PrPc*) molecule is in equilibrium with higher order complexes of PrPc under physiological conditions. This higher order structure (which may contain as little as 2 or 4 molecules of PrPc as well as other molecules) acts as a receptor for the infectious agent (*X* is the strain-coding molecule of the prion or virino). The binding of X to the PrPc (or *virino precursor*) complex alters its properties and causes it to accumulate in and around cells as amorphous and fibrillar aggregates of PrP: these aggregates (PrPSc) are resistant to complete proteolysis under conditions which destroy PrPc. This abstract model was devised in 1986 to explain some of our data on the molecular composition of *SAF* [24]. There are several ideas implicit in this process, some of which remain to be verified experimentally: (i) monomeric PrP should not be infectious, (ii) PrPc and PrPSc may only differ in conformation, (iii) amorphous and fibrillar PrPSc should contain both infectivity and the strain determinant molecule (*X*), and (iv) a higher order complex of PrPc, the virino precursor complex, should exist in normal brain and other tissues

to host protein (possibly PrP), and this protein coat provides protection from the host immune system and degradative processes. In this virino complex, it is the second molecule, not PrP, which determines the strain of the infectious particle and the phenotype of the infection. These ideas are summarised graphically in Fig. 2.

This interaction of agent and receptor (PrP) might be expected to disrupt the normal turnover of the protein and lead to its accumulation. It might also be expected to impair its function, about which little is yet known. Its location on the cell surface and other biochemical evidence has provoked speculation that the protein acts as a cell-surface receptor, a glial cell growth factor [40], an acetyl-choline receptor inducer (ARIA) [19], a cell adhesion molecule or (on lymphocytes) as a lymphoid cell activator [14]. However, the protein's function may be non-essential or shared by other molecules because, even though it is widely expressed in embryonic and adult tissues [32, 33], the deletion of the murine PrP gene from embryonic stem cells by homologous recombination does not prevent these cells developing into fertile chimeric mice whose offspring, PrP null mutants, appear normal [13]. In the near future, challenge of these mice with different strains of infectivity will show if other molecules can substitute for PrP during the development of disease (Weissmann, this volume), and will provide a test of its putative role as the precursor of the infectious particle – a crucial test of the prion and virino models of the infectious particle.

The search for other host molecules which might effect the kinetics of PrPSc accumulation (and hence observed phenotype) continues [30, 38, 39] and these ideas about prions and virinos have been melded together in a unified hypothesis of agent structure [55]: prions have been called apo-prions, virinos renamed holo-prions and the second component of the virino has been dubbed a co-prion: but, whatever the nomenclature, it is clear that the application of molecular biological techniques to the structure and function of PrP and the spongiform encephalopathies has revolutionised our thinking and brought us close to a molecular understanding of these enigmatic diseases.

Cell Biology

Where does the conversion of PrPc to PrPSc take place?

The biosynthesis of PrP and the kinetics of its post-translational modification and translocation to and from the plasma membrane have been investigated by pulse-chase experiments in cell cultures. The half-life of PrPc in mouse neuroblastoma cells (N2A) was calculated as 3–5 h, in stark contrast to that of PrPSc which was estimated at >15 h in persistently infected N2A cells (sc^+-MNB's). Hence, in cell culture, as well as in the central nervous system, infection can specifically interfere with the metabolic turnover of this protein. While the exact mechanism of interconversion and identity of these forms of N2A-cell protein with PrPc/PrPSc of brain and other tissues remains to be established, these in vitro experiments have provided insights into how and where in the cell this pathological process may occur. Although polarised cells such as neurons may have specialised mechanisms for targeting proteins to their functional sites, the traffic of membrane components in N2A cells may be represented by Fig. 3.

The conversion of pulse-labelled PrPc is sensitive to exogenous proteases in the culture medium but is not inhibited by brefeldin A, a fungal antibiotic which reversibly disrupts the endoplasmic reticulum-cis Golgi network [42, 53]. While the development of protease-resistance of PrPc, a key difference between PrPc and PrPSc, does not seem to be influenced by monensin (an inhibitor of mid-Golgi glycosylation) or lysosomotropic amines, these reagents prevent the subsequent conversion of PrPSc to PrP27–30 [7, 53]. These, and other studies [15, 16],

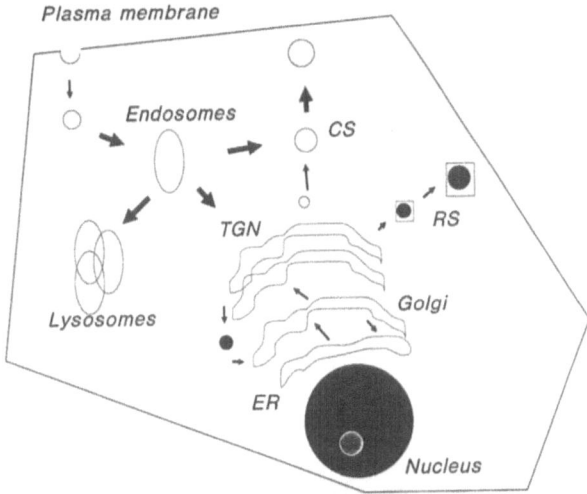

Fig. 3. Trafficking of membrane proteins in the cell. The typical life cycle of a membrane glycoprotein, such as PrPc, involves biosynthesis on the ribosome-endoplasmic reticulum, vesicular transport to the sites of N-glycan-glycolipid addition and processing in the cis-Golgi-trans-Golgi system. For cell surface proteins, this is followed by targetted fusion of vesicles into the plasma membrane by a regulated (*RS*) or constitutive (*CS*) (secretory) pathway. The membrane protein has a variable functional lifetime (for PrPc in cultured cells, t1/2 is about 5 h) before it is engulfed in a coated (or non-coated) vesicle and either re-cycled through to the Golgi system or dispatched to lysosomes for degradation. This endocytic pathway has been implicated as the cellular site of conversion of PrPc to PrPSc [7, 16, 52]

implicate the plasma membrane or the endosome/lysosome system as the site of conversion of PrPc to PrPSc with subsequent N-terminal proteolysis to PrP27–30 occuring in an acidic, endosome/lysosome compartment.

Knowing the site and mechanism of PrPc/PrPSc conversion will aid the development of a rational approach to therapy [9, 43] and the lysosome has long been recognised as a key organelle (and potential therapeutic target) in the development of the spongiform pathology of these enigmatic diseases [34]. While lysosomal involvement (glial or neuronal) in the development of pathology would not be surprising, claims based on ultrastructural studies that the lysosome is a "bioreactor" of infectivity, THE organelle replicating ALL strains of scrapie, and THE site of conversion of PrPc to PrPSc [35] are premature – and probably wrong. This "bioreactor" model also seems to disregard the fact that replication of infectivity (and death) can occur in the absence of spongiform pathology at the level of the light microscope.

Practical issues

Key problems on the cause, structural origin and mechanism of degeneration in these transmissible encephalopathies remain to be solved,
and development of therapeutic approaches to these diseases has been
severely hindered by the long, pre-clinical phase of disease and the lack
of a diagnostic test for the causative agent.

Preclinical diagnosis

The development of more specific and sensitive assays for the infection-
related form of PrP (PrPSc) is a major aim of those trying to improve
methods for the pre-clinical detection of disease. Modified forms of PrP
(SAF/PrPSc) accumulate in the brain and some peripheral tissues
(spleen, lymph nodes) during the development of scrapie and related
diseases. This accumulation parallels increases in the titre of the
infectious agent in these tissues, and some workers have claimed
co-purification of infectivity with a 30 kDa PrP fraction on SDS-poly-
acrylamide gel electrophoresis [10] and gel filtration chromatography
[48]. We have tried to confirm these observations that monomeric PrP is
infectious: essentially, SAF were purified from 263K-infected hamster
brain [24] and extracted in 1%-SDS, 1%-mercaptoethanol, 0.1M-Tris
HCl, pH 7.4 by heating at 80°C for 5 min. The extract was fractionated
using a 6%-PAGE gel (fitted with Zitex filters, 30–60 μm) and an ABI
A230 HPEC. Protein eluting in the 20–35 KDa range was pooled, and
these pools, SAF and the SDS/ME extract were tested for infectivity by
i.c. injection of hamsters. Hamsters (n = 4) injected with SAF died after
72 +/− 1 days, those injected with the SDS/ME extract died after 79
+/− 3 days while animals injected with the 20–35 KDa PrP fraction have
survived for over 300 days. Hence, we were unable to confirm recovery
of infectivity in a fraction of molecular size equivalent to PrP monomer
by PAGE gel electrophoresis. Experiments to identify the fate of
infectivity in this type of experiment are in progress.

While immuno-detection of PrPSc may greatly enhance the post-
mortem diagnosis of pre-clinical disease [45], the application of this
technique to living animals or man has been very limited (but see [29]).

In vivo studies using non-invasive methods of diagnosis

Various non-invasive methods, such as electro-encephalography (EEG),
X-ray computer-assisted tomography (CT), single photon emission com-

puted tomography (SPECT) and positron emission tomography (PET), have been used to monitor neurodegenerative diseases of this type in man [4] but their systematic use to detect early changes in the brain and to follow the development of disease has been limited. Recently, much interest has focused on the use of in vivo proton magnetic resonance spectroscopy (MRS) and imaging (MRI) to monitor virus-induced damage in the brains of man and animal. Over the years, rodent models of the spongiform encephalopathies have proved invaluable in elucidating the biology, molecular biology and pathogenesis of these diseases and we have used in vivo MRS of the brain of scrapie-affected mice to evaluate the potential of this technique for non-invasive, pre-clinical diagnosis of infection in the living host.

In vitro and in vivo MRS of mice

Following a screen of various mouse models by in vitro proton NMR of perchloric acid extracts of brain, we selected the VM mouse infected with the ME7 strain of scrapie for in vivo studies. In this mouse model the clinical course of disease is predictable, characterised by a long incubation time of 340–370 days and the detection of the disease isoform of PrP in the brain by immunoblotting at 100–110 days post-inoculation. Vacuoles can be seen in the hypothalamus and the septal nuclei of the para-terminal body at around 170–190 days, and severe vacuolation is present throughout the brain (but notably in the cortex) during the clinical phase of disease.

By in vitro analysis, levels of N-acetyl aspartate (NAA), an amino-acid believed to be localised to neurons in the brain, were found to be greatly reduced compared to levels of choline-containing compounds (Cho), creatine and phosphocreatine (Cr) in ME7 scrapie-affected, VM mice. Changes in the absolute levels of NAA, and in NAA/Cr and NAA/Cho ratios were observed from midway through the incubation period of the disease, and they correlate in magnitude and timing of appearance with the incidence and severity of vacuolar degeneration in this murine scrapie model. This fall in NAA/Cr or NAA/Cho ratio could be detected in vivo during the later stages of disease. Intriguingly, a novel in vivo signal at 1.75 ppm was observed in test animals at an early stage of infection, and this signal increased during the incubation period of the disease (Figs. 4, 5). Observation of this signal precedes in vitro detection of PrP and vacuolar histopathology by weeks and clinical signs by several months. The chemical identity giving rise to this signal or its significance is not yet known, although its spectral properties indicate the compound may be a lipid [5]. Notably, an MRS study in a patient with

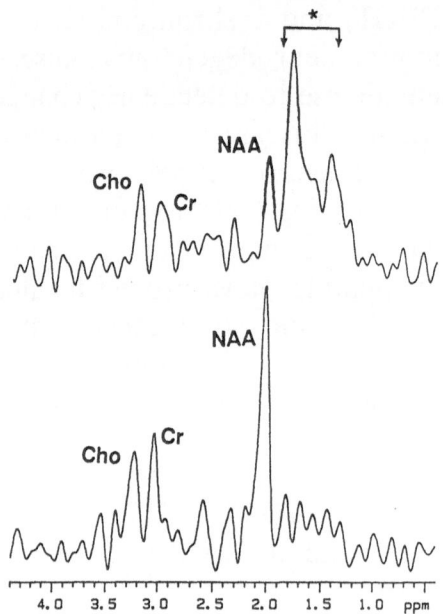

Fig. 4. In vivo proton NMR spectrum in scrapie-infected (above) and control (below) mice [5]

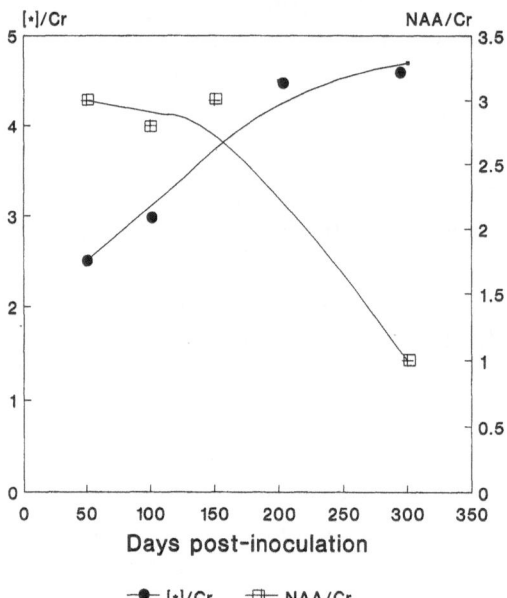

Fig. 5. Fall of N-acetyl aspartate and rise of an unknown lipid component (*) detected by proton NMR during the pathogenesis of murine scrapie [5]

Creutzfeldt-Jakob disease also found a decrease in NAA levels compared to controls, but no change in signal equivalent to that observed in mice at 1.75 ppm was observed [12].

These results, together with imaging data not presented here, suggest that the combined use of MRS and MRI may be useful for the study of this type of neurodegenerative disease in animal and man.

DNA typing: screening for susceptibility to disease

Another practical advance made in the wake of the studies on PrP and its gene has been the use of DNA linkage studies to show PrP may be a product of the incubation time control loci in man and animal. There are linkages of the PrP gene to the alleles of *Sinc* in *Sinc* congenic mouse strains [28], *Sip* in sheep [21] and to the incidence of the human familial diseases (Kretschmar, this volume). This linkage is supported by transgenic mouse studies [49]. Practically, PrP DNA analysis may be used to aid genetic counselling of those families in which the incidence of human spongiform encephalopathy is high, and similar gene analysis in sheep may aid the control and management of scrapie in sheep [21, 27].

Acknowledgements

To all my colleagues at the Neuropathogenesis Unit, Edinburgh, and to my collaborators in the University of London, especially Jane Cox, Jimmy Bell and Steve Williams.

References

1. Alper T, Cramp WA, Haig DA, Clarke MC (1967) Does the agent of scrapie replicate without nucleic acid? Nature 214: 764–766
2. Alper T (1992) The infectivity of spongiform encephalopathies: does a modified membrane hypothesis account for lack of immune response? FEMS Microbiol Immunol 89: 235–242
3. Baker HF, Ridley RM (1992) The genetics and transmissiblity of human spongiform encephalopathy. Neurodegeneration 1: 3–16
4. Bastian FO (ed) (1991) Creutzfeldt-Jakob disease and other transmissible spongiform encephalopathies. Mosby-Year Book, St. Louis
5. Bell JD, Cox IJ, Williams SCR, Belton PS, McConnell I, Hope J (1991) *In vivo* detection of metabolic changes in a mouse model of scrapie using nuclear magnetic resonance spectroscopy. J Gen Virol 72: 2419–2423
6. Bolton DC, McKinley MP, Prusiner SB (1982) Identification of a protein that purifies with the scrapie prion. Science 218: 1309–1311

212   J. Hope

 7. Borchelt DR, Taraboulos A, Prusiner SB (1992) Evidence for synthesis of scrapie
    prion proteins in the endocytic pathway. J Biol Chem 267: 16188–16199
 8. Brown HR, Goller NL, Rudelli RD, Merz GS, Wolfe GC, Wisniewski HM,
    Robakis NK (1990) The mRNA encoding the scrapie agent protein is present in a
    variety of non-neuronal cells. Acta Neuropathol 80: 1–6
 9. Brown P (1990) A therapeutic panorama of the spongiform encephalopathies.
    Antiviral Chem Chemother 1: 75–83
10. Brown P, Liberski PP, Wolff A, Gajdusek CD (1990) Conservation of infectivity
    in purified fibrillary extracts of scrapie-infected hamster brain after sequential
    enzymatic digestion or polyacrylamide electrophoresis. Proc Natl Acad Sci USA
    87: 7240–7244
11. Bruce M, McBride PA, Farquhar CF (1989) Precise targeting of the pathology of
    the sialoglycoprotein, PrP, and vacuolar degeneration in mouse scrapie. Neurosci
    Lett 102: 1–6
12. Bruhn H, Weber T, Thorwirth V, Frahm J (1991) *In-vivo* monitoring of neuronal
    loss in Creutzfeldt-Jakob disease by proton magnetic resonance spectroscopy.
    Lancet 337: 1610–1611
13. Buehler H, Fischer M, Lang Y, Bluethmann H, Lipp H-P, DeArmond SJ, Prusiner
    SB, Aguet M, Weissmann C (1992) Normal development and behaviour of mice
    lacking the neuronal cell-surface PrP protein. Nature 356: 577–582
14. Cashman NR, Loertscher R, Nalbantoglu J, Shaw I, Kascsak RJ, Bolton DC,
    Bendheim PE (1990) Cellular isoform of the scrapie agent protein participates in
    lymphocyte activation. Cell 61: 185–192
15. Caughey B, Neary K, Butler R, Ernst D, Perry L, Cheseboro B, Race R (1990)
    Normal and scrapie-associated forms of prion protein differ in their sensitivities
    to phospholipase and proteases in intact neuroblastoma cells. J Virol 64: 1093–
    1101
16. Caughey B, Race RE (1992) Potent inhibition of scrapie-associated PrP accumu-
    lation by Congo Red. J Neurochem 59: 768–771
17. Dickinson AG, Outram GW (1988) Genetic aspects of unconventional virus infec-
    tions: the basis of the virino hypothesis. In: Novel infectious agents of the central
    nervous system, Ciba Symp. 135. Wiley, Chichester, pp 63–83
18. Diringer H, Gelderbolom H, Hilmert H, Ozel M, Edelbluth C, Kimberlin RH
    (1983) Scrapie infectivity, fibrils and low molecular weight protein. Nature 306:
    376–378
19. Harris DA, Falls DL, Walsh W, Fischbach GD (1991) A prion-like protein from
    chicken brain co-purifies with an acetyl-choline receptor-inducing activity. Proc
    Natl Acad Sci USA 88: 7664–7668
20. Gabizon R, Prusiner SB (1990) Prion liposomes. Biochem J 266: 1–14
21. Goldmann W, Hunter N, Benson G, Foster JD, Hope J (1991) Different scrapie-
    associated fibril proteins (PrP) are encoded by lines of sheep selected for different
    alleles of the *Sip* gene. J Gen Virol 72: 2411–2417
22. Hecker R, Taraboulos A, Scott M, Pan K-H, Yang S-L, Torchia M, Jendroska K,
    DeArmond SJ, Prusiner SB (1992) Replication of distinct scrapie prion isolates
    is region specific in brains of transgenic mice and hamsters. Genes Dev 6: 1213–
    1228
23. Hope J, Manson J (1991) The scrapie fibril protein and its cellular isoform. Curr
    Top Microbiol Immunol 172: 57–74
24. Hope J, Morton LJD, Farquhar CF, Multhaup G, Beyreuther K, Kimberlin RH
    (1986) The major protein of scrapie-associated fibrils (SAF) has the same size,

charge distribution and N-terminal protein sequence as predicted for the normal brain protein (PrP). EMBO J 5: 2591–2597

25. Hope J, Multhaup G, Reekie LJD, Kimberlin RH, Beyreuther K (1988) Molecular pathology of scrapie-associated fibril protein (PrP) in mouse brain affected by the ME7 strain of scrapie. Eur J Biochem 172: 271–277

26. Hope J, Reekie LJD, Hunter N, Multhaup G, Beyreuther K, White H, Scott AC, Stack MJ, Dawson M, Wells GAH (1988) Fibrils from brains of cows with new cattle disease contain scrapie-associated protein. Nature 336: 390–392

27. Hunter N, Hope J (1991) The genetics of scrapie susceptibility in sheep (and its implications for BSE). In: Owen JB, Axford RFE (eds) Breeding for disease resistance in farm animals. C.A.B.I., Wallingford, pp 329–344

28. Hunter N, Dann JC, Bennett AD, Somerville RA, McConnell I, Hope J (1992) Are *Sinc* and the PrP gene congruent? Evidence from PrP gene analysis in *Sinc* congenic mice. J Gen Virol 73: 2751–2755

29. Ikegami Y, Ito M, Isomura H, Momotani E, Sasaki K, Muramatsu Y, Ishiguro N, Shinagawa M (1991) Pre-clinical and clinical diagnosis of scrapie by detection of PrP protein in tissues of sheep. Vet Rec 128: 271–275

30. Kellings K, Meyer N, Mirenda C, Prusiner SB, Riesner D (1992) Further analysis of nucleic acids in purified scrapie prion preparations by improved return refocusing gel electrophoresis. J Gen Virol 73: 1025–1029

31. Kretszchmar HA, Prusiner SB, Stowring LE, DeArmond SJ (1986) Scrapie prions are synthesized in neurones. Am J Pathol 122: 1–5

32. Manson J, McBride P, Hope J (1992) Expression of the PrP gene in the brain of *Sinc* congenic mice and its relationship to the development of scrapie. Neurodegeneration 1: 45–52

33. Manson J, West JD, Thomson V, McBride P, Kaufman MH, Hope J (1992) The prion protein gene: a role in mouse embryogenesis? Development 115: 117–122

34. Marsh RF, Sipe JC, Morse SS, Hanson RP (1976) Transmissible mink encephalopathy: reduced spongiform degeneration in aged mink of the Chediak-Higashi genotype. Lab Invest 34: 381–386

35. Mayer RJ, Landon M, Laszlo L, Lennox G, Lowe J (1992) Protein processing in lysosomes: the new therapeutic target in neurodegenerative disease. Lancet 340: 156–159

36. Merz PA, Somerville RA, Wisniewski HM, Iqbal K (1981) Abnormal fibrils from scrapie-affected brain. Acta Neuropathol 65f: 63–74

37. Merz PA, Rowher RG, Kascsak R, Wisniewski HM, Somerville RA, Gibbs CJ, Gajdusek DC (1984) Infection-specific particle from the unconventional slow virus diseases. Science 225: 437–440

38. Meyer N, Rosenbaum V, Schmidt B, Gilles K, Mirenda C, Groth D, Prusiner SB, Riesner D (1991) Search for a putative scrapie genome in purified prion fractions reveals a paucity of nucleic acids. J Gen Virol 72: 37–49

39. Oesch B, Teplow DB, Stahl N, Serban D, Hood LE, Prusiner SB (1990) Identification of cellular proteins binding to the scrapie prion protein. Biochemistry 29: 5848–5855

40. Olescak EL, Murdoch G, Manuelidis L, Manuelidis EE (1988) Growth factor production by Creutzfeldt-Jakob disease cell lines. J Virol 62: 3103–3108

41. Outram GW (1976) The pathogenesis of scrapie in mice. In: Kimberlin RH (ed) Slow virus diseases of animals and man. North Holland Research Monographs, vol 44. North-Holland, Amsterdam, pp 325–357

42. Pelham H (1992) Multiple targets for brefeldin A. Cell 67: 449–451

43. Pocchiari M, Salvatore M, Ladogana A, Ingrosso L, Xi YG, Cibati M, Masullo C (1991) Experimental drug treatment of scrapie: a pathogenetic basis for rationale therapeutics. Eur J Epidemiol 7: 556–561
44. Prusiner SB (1991) Molecular biology of prion disease. Science 252: 1515–1522
45. Race R, Ernst D, Jenny A, Taylor W, Sutton D, Caughey B (1992) Diagnostic implications of detection of proteinase-K resistant protein in spleen, lymph nodes and brain of sheep. Am J Vet Res 53: 883–889
46. Raeber AJ, Borchelt DR, Scott M, Prusiner SB (1992) Attempts to convert the cellular prion protein into the scrapie isoform in the cell-free systems. J Virol 66: 6155–6163
47. Rohwer R (1991) The scrapie agent: A virus by any other name. Curr Top Microbiol Immunol 172: 195–232
48. Safar J, Wang W, Padgett MP, Ceroni M, Piccardo P, Zopf D, Gajdusek DC, Gibbs CJ (1990) Molecular mass, biochemical composition, and physico-chemical behavior of the infectious form of the scrapie precursor protein monomer. Proc Natl Acad Sci USA 87: 6373–6377
49. Scott M, Foster D, Mirenda C, Serban D, Coufal F, Walchli M, Torchia M, Groth D, Carlson GA, DeArmond SJ, Westaway D, Prusiner SB (1989) Transgenic mice expressing hamster prion protein produce species-specific scrapie infectivity and amyloid plaques. Cell 59: 847–857
50. Stahl N, Borchelt DR, Hsiao K, Prusiner SB (1987) Scrapie prion protein contains a glycosylinositol phospholipid. Cell 51: 229–240
51. Stahl N, Baldwin MA, Hecker R, Pan K-M, Burlingame AL, Prusiner SB (1992) Glycosylinositol phospholipid anchors of the scrapie and cellular prion proteins contain sialic acid. Biochemistry 31: 5043–5053
52. Taraboulos A, Serban D, Prusiner SB (1990) Scrapie prion proteins accumulate in the cytoplasm of persistently infected cultured cells. J Cell Biol 110: 2117–2132
53. Taraboulos A, Raeber AJ, Borchelt D, Serban D, Prusiner SB (1992) Synthesis and trafficking of prion proteins in cultured cells. Mol Biol Cell 3: 851–863
54. Weissmann C (1991) Spongiform encephalopathies: The prion's progress. Nature 349: 569–571
55. Weissmann C (1991) A unified theory of "prion propagation". Nature 352: 679–683

Author's address: Dr. J. Hope, AFRC & MRC Neuropathogenesis Unit, Ogston Building, West Mains Road, Edinburgh, EH9 3JF, U.K.

Arch Virol (1993) [Suppl] 7: 215–225

Analysis of nucleic acids in purified scrapie prion preparations

K. Kellings[1], **N. Meyer**[1], **C. Mirenda**[2], **S.B. Prusiner**[2], and **D. Riesner**[1]

[1] Heinrich-Heine-Universität Düsseldorf, Institut für Physikalische Biologie,
Düsseldorf, Federal Republic of Germany
[2] University of California, Department of Neurology, San Francisco,
California, U.S.A.

Summary. Amount, type, and size of nucleic acid molecules associated with purified prion preparations were analyzed. Return refocusing gel electrophoresis (RRGE) was developed to detect homogeneous and heterogeneous nucleic acids extracted from highly purified scrapie prion preparations. With this method all types of nucleic acids in the size range from 13 to several thousand nucleotides could be analyzed. The recovery of all nucleic acids, after deproteinization and two-phase extraction was higher than 90%. Despite extensive nuclease digestions some small polynucleotides remained. Although a scrapie-specific nucleic acid cannot be excluded, the results further define the possible characteristics for such a hypothetical molecule. If it was homogeneous in size, then it would be <80 nt in length at a particle-to-infectivity ratio (P/I) near unity; if the other extreme, i.e. totally heterogeneous scrapie-specific nucleic acids were assumed, then scrapie-specific nucleic acids would have to include molecules smaller than 240 nt. In order to exclude the possibility that unspecific background nucleic acid is entrapped in prion-rods, infectious material has to be prepared without a proteolysis and rod formation, and the analysis of nucleic acids performed with those preparations.

Introduction

A quarter of a century after Alper and colleagues [1, 2] first proposed the heretical idea that the causative agent of scrapie does not depend on an intrinsic nucleic acid moiety for replication, it has not been satisfactorily established whether this is indeed the case or not. A wealth of data indicate that a protein, designated prion protein (PrP), is required for scrapie infectivity. Protease treatment and protein denaturing agents affect infectivity, whereas procedures that modify or hydrolyse nucleic acids do not alter infectivity [3]. In spite of many attempts, no physical or chemical evidence for a scrapie-specific nucleic acid has been found to date. A recent review [4] was devoted to a critical assessment of the different approaches, which have not as yet provided unequivocal evidence for or against a nucleic acid component. The existence of multiple isolates or "strains" with different biological properties [5, 6] has offered the strongest argument for a scrapie-specific nucleic acid. The failure to explain such strain variation in terms of molecular variation in PrP continues to stimulate the search for a scrapie-specific nucleic acid [7, 8].

In the studies reported here, physico-chemical methods were used to search for a scrapie-specific nucleic acid. These studies, like other approaches, neither establish or exclude a nucleic acid component to infectivity, but they do put constraints on the size of such a component, which cannot be larger than about 100 nucleotides.

Materials and methods

Purification of prions

Hamster prions were purified by sedimentation and treated with DNase and Zn^{2+}, treatments which do not reduce infectivity, but do reduce contamination by extraneous host nucleic acids. In order to improve the recovery of prion infectivity during the collection of prion rods from sucrose gradient fractions, ultracentrifugation steps (2×4 h, 100,000 g) were substituted for the ethanol precipitation previously used. Furthermore, in some preparations, DNase digestion of rods was omitted because the material was treated with Bal31, micrococcal nuclease and RNaseA after formation of detergent-lipid-protein complexes (DLPC). The prion rods were dispersed into DLPCs which resulted in retention of infectivity but made nucleic acids, formerly protected in rods, accessible to degradation. DLPCs were digested either with a mixture of DNase, Bal31, and RNase A or a mixture of micrococcal nuclease, alkaline phosphatase, RNase A and phosphodiesterase. After dissociation in SDS and deproteinization by proteinase K treatment and twofold phenol-chloroform-isoamyl alcohol (25:24:1) extraction, the preparations were analyzed by PAGE and return-refocusing gel electrophoresis, respectively. The details are described elsewhere [9, 10].

Return refocusing gel electrophoresis

Because heterogeneous nucleic acids would migrate in PAGE as many bands and would be unresolved in a background smear, the method of return refocusing gel electrophoresis (RRGE) was developed. The method is depicted in Fig. 1. After conventional PAGE heterogeneous nucleic acids are dispersed over the whole length of the lane. The lane is cut into a few segments, each corresponding to a well defined range of M_r. The segments are repolymerized into the bottom of new gel matrices and a second electrophoresis is performed with reversed polarity so that the nucleic acids migrate into the new gel matrix. Because all nucleic acids in a gel segment begin

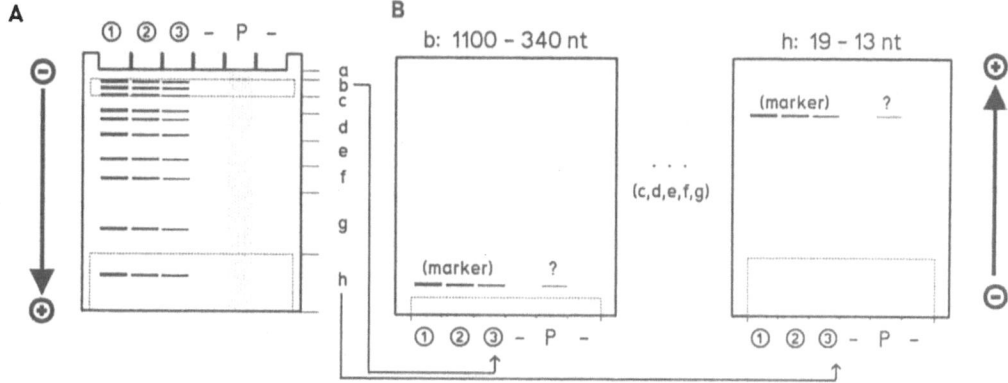

Fig. 1. Principle of the return refocusing gel electrophoresis (RRGE). After conventional PAGE (e.g. 100 min, 250 V) heterogeneous nucleic acids are dispersed over the whole length of the lane (lane *P* in **A**). The lane is cut into a few segments (*a–h*), each corresponding to a well defined range of M_r. The segments are repolymerized into the bottom of new gel matrices (**B**) and a second electrophoresis (250 V) is performed with reversed polarity so that the nucleic acids migrate into the new gel matrix. Because all nucleic acids in a gel segment begin migration from the same position at the beginning of the first PAGE, they meet again after reversal of the polarity of the second electrophoresis if the second run is stopped at a definite time. By adding SDS to the second PAGE, the focusing effect still works for nucleic acids, while other substances such as proteins and polysaccharides remain dispersed. This is a significant advantage since proteins, like nucleic acids, stain with silver. The times of refocusing of different gel segments are chosen to be optimal for the different segments (between 42 and 48 min). The unknown nucleic acid amount (? in **B**) of the prion sample is determined by conparison with the nucleic acid markers of known concentrations (markers 1, 2, 3). Only the two gel segments *b* and *h* are given as an example for the refocusing step; gel segment "a" is not used for refocusing. Figure reproduced from Kellings et al. [10]

migration from the same position at the beginning of the first PAGE, they meet again after reversal of the polarity of the second electrophoresis if the second run is stopped at a definite time. By adding SDS to the second PAGE, the focusing effect still works for nucleic acids while other substances such as proteins and polysaccharides remain dispersed. This is a significant advantage since proteins, like nucleic acids, stain with silver. The unknown nucleic acid amount of the prion sample is determined by comparison with nucleic acid markers of known concentrations. The details were described before [9, 10].

Results and discussion

A direct assault on the problem by physico-chemical methods must currently satisfy two basic requirements: (i) It has to be capable of detecting the hypothetical scrapie nucleic acid that might be DNA or

RNA, single- or double-stranded, circular or linear, capped, chemically modified or covalently bound to proteins, and homogeneous or heterogeneous in size; it should be emphasized, however, that only conventional nucleic acid chemistry can be considered. Chemically modified nucleic acid, which have not been described as natural compounds so far, and would have unknown chemical properties will not be considered. (ii) It has to be capable of detecting unlabeled nucleic acids in the pg range, calculated as follows: if we take 100 pg as the current limit of nucleic acid detection, 2×10^9 ID_{50} with a hypothetical nucleic acid of 100 nt would contain this amount of nucleic acid, assuming a particle (nucleic acid molecule)-to-infectivity ratio (P/I) of at least one. A smaller scrapie-specific nucleic acid would require starting with more ID_{50} units, a larger nucleic acid material with fewer ID_{50} units to yield the same amount of the nucleic acid in pg. The requirement for sensitive detection of unlabeled nucleic acids is imposed by the inefficiency of radiolabelling nucleic acids in animals and the low titers of the agent in cell culture.

Search for homogenous nucleic acids by PAGE

Polyacrylamide gel electrophoresis (PAGE) combined with silver staining satisfies all of these requirements if the hypothetical nucleic acid is a discrete species, but is not sufficiently sensitive for heterogeneous acids which must be investigated by the refocusing technique discussed in a later section.

In the PAGE studies [9] hamster prions were purified and deproteinized as described under Materials and Methods. Some background smearing as well as distinct bands migrating near the dye front were evident in DNase- and Zn^{2+}-treated material, whereas omission of these treatments resulted in a prion fraction with a large number of stained bands in addition to the rapidly migrating molecules in the size range between 4 and 15 bases. It could be shown that these bands are likely either complex Asn-linked oligosaccharides released from PrP 27–30 during proteinase K digestion or non-covalently bound sugar polymers which copurified with the prion rods. No bands above 15 nt could be detected.

If control nucleic acids (oligo RNA; 51 nt; oligo DNA, 29 nt; oligo DNA, 54 nt; tRNA, 80 nt; viroid RNA, 359 nt) were analyzed in the same way as infectious scrapie prions, 3×10^{10} molecules of each of the control nucleic acids were readily detected. If the control nucleic acids were hydrolyzed with DNase I and Zn^{2+}, no silver stained bands were detected. If the control nucleic acid molecules were added to the scrapie sample after DNase and Zn^{2+} treatment but before destroying the infectivity by SDS, all of them were visible except those of 10 nt and 11 nt in length because these were hidden by the non-nucleic acid molecules co-purifying with the prions.

The prion infectivity was determined to be 1.2×10^{10} ID_{50} prior to deproteinization. Taking 170 pg nucleic acid in a single band as a conservative limit of detection, 1.2×10^{10} nucleic acid molecules of at least 25 nt in length would have been detected. Within the limits of the bioassay, with variation in titers by a factor of 10, these investigations supported the conclusion that the infectious particle does not contain a homogeneous nucleic acid component. Since the analysis utilized 20% polyacrylamide gels, nucleic acids >300 nt in length also could not be analyzed, but were assessed in subsequent experiments.

Analysis of heterogeneous nucleic acids by RRGE

The possibility remained that prions contain nucleic acid molecules of non-uniform length. In such a case, the nucleic acid would migrate during PAGE as many bands and each band might either be below the threshold for detection or not resolved from neighbouring bands, resulting in a smear of staining. To evaluate this unprecedented but formal possibility, we developed a technique to measure nucleic acid molecules of variable size. With RRGE, nucleic acids can be separated from other stainable molecules and focused into one sharp band. The heterogeneous nucleic acid can then be detected with a sensitivity close to that of the detection of a homogeneous nucleic acid.

In a first series of experiments [9] prion samples were analyzed by RRGE after intensive nuclease digestion but without DLPC formation. After RRGE, clear bands were obtained in size range from 50 to 300 nt. The total amount of heterogeneous nucleic acid in the prion sample was estimated to be about 20 ng.

Nuclease digestion studies were carried out prior to RRGE in order to confirm the nucleic acid nature of the bands and to differentiate between RNA and DNA. It was found that the RNA and DNA detected differed in size distribution: nucleic acids below 50 nt turned out to be mainly RNA, while longer molecules were mainly DNA.

When prion rods were dispersed into DLPCs which resulted in retention of infectivity [11, 12], it could be expected that nucleic acids that were possibly protected from degradation by inclusion in the rod-shaped aggregates would become accessible to nucleases upon formation of the DLPCs. Purified prion samples were dispersed into DLPCs and digested either with a mixture of DNase Bal31 and RNase A or a mixture of micrococcal nuclease, alkaline phosphatase, RNase A and phospho-

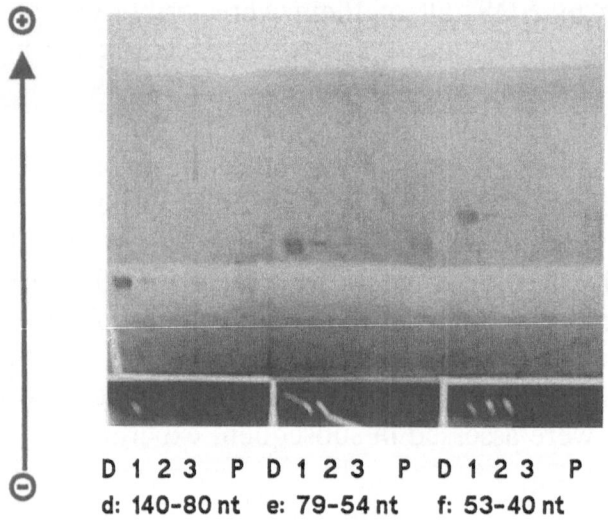

⊕

⊖

D 1 2 3   P   D 1 2 3   P   D 1 2 3   P
d: 140–80 nt   e: 79–54 nt   f: 53–40 nt

Fig. 2. RRGE of a prion sample after DLPC formation and nuclease digestion. The refocused gel segments *d* to *f* (cf. Fig. 1) are shown. Each gel segment consists of six lanes: (*D*) sonicated calf thymus DNA, 300 ng (total); (*1*), (*2*), (*3*) nucleic acid marker, 1, 2, 3, (cf. Fig. 1); empty; and (*P*) prion sample. The comparison of band intensities had to be carried out on the original gel, not on the photographic reproduction. The intensity as well as the width of a band were considered for comparison between prion and marker bands. Bands from adjacent gel segments were also included in the comparison. The sonicated ctDNA was used as an internal control of refocusing and not for quantification. The amounts of nucleic acids in the prion sample were estimated as 300 pg (*d*), 450 pg (*e*), and 600 pg (*f*). Figure modified from Kellings et al. [10]

diesterase. Due to this treatment the nucleic acid content of prions was reduced 10-fold. Control nucleic acids, which were added to the DLPC fraction and treated as described above, were degraded to mono- and di-nucleotides. Analysis of this material by state-of-the-art method of RRGE (Fig. 2) which extend the size range to 13–1,100 nucleotides documented nucleic acids present in the prion sample at concentrations estimated by comparison with M_r markers. The estimation of nucleic acid content of the prion sample was carried out by comparison with the marker bands (cf. Fig. 2).

Alltogether seven independent prion samples were analyzed by the procedure described above, with slight modifications. Before the de-proteinization/nucleic acid extraction step the infectivities of these pre-parations were log ID_{50} 7.8, 8.5, 8.7, 8.3, 8.3, 8.5, 8.7. These infectivities were used *in toto* for one gel electrophoretic analysis. The scrapie infectivity was monitored by an incubation time interval procedure (Prusiner et al. [13]) [13] at all steps of the preparation. Separation of the prion rods from the sucrose used for gradient centrifugation resulted in some experiments in a loss of infectivity of 1–3 orders of magnitude, probably as a result of aggregation and denaturation. It is worth noting that the dispersion of ethanol-precipitated prion rods into DLPCs frequently increased the titer more than 10-fold.

The yield of nucleic acid after deproteinization was estimated quantitatively by several radiolabelled nucleic acids and amounted in all cases to approximately 90% [10].

Ratio of nucleic acid molecules per infectious unit

Based on the amount of nucleic acid estimated from RRGE and the titers of the prion fractions prior to boiling in SDS, the ratio of nucleic acid molecules to ID_{50} units (P/I) was calculated. If the nucleic acids detected were related to scrapie infectivity, one of two alternative paradigms would be correct. First, a putative scrapie-specific nucleic acid of uniform length might be hidden amongst an ensemble of background nucleic acid. Such a scrapie-specific nucleic acid would not have been detected by conventional PAGE, even if it was present in sufficient amounts. Second, one must consider the possibility of a scrapie-specific polynucleotide which is heterogeneous in length. In Fig. 3 the numbers of nucleic acid molecules per ID_{50} unit are plotted as a function of their length as estimated from the individual gel sections. In this plot, the calculation was based upon the first paradigm, i.e. a well-defined scrapie-specific nucleic acid among the heterogeneous background nucleic acids. Data from all published experiments [9, 10] are presented in the plot.

A significant decrease in the number of nucleic acid molecules per ID_{50} unit was found as the size of the polynucleotide increased. For small nucleic acid molecules (20 nt), about 10 to several hundred molecules/ID_{50} were estimated. If the scrapie-specific nucleic acid were longer (>76 nt), the particle-to-infectivity ratio would fall below unity and continues to drop several orders below unity. Thus, one can safely conclude, that a unique nucleic acid of this size range cannot be scrapie-specific. The straight line in Fig. 3 is an interpolation of the experimental data by linear regression in order to determine an average nucleic acid size at a P/I ratio of 1.

A discussion of the maximum error [10] assumed one log ID_{50} lower than measured and twofold higher nucleic acid content; this estimation shifts P/I ratio of unity from 76 nt to 165 nt. It is worth noting that the limit of one nucleic acid molecule per ID_{50} (at a P/I ratio of 1) is an extreme assumption [cf. 9], for comparison 10^4 to 10^5 PrP molecules are necessary for one infectious unit [13].

If heterogeneous scrapie-specific nucleic acids were assumed, all molecules of a certain heterogeneity class would have to be added up

222   K. Kellings et al.

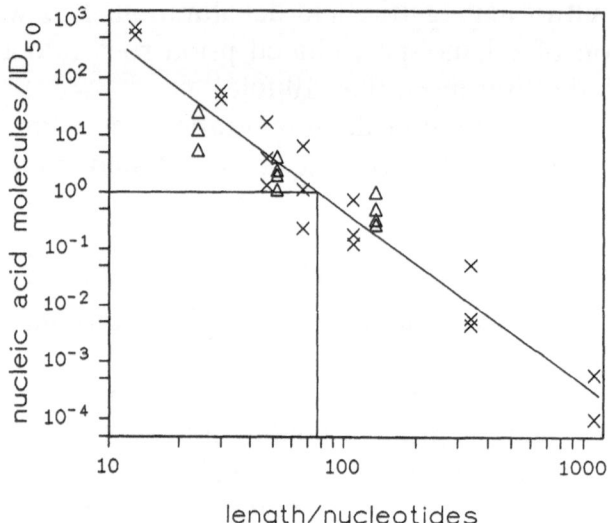

length/nucleotides

Fig. 3. Relationship of P/I ratio to average length of nucleic acid species from determinations of the kind illustrated in Fig. 2 on seven independent prion samples. The relationship is linear over the size range of 10–1,100 nucleotides with an intercept of about 80 nt for a P/I of unity. Only small nucleic acids <80 nt have P/I >1. Data ($\triangle$) were taken from Meyer et al. [9], ($\times$) from Kellings et al. [10]. The relationships were calculated as follows, using fragment e in Fig. 2 as an example: this fragment contained 450 pg of nucleic acids in the size range of 54–79 nt. Assuming a continuous distribution of different sizes, the 26 species in this class will have an average MW of 22×10^3 and there will be:

$$450 \times 10^{-12} \text{g} \times \frac{6 \times 10^{23}\,\text{mol}}{26 \times 22 \times 10^3\,\text{g}} = 4.7 \times 10^8\,\text{mol}$$

of a particular size in this ensemble. Since the starting sample contained $10^{8.7}$ (5×10^8) ID_{50}, the P/I is approximately 1 for a hypothetical discrete scrapie specific nucleic acid in the ensemble. Figure from Riesner [4]

to account for the corresponding P/I ratio. Such heterogeneity would be an intrinsic property of the scrapie-genome and not an artefact from nuclease digestion, because the bioassays yielding the ID_{50} value were carried out after nuclease digestion. One might discuss several cases of heterogeneity, e.g. that all molecules of sizes between 200 nt and 300 nt could act as scrapie-genome. It is most interesting, however, to discuss an extreme hypothetical situation, i.e. total heterogeneity. In that case it was estimated [10] that all nucleic acid molecules from 239 nt to infinite lengths, which were detected in prion samples, add up to a P/I ratio of unity [10]. Consequently, even in this unprecedented case, nucleic acid molecules as small as 239 nt act as scrapie genome.

Conclusions and prospects

In spite of the present and many other studies [4] the question of whether scrapie infectivity depends on a nucleic acid component or not has not been settled. No essential nucleic acid has been isolated, but neither have all been excluded. The argument that the physical, chemical and enzymatic treatments applied to infectious material constitute evidence against a nucleic acid component is greatly weakened by the demonstration that small nucleic acids (up to several hundred nucleotides) in prion preparations withstand these treatments; and by the tacit assumption in these arguments that a hypothetical scrapie nucleic acid will have similar properties to nucleic acids free in solution, in contact with viral proteins or components of cellular extracts. This assumption may well be invalid as PrP clearly differs from viral and cellular proteins with respect to solubility, resistance against proteinase K and tendency to self-aggregate.

Our approach to exclude a nucleic acid component to infectivity by largely structure-independent methods for detection and quantitation has failed to do so but has defined constraints on acceptable models:

(i) The hypothetical scrapie-specific nucleic acid might be a well-defined molecular species hidden in the smear of heterogeneous nucleic acids which represent preparative impurities of the sample. In that case, the "genome" must be about 76 nt or smaller, as longer molecules were not present in concentrations above one molecule per infectious unit. If we assume an order of magnitude error in the bioassay for determining the infectious units and a twofold underestimation of nucleic acid, the limit would be 165 instead of 76 nucleotides chain length. The one sure conclusion is that larger nucleic acids can be excluded. These smaller molecules are nonetheless of uncertain functional significance since their co-purification with infectivity does not infer that they are essential for infectivity. Speculations about possible functional modes of a small scrapie-specific nucleic acid were outlined in a recent review [4].

(ii) The scrapie-specific nucleic acid itself might be heterogeneous in size. In this case, more of the detected nucleic molecules would serve as candidate "genomes" and all the molecules detected in a particular size range have to be counted. In this model, the limit size of the ensemble is ≤240 uncleotides in a P/I of one. Some observations on viroids raise the possibility that a heterogeneous ensemble of nucleic acids might be capable of information storage and replicative functions. For example, cadang-cadang viroid molecules have been described with variable lengths from 287 to 301 nucleotides, and all were infectious [14]. Circular viroids could be converted within the host cell into linear molecules which retain infectivity by cleavage at several different sites in the molecule [15]. Infectious viroids could be recovered also from those plants inoculated with two cDNA clones each encoded by different portions of the viroid sequence [16].

(iii) The relationship shown in Fig. 3 might simply reflect the decreasing efficiency of nuclease degradation with decreasing size of the nucleic acid in combination with the known protective effects of PrP. In this case, the nucleic acids would be mere impurities in the highly purified preparations without any functional relevance for scrapie infectivity. At present, this possibility appears to us most probable, particularly if all other information on the scrapie infectious agent, for example also that from experiments with transgenic animals is included [17, 18]. Recent experiments (Kellings et al., unpubl.) in which even more harsh nucleic acid degradation procedures were applied, did not result in a reduction of the P/I ratio. One has to assume, that the nucleic acids found in this work as associated with purified prions are intimately bound and protected by the prion protein, probably in an aggregated form. Since these aggregates are formed during the preparation as a result of proteinase K treatment and solubilization in detergents [19] one could speculate that the nucleic acids are entrapped in the aggregates during the preparation. They are partially but not completely released during DLPC formation as may be concluded from the finding that the amount of nucleic acids in prion preparations is reduced after DLPC formation by more than an order for magnitude. It can be inferred from these considerations that the amount of nucleic acids in prion preparations might be lowered only, if degradation of nucleic acids is carried out before formation of prion protein aggregates. Those investigations are presently carried out.

Acknowledgements

This work was supported by research grants from the National Institutes of Health (NS14069, NS22786 and AG08967), from the Minister für Wissenschaft und Forschung von Nordrhein-Westfalen, Germany and the Fonds der Chemischen Industrie.

References

1. Alper T, Haig DA, Clarke MC (1966) The exceptionally small size of the scrapie agent. Biochem Biophys Res Commun 22: 278–284
2. Alper T, Cramp WA, Haig DA, Clarke MC (1967) Does the agent of scrapie replicate without nucleic acids? Nature 214: 764–766
3. Prusiner SB (1982) Novel proteinaceous infectious particles cause scrapie. Science 216: 136–144

4. Riesner D (1991) The search for a nucleic acid component to scapie infectivity. Semin Virol 2: 215–226

5. Bruce ME, Dickinson AG (1987) Biological evidence that scrapie agent has an independent genome. J Gen Virol 68: 79–89

6. Kimberlin RH, Cole S, Walker DA (1987) Temporary and permanent modifications to a single strain of mouse scrapie on transmission to rats and hamsters. J Gen Virol 68: 1875–1881

7. Prusiner SB (1991) Molecular biology of prion diseases. Science 252: 1515–1522

8. Weissmann C (1991) A "unified theory" of prion propagation. Nature 352: 679–683

9. Meyer N, Rosenbaum V, Schmidt B, Gilles K, Mirenda C, Groth D, Prusiner SB, Riesner D (1991) Search for a putative scrapie genome in purified prion fractions reveals a paucity of nucleic acids. J Gen Virol 72: 37–49

10. Kellings K, Meyer N, Mirenda C, Prusiner SB, Riesner D (1992) Further analysis of nucleic acids in purified scrapie prion preparations by an improved return refocusing gel electrophoresis (RRGE). J Gen Virol 73: 1025–1029

11. Gabizon R, McKinley MP, Prusiner SB (1987) Purified prion proteins and scrapie infectivity copartition into liposomes. Proc Natl Acad Sci USA 84: 4017–4021

12. Gabizon R, Groth DF, McKinley MP, Prusiner SB (1988) Immunoaffinity purification and neutralization of scrapie prion infectivity. Proc Natl Acad Sci USA 85: 6617–6621

13. Prusiner SB, Cochran SP, Groth DF, Downey DE, Bowman KA, Martinez HM (1982) Measurement of the scrapie agent using an incubation time interval assay. Ann Neurol 11: 353–358

14. Haseloff J, Mohamed NA, Symons RH (1982) Viroid RNAs of cadang-cadang disease of coconuts. Nature 299: 316–322

15. Palukaitis P, Zaitlin M (1987) The nature and biological significance of linear potato spindle tuber viroid molecules. Virology 157: 199–210

16. Tabler M, Sänger HL (1984) Cloned single- and double-stranded DNA copies of potato spindle tuber viroid (PSTV) RNA and co-inoculated subgenomic DNA fragments are infectious. EMBO J 3: 3055–2062

17. Scott M, Foster D, Mirenda C, Serban D, Coufal F, Wälchli M, Torchia M, Groth D, Carlson G, DeArmond SJ, Westaway D, Prusiner, SB (1989) Transgenic mice expressing hamster prion protein produce species-specific scrapie infectivity and amyloid plaques. Cell 59: 847–857

18. Hsiao KK, Scott M, Foster D, Groth D, DeArmond SJ, Prusiner SB (1990) Spontaneous neurodegeneration in transgenic mice with mutant prion protein. Science 250: 1587–1590

19. McKinley MP, Meyer R, Kenaga L, Rahbar F, Cotter R, Serban A, Prusiner SB (1991) Scrapie prion rod formation in vitro requires both detergent extraction and limited proteolysis. J Virol 65: 1440–1449

Authors' address: Dr. D. Riesner, Heinrich-Heine-Universität Düsseldorf, Institut für Physikalische Biologie, Universitätsstrasse 1, D-40225 Düsseldorf, Federal Republic of Germany.

Arch Virol (1993) [Suppl] 7: 227–243

Prions and molecular chaperones

J.-P. Liautard

INSERM U-65, Departement Biologie Santé, Université de Montpellier II,
Montpellier, France

Summary. Molecular chaperones are proteins involved in the folding of other proteins. Among these chaperones, some are involved in their own folding (auto-chaperones). A question arises: what is the mechanism of the chaperone folding catalysis? A model for protein folding that uses the thermodynamics of irreversible processes and statistical mechanics to describe the phenomenon is proposed; the analysis presents a clear link between these two aspects. A consequence of this model is the possible existence of misfolded proteins. This point is discussed and some experimental results arguing in this direction detailed. This thermo-kinetic model is applied to protein folding driven by a molecular chaperone. Analysis of folding shows that a misfolded chaperone can induce misfolding in protein and, in the case of autofolding (auto-chaperone), may lead to new misfolded chaperones. The consequences are explored by computer simulations. They show that such an auto-chaperone could behave as a new kind of informative molecule and replicate a misfolded structure by a process similar to infection. A quantitative model, displaying the epidemiologic characters of prion infections, is derived from this hypothesis. This hypothesis satisfactorily explains the three manifestations (infectious, genetic and sporadic) that are the characteristic features of all prion diseases. Are prions really molecular chaperones required for their own assembly? Analysis of the structure of prions revealed some features shared by true molecular chaperones. This analysis suggests the positions of the mutations likely to lead to the characteristic early onset of encephalopathy. They are in good agreement with experimental results.

Introduction

Proteins that appear to be essential for the correct cellular folding and assembly of many proteins have been identified; they are called "molecular chaperone", reflecting their role in guiding nascent polypeptides into their correct, biologically active conformation [38]. Analysis of the chaperone assited folding is performed at the phenomenological level. Very little is known on the mechanisms involved at the submolecular level. To explain the folding guided by a chaperone the need of an accurate model of protein folding is obvious.

The understanding of protein folding is still one of the main challenges of present-day biology [18]. The models presently used to try to calculate the protein structure always begin with the hypothesis that the native conformation is the most stable [23]. This is probably true for the majority of proteins but is false in its generality. Indeed, the existence, under physiological conditions, of misfolded proteins implies that the protein folding process is not entirely governed by thermodynamic stability. Furthermore, examples of native proteins that are not under the most stable conformation has been recently reported: the subtilisin [45] and the alpha lytic protease [2]. Consequently these proteins have only a kinetic stability. Finally, the existence of molecular chaperones is an argument for the reality of such metastable conformation. A protein that needs a chaperone would probably be in an incorrect conformation for a long time. This means that the kinetics of correct folding is impaired by an energetic barrier. This barrier could be removed by the molecular chaperone in a way comparable to the enzymatic mechanism that accelerates a chemical reaction.

It is striking to note that in terms of thermodynamics, under physiological conditions, protein folding is an irreversible process [26]. Indeed, starting with a 100% unfolded protein the folding process leads to 100% folding and it is necessary to change the conditions to denature the protein. I propose a model of protein folding that takes these observations into account. This model has been used to analyse chaperone directed protein folding [27]. The consequences of this model are analysed, they reveal that a molecular chaperone can be misfolded and propagate misfolding in the cell. In such a circumstance the molecular chaperone behaves as an infectious agent that uses the cell components to replicate a misfolded conformation.

Thermo-kinetic analysis of protein folding

A. Analysis of the starting hypothesis

The thermo-kinetic model I propose is based on three implicit hypotheses. I think it was better to analyze these hypotheses at the light of present experimental knowledge before developing the model.

Hypothesis I

The folding is a kinetic process.

Correct folding is generally believed to lead to the most stable protein [23], however there is no real proof that this is always the case [18]. On the contrary, it has been shown that some proteins have only kinetic stability [23]. For example, the conformation of folded prosubtilisin is unchanged by removal of a 77-residue N-terminal segment, but if this segment is removed before folding a new conformation is adopted [45]. Thus mature subtilisin has only kinetic stability. A comparable result has been reported recently for the alpha-lytic protease; Baker et al. [2] have shown that this protein folds correctly only in the presence of a pro-region (which may be removed after folding) but stays stable with no detectable interconversion in its abscence. The authors conclude that either the native state or an intermediate state is kinetically trapped [2]. Metastable conformations do not violate kinetic theories of protein folding [4], and the very existence of molecular chaperones is a strong argument for such metastable conformation. Proteins which require a molecular chaperone fold to an off-pathway and precipitate, in the abscence of a chaperone [13]. This implies that the true pathway is a kinetics of folding blocked by an energy barrier which is removed by the molecular chaperone, much like enzyme catalysis. Furthermore, the very existence of intermediates of folding [23] shows that at least the approach to the native structure takes a path that is kinetically driven. These observations make legitimate use of kinetic theories of protein folding to analyze the mechanisms involved in protein folding. However, it is clear that most proteins are in their most stable conformation. Thus, a good model based on the hypothesis of kinetic folding should lead to the most stable conformation in the majority of cases. This will be discussed below, after the development of the thermo-kinetic model.

Hypothesis II

Secondary structures fold more rapidly than tertiary ones.

It is generally accepted that substantial secondary structure appears during the very early stage of folding (perhaps in less than one millisecond), while tertiary structure emerges during the middle stage of folding [29]. Very conclusive results have been obtained by Bycroft et al. [7] that have shown the formation of early native-like structure in the folding of barnase. Data from stopped-flow circular dichroism (CD) are also in agreement with this hypothesis [40]. Further, Ptitsyn et al. [36] proposed that the molten globule forms after the secondary structure.

Finally, this hypothesis is assumed by the famous diffusion-collision model of M. Karplus [4]. Secondary structure formation is so rapid when compared with tertiary structure interaction that equilibrium is reached thus allowing calculation according to statistical mechanics. This may be done according to the model developed by Zimm and Bragg [44].

Hypothesis III

The folding is an irreversible process.

Denaturation or renaturation of most proteins may occur under reversible conditions; by a slow increase (or decrease) of a denaturant. However, when a jump in denaturing condition is applied to the system, or under physiological conditions, the reality of reversibility is not so evident. The criterion of reversibility is the conservation of entropy in a system that does not exchange heat [22]. What happens when a protein folds spontaneously? According to Prigogine [33], the change in entropy of the system dS per time unit dt will be:

$$dS/dt = J_k X_k$$

(Jk represents the generalized flux, Xk the forces). For a chemical reaction this equation becomes, according Prigogine [33]:

$$dS/dt = v(A/T)$$

(v is the speed of folding and A the chemical affinity). Clearly v should be positive (otherwise folding does not occur) and the affinity should be positive to promote this folding. Thus $dS/dt > 0$ implying that the folding is an irreversible process.

I also suppose for the model that this process although irreversible is not too distant from the equilibrium state and thus the analysis may be performed using the linear equation. It is accepted (see Ref. 33 for a discussion) that a chemical reaction could be analyzed by linear approximation if the affinity A is very small compared to RT (i.e., $A \ll RT$; R is gaz constant and T the absolute temperature). In the present model I consider the affinity between two amino acids as a fundamental driving force; this affinity is very low for all pairs of amino acids and thus this criterium of linearity is fulfilled. Furthermore, it is known that linearity often corresponds to maximum efficiency in biological systems [21].

B. Thermodynamic analysis

According the kinetic hypothesis (hypothesis I) discussed above, the protein folding could be analyzed by taking the association between two amino acids as a fundamental process. Thus, one should calculate the speed of interaction between all pairs of amino acids along the protein chain.

Starting with two amino acids (a_i and a_j), according to hypothesis III, one can calculate the association flux (i.e. speed) using the linear thermodynamics of irreversible processes [21, 22, 33] with equation (1).

$$J_a = \sum_b L_{ab} \cdot X_b \qquad (1)$$

This equation is very general. It is therefore necessary to explain the meaning of the generalized forces X_b and the phenomenologic coefficients L_{ab} in the case of a protein.

We are dealing with a chemical reaction so the folding forces are proportional to the affinity A_b and the flux is the association speed V_a [21, 22, 33]. Equation (1) becomes:

$$V_a = \sum_b L_{ab} \cdot A_b \qquad (2)$$

The phenomenological coefficients L_{ab} are very important in this model and have to be defined.

In order to simplify the reasoning, one considers the association of two parallel beta-strands, each containing respectively the amino acids a_i and a_j. Thus amino acid a_i associates with amino acid a_j, the amino acid a_{i+1} will react with amino acid a_{j+1} only if they are in the right conformation. Under these conditions the association speed between a_i and a_j will be increased by that between a_{i+1} and a_{j+1}. This implies that the interference coefficients are proportional to the probability of finding the amino acids in a configuration that allows the interaction. Taking into account hypothesis II, this probability has a physical meaning, it is the probability defined by Boltzman's law. This probability can be calculated using the model of Zimm and Bragg [44]. Thus the reaction flux can be written according to equation:

$$V_{ai \to aj} = \sum_n \left[\prod_m \left(P^{ss}_{ai+m} \cdot P^{ss}_{aj+m} \right) \left(A_{ai+n \to aj+n} \right) \right] \qquad (3)$$

where p^{ss}_{ai} is the probability (according Boltzman's law) that amino acid a_i is in the secondary structure ss allowing interaction; n is the number of amino acids around a_i that influence the speed of association and m should be taken between 0 and n.

It is very easy to calculate that the coefficients, thus defined, satisfy the properties of phenomenological coefficients i.e.:

– $L_{ii} > 0$
– $(L_{ij} + L_{ji})2 < 4(L_{ii} \cdot L_{jj})$
– $L_{ij} = L_{ji}$

Mechanism of chaperone catalyzed folding

One of the major consequences of the thermo-kinetic model is the possibility that proteins fold incorrectly. Indeed the kinetic folding is clearly a probabilistic process giving rise to the possibility of misfolded molecules. These proteins are not necessarily in the most thermodynamically stable conformation, and are not biologically active. Thus, it is necessary for the cell to identify these misfolded proteins in order to try to drive them in the correct conformation. This is the role of the molecular chaperone (such as the hsp70 class) which can recognize a protein in the denatured state [16].

The model presented also suggests that interactions between proteins could compete with protein folding. In this hypothesis, the protein cannot be considered as an isolated system. The major consequence arising from this hypothesis is that a protein can use an other protein to help its own folding: the chaperone catalysed folding.

Among the molecular chaperons, chaperonins (proteins from the hsp60 class) are involved in guiding the folding of other proteins [32, 38, 41]. There is now firm evidence that chaperones – at least of the hsp60 type – facilitate the productive folding of several proteins. Initially, refolding intermediates associate with the chaperonin. The rapid binding competes with non-productive reactions, such as aggregation [5].

I have analyzed this phenomenon using the thermo-kinetic model of protein folding. The thermodynamics of irreversible processes indicate that the speed of folding determines the final interactions between the amino acids. Generalization to folding catalyzed by molecular chaperones is straightforward. Let us consider the case of an interaction between two molecules (an unfolded molecule I, and a folded molecule J: the molecular chaperone). The speed of interaction between two amino acids, one (a_i) from molecule I and the other (a_j) from the molecule J is governed by the relation (3). But under these hypotheses, since molecule J (the molecular chaperone) is folded, this implies that $p_{aj+m}^{ss} = 1$ for all m. Because molecule I cannot fold without a molecular chaperone, its p_{ai+m}^{ss} is very low for all m. Therefore, spontaneous correct folding does not occur, and the folding speed is governed by the molecule J. On the other hand, if there is no molecular chaperone the folding could be driven in another direction leading to misfolding or aggregation. Thus the molecular chaperone is present to avoid a misfolding of the protein. This explains the results obtained by Landry and Gierasch [24] that a polypeptide which is unstructured if free in aqueous solution, adopts an alpha-helical conformation upon binding to GroEL. Clearly the peptide presents a p^{ss} very low for alpha-helical structure but is stabilized when it interacts with the chaperonin. It has been proved that an oligopeptide, otherwise found in random coil conformation in solution, assumes an alpha-helical structure upon binding to GroEL (hsp60) and is maintained in extended conformation upon binding to DnaK (hsp70) [25].

The role of ATP in this reaction is probably to change the conformation of chaperone to detach it.

Consequences of misfolded chaperone

As stressed above, a major consequence of the thermo-kinetic model is the possibility that misfolded proteins which arise accidentally may exist. This means that in a population of protein a tiny part could be misfolded. For a common protein the only consequence will be a loss of the biological activity. The protein will probably be rapidly eliminated by the cell. On the other hand, if this misfolded protein is a molecular chaperone that catalyses the folding of other proteins, the outcome could be of importance. I have analyzed the consequences of an accidental misfolding of a chaperone on the folding of the forthcoming proteins.

The thermodynamics of irreversible processes indicates that the speed of folding determines the final interactions between the amino acids (see relation (3)). Consider the case of an interaction between two mlecules (an unfolded molecule I, and a folded molecule J: the molecular chaperone) of the same kind of protein. The speed of interaction $V_{ai \rightarrow aj}$ between two amino acids, one (a_i) from molecule I and the other (a_j) from the molecule J is governed by the relation:

$$V_{ai \rightarrow aj} = \sum_n \left[\prod_m \left(P^{ss}_{ai+m} \cdot P^{ss}_{aj+m} \right) \left(A_{ai+n \rightarrow aj+n} \right) \right] \qquad (4)$$

As molecule J, the molecular chaperone, is folded, $p^{ss}_{aj+m} = 1$ for all m but; as molecule I cannot fold without a molecular chaperone, its p^{ss}_{aj+m} is very low for all m. Hence, spontaneous correct folding does not occur, and the folding speed is governed by the molecule J. Thus, the conformation of a_i and its neighbors should be that of the corresponding residues a_j and its neighbors, otherwise the affinity $A_{ai \rightarrow aj}$ is very low and interaction cannot occur. Therefore, if molecule J is misfolded, then molecule I will also be misfolded so that interaction between amino acids occurs, and the new misfolded molecule should have the same structure as its molecular chaperone. This result implies that a structural information can be transmitted by proteins. Thus a kind of biological information can be transmitted, under specific circumstances, without the intervention of nucleic acid. Therefore, the structure of a protein is informative in itself.

Furthermore, a chaperone can catalyze its own folding (I call these molecules auto-chaperone). Such chaperone folding depending on itself has been described [9]. Under these conditions the misfolded chaperone should invade the cell. This situation has been analyzed.

Characteristics of the auto-chaperone misfolding invasion

According to the deductions presented above, a model can be developed to simulate auto-chaperone misfolding invasion. The conditions are that the chaperone is involved in its own folding (auto-chaperone) as it is the case for chaperonin [9].

Consider the number Mt of misfolded molecular chaperones at time t when a new molecule is synthesized. This number will depend on the number of misfolded chaperones at time t − 1 (called M_{t-1}). Furthermore, there is a probability (very low but not zero) of spontaneous misfolding (if folding of the chaperone is not aided) noted P_i (it is easy to demonstrate that P_i is directly proportional to p^{ss} of equation (3)). Otherwise, the folding will be performed on a molecular chaperone and the probability of misfolding is $(M_{t-1}/(F_{t-1} + M_{t-1}))$, where F_{t-1} is the number of correctly folded molecules ($F_0 = 100\%$ at time t0). The elimination of the misfolded chaperone, as a physiologic degradation of the protein, may also be calculated. The simplest relationship of this degradation is kM_{t-1} (k is a constant). This analysis can be summarized by:

$$M_t \leftarrow M_{t-1} + P_i + [M_{t-1}/(F_{t-1} + M_{t-1})] - (kM_{t-1}) \qquad (5)$$

The simulation was performed with a computer by iteration. Ten different simulations were performed in order to show the probabilistic process equivalent to a sporadic apparition (Fig. 1). The number of iterations is proportional to time. Figure 1A depicts this sporadic apparition of misfolded chaperones after a more or less long lag period. According to the model, point mutations change the value of p^{ss} and thus can favor misfolding by increasing P_i. The calculations simulate this phenomenon (Fig. 1B). Increasing the probability P_i four times results in reduction of the invasion lag period of about a third. External addition of M (misfolded auto-chaperone) is equivalent to an infection. The results of this simulation are presented in Fig. 1C model such an infection. After injection of M (arrow) the invasion appears very rapidly, with a short lag period.

These results demonstrate that, under specific circumstances, a kind of biological information can be transmitted directly by proteins without the intervention of nucleic acid. Hence, the structure of a protein is

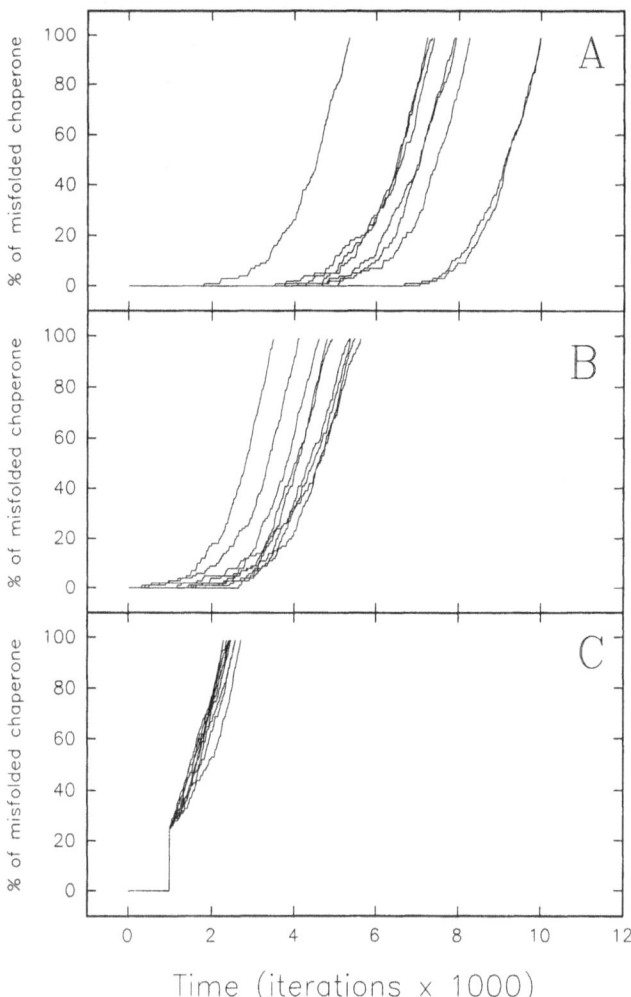

Fig. 1. Computer simulation of the invasion by a misfolded chaperone. According to the model presented (equation (5)) the appearance of the misfolded molecule depends first of a spontaneous misfolding probability (Pi) and second from the probability to meet a misfolded chaperone instead of a correctly one. The invasion is simulated by analysing the folding behaviour of each new synthesised molecule with computer. Time is proportional to the number of iterations ×1,000. Invasion is given as a percentage of misfolded protein. **A** Typical results for the simulation of the sporadic invasion. In **B** the value of Pi is four times the one used in **A** in order to simulate genetic susceptibility. Under these conditions invasion occurs earlier. In **C** misfolded molecular chaperones are added at time 1 iteration (×1,000) and simulation performed using the parameter of **A**. Complete invasion is very rapid after the addition of misfolded chaperone

informative by itself. Such proteins could be regarded as infectious organisms, replicating within a cell using, as do viruses, the cell's own components. Misfolding can be considered as a "structural mutation" that propagates in the cell.

Features of prion diseases that tally with the model

The characteristics of the invasion by a misfolded auto-chaperone resemble those exhibited by prion disease. A direct comparison shows that the model could explain most of the feature of this disease.

The features of the diseases resulting from the chaperone model are the following:

– Localization is mainly in non-dividing cells (brain) or in organs that cannot eliminate the rogue chaperone (k must be small).
– Sporadic appearance, this is a consequence of the Boltzman's law governing the thermodynamical probability of a conformation in the protein folding (parameter P_i).
– Genetic transmission of succeptibility results from the increased probability of misfolding associated with a different amino acid (parameter P_i).
– Punctual mutation of the chaperone which may be at the origin of the heritability by increasing the probability of spontaneous misfolding (parameter P_i). Different mutations will give different speeds of invasion (governed by P_i).
– It exists a possibility of infection if the rogue chaperones manage to enter the cells or the organ considered (the seeding effect that increases M_t).
– All increases in synthesis will decrease the length of the sporadic appearances of the disease (the number of "iterations" increasing per time unit).

Are prions molecular chaperones?

To validate the model, it is necessary to prove that prion proteins are really chaperones. The best known molecular chaperones are the heat-shock proteins. The evidence that prions could be molecular chaperones is two-fold. First, molecular chaperones, as proteins, can be denatured, and prions lose their infectivity after denaturation [34] implying that the tertiary structure is of importance. Second, sequences and structure homology between prion molecules and genuine molecular chaperones can be shown by sequence analysis.

1. Sequence similarity

There are some structural similarities between prions and molecular chaperones. Sequences have been examined for convergence by studying

```
Glycine        **    **   **   **   **   **   **   **
SSA1           KLYQGAGGAPGGAAGGAPGGFPGGAPPEAEGPTVEEV
                 **    **   **   **   **   +-   *+   --
hsc70            GGMPGGMPGGFPGGGAPPS-GGASSGPTIEE
                 **    **   **   **   ++   **   -*   --
hsp70          MGGGDGPGGMPEGMPGGMPGGMPGGMGGGM-GGASSPK
                 **    **   -*   **   **   **   **   **   -+
prion          RPKGGWNTGGSRYPGQG-SPGGNRYPPQGGTWGQPHGGG
                 **    **   -+   *-   **   -+   **   *-   **
prion          WGQPHGGW-GQPHGGSWGQPHGG-GWGQ-GGTHNQW
                 -*    **   *-   **   *-   **   -*   **   --
prion          KHVAGAAAAGAVVGG-LGGYMLGSA-MSRPMIHFGNDWE
                 *-    *-   **   **   -*   --   +-   -*
```

Fig. 2. Comparison of prion and heat shock proteins of the hsp70 family. Evidence for the same repetition of glycines. When prions are analysed for repetitions by Fourier transform analysis [31], a striking repetition of glycines is evident. Analysis of hsp reveals comparable repeats in some of them. The figure presents the alignment of three consecutive repetitions of mouse prion with those found in hsp70, hsc70 and SSA1

the periodicities in the sequences by Fourier transform analysis [31]. The results show that prion, hsp70, hsc70 and SSA1 all have strikingly similar glycine repeat sequences (Fig. 2). Furthermore, glycine is a very mobile amino acid, with all its conformations having about the same energy [18]. Hence these sequences have a low propensity to adopt a definite secondary structure. It is clear, however, that this similarity is not statistically significant.

2. Secondary structure analysis

Secondary structure analysis has been performed on the prion sequence. Three methods have been used. The GOR method [15] and the GCR method [19] give a probability of structure and the hydrophobic moment (mH) [12] is more related to the actual structure of the protein. The results of this analysis are presented in Fig. 3. One can conclude that, besides some unstructured regions, the protein exhibits two alpha-helix and seven beta-sheets (Fig. 3).

3. Structural similarity with hsp70

Because the comparisons of sequences have revealed a similarity between hsp70 (a genuine chaperone) and the prion, I have compared the predicted secondary structure of prion with the one of hsp70. Hsp70 comprises two domains. The first has crystalised and its 3D structure determined, it is homologous to actin [14]. The second has been carefully examined and found to be homologous to HLA of which 3D structure

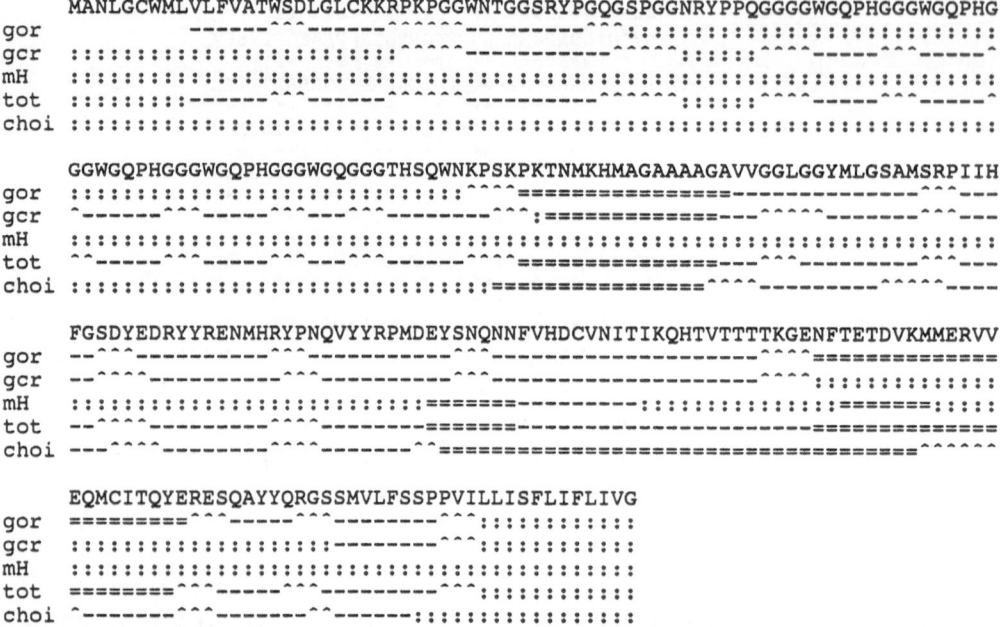

```
     MANLGCWMLVLFVATWSDLGLCKKRPKPGGWNTGGSRYPGQGSPGGNRYPPQGGGGWGQPHGGGWGQPHG
gor        -------^^^-------^^^^----------^^^:::::::::::::::::::::::::::::::::
gcr  :::::::::::::::::::::::::^^^^-----------^^^^^:::::::^^^_____^^^_____^
mH   :::::::::::::::::::::::::::::::::::::::::::::::::::::::::::::::::::::::::::
tot  :::::::::::------^^^-------^^^^----------------^^^^^:::::::^^^_____^^^_____^
choi :::::::::::::::::::::::::::::::::::::::::::::::::::::::::::::::::::::::::::::
```

```
     GGWGQPHGGGWGQPHGGGWGQGGGTHSQWNKPSKPKTNMKHMAGAAAAGAVVGGLGGYMLGSAMSRPIIH
gor  :::::::::::::::::::::::::::::^^^^===============---------------^^^---
gcr  ^-------^^^-----^^^---^^^--------^^^:===============---^^^^------^^^---
mH   :::::::::::::::::::::::::::::::::::::::::::::::::::::::::::::::::::::::::
tot  ^^-----^^^-----^^^---^^^------===============---^^^---------^^^---
choi :::::::::::::::::::::::::::::;===============^^^^---------^^^^----
```

```
     FGSDYEDRYYRENMHRYPNQVYYRPMDEYSNQNNFVHDCVNITIKQHTVTTTTKGENFTETDVKMMERVV
gor  --^^^-------^^^-----------^^^------===============
gcr  --^^^----------^^^----------^^^----------------^^^^:::::::::::::::
mH   :::::::::::::::::::::=======---------:::::::::::::::=======:::::
tot  --^^^-------^^^-----======---------------=============
choi ---^^^--------^^^------^^=====================================^^^^
```

```
     EQMCITQYERESQAYYQRGSSMVLFSSPPVILLISFLIFLIVG
gor  =========^^^-----^^^---------^^^::::::::::::::
gcr  ::::::::::::::::::::------^^^:::::::::::::
mH   ::::::::::::::::::::::::::::::::::::::::::::
tot  ========^^^-----^^^---------^^^::::::::::::::
choi ^--------^^^-------^^------:::::::::::::::::
```

Fig. 3. Secondary structure analysis of the prion molecule. The methods used are those of Garnier et al. [15] (gor), Gibrat et al. [19] (ggr) and Einsenberg et al. [12] (mH). choi represents the selected structure for model building. ---: beta-sheet; ^^^: turn or coil; ===: alpha-helix; ::::: non predicted

has been determined [3]. The 3D structure has been deduced [37]. This structure comprises two alpha-helix and eight beta-sheets. The distribution of secondary structure resembles greatly to that found for prions. We (with D. Gregut and L. Chiche) are using this observation to build a 3D computer model of the prion protein (complete analysis of this molecular model will be presented elsewhere).

4. Analysis of the consequences predicted by this model

Prediction of the position of the mutations. Amyloid deposits specific of prion diseases contain the prion in the beta-sheet conformation. According to the chaperone model, the change in secondary structure would be the origin of this misfolding implying that the alpha-helix are destroyed. In the genetic appearance of disease this hypothesis implies that the mutations should be in the alpha-helix and that they favor other conformations (i.e., beta-sheet or random coil).

There are numerous mutations associated with prion disease [8]. Mutations M109 → F, A117 → V, D177 → N, F197 → S, E 199→ K fall in one of the two alpha-helix and each mutation decreases the alpha-

Table 1. Prion mutations that lead to disease are less favourable for alpha-helix formation

	First alpha-helix (102–118)	Second alpha-helix (168–203)
Mutation	M109 → F	D177 → N
P(alpha)[a]	1.2 → 1.12	0.98 → 0.73
P(beta)[a]	1.67 → 1.28	0.80 → 0.65
P(coil)[a]	0.61 → 0.81	1.09 → 1.33
Mutation	A117 → V	F197 → S
P(alpha)[a]	1.45 → 1.14	1.12 → 0.79
P(beta)[a]	0.97 → 1.65	1.28 → 0.72
P(coil)[a]	0.66 → 0.66	0.81 → 1.27
Mutation	P102 → L	E109 → K
P(alpha)[a]	0.59 → 1.34	1.53 → 1.07
P(beta)[a]	0.62 → 1.22	0.26 → 0.74
P(coil)[a]	1.45 → 0.66	0.87 → 1.05

[a] The amino acid propensity to form alpha-helix is analysed by comparing the value proposed by Chou and Fassman [10]

helix formation (Table 1). However, there are two exceptions the P102 → L mutation that is in the alpha-helix but the mutation should favor alpha-helix (Table 1) and A136 → V that is in the beta-sheet domain. Concerning P102 → L it should be specified that this P102 is directly in contact with the solvent (water) at the outer part of the helix. The presence of a very hydrophobic residue (such as leucine) in this position should greatly disturb the folding. Furthermore, the presence of a proline at the beginning of the first alpha-helix seems to be constant since it is found in the HLA as well as in the hsp70. The case of A136 is not presently explained by the model; however it seems that this mutation is different from the other because only the homozigots show symptoms.

Experimental results and the chaperone model of prion invasion

Prusiner et al. [35] have demonstrated that transgenic mice expressing prion proteins exhibited distinct incubation times depending of the quantity of mRNA synthetized. Furthermore, they showed that the prion inoculum dictates which prion is synthetized de novo. These results are well explained by the chaperone model that correlates the synthesis with the early appearance and that show that a direct interaction between the proteins is necessary.

Unlike several other well known molecular chaperones there is no evidence of an ATP binding site in the prion molecule. However, at least two genuine chaperones function without ATP: the prokaryotic SecB [17] and the eukaryotic hsp90 [43].

Prion proteins have been shown to be membrane proteins [39]. This is not incompatible with a chaperone function since membrane-bound molecular chaperones have been described [11].

Recently, Büeler et al. [6] have shown that mice lacking the cell-surface protein PrP (prion) exhibit normal development and behavior. This finding is not incompatible with a chaperone activity, indeed HtpG (hsp90) can be deleted without noticeable effects on bacterial growth [17]. If one supposes that the major role of chaperone is to increase the speed of folding or/and the rate of correct folding, it seems natural that animals survive without certain chaperones.

The dogma of molecular biology implies that the only molecule able to transport information is nucleic acid. A recent review from Weissmann [42] tries to reconcile the experimental results and the dogma by hypothysing a nucleic acid involved in the prion diseases. In the chaperone model presented above there is no need to suppose the existence of a nucleic acid, and the explanation does not violate the laws of biology or of physics. However, they evoke a new concept: chaperone are proteins able to transport a kind of information.

Are there other diseases of which etiology could be explained by the misfolded chaperone model?

There are numerous different chaperones, existing in all compartments of the cells, that still remain to be identified and characterised. The model presented implies that they may all be targets of chaperone diseases. The characteristics of these diseases can be predicted by the model. One of these is the formation of insoluble aggregates if the misfolded chaperone cannot be eliminated. The amyloid diseases should be considered. They appear generally after an accident that results in an increase in synthesis [30], but they are often sporadic [20] or genetic [1, 20].

A special attention should be paid to Alzheimer's disease. This widely spread disease shares most of the feature depicted by chaperone model. In a recent paper I have shown that the β-amyloid precursor depicts characteristics of molecular chaperone [28].

References

1. Almeida MR, Longo AI, Sakaki H, Costa PP, Saraiva MJM (1990) Prenatal diagnostic of familial amyloidotic neuropathy. Hum Genet 85: 623–626
2. Baker D, Sohl J, Agard DA (1992) A protein-folding reaction under kinetic control. Nature 356: 263–265
3. Bjorkman PJ, Saper MA, Samraoui B, Bennett WS, Strominger JL, Wiley DC (1987) Structure of the human class I histocompatibility antigen, HLA-A2. Nature 329: 506–512
4. Brooks C, Karplus M, Pettitt M (1988) Proteins: a theoretical perspective of dynamics, structure, and thermodynamics. Wiley, New York
5. Buchner J, Schmidt M, Fuchs M, Jaenicke R, Rudolph R, Schmid FX, Kiefhaber T (1991) GroE facilitates refolding of citrate synthase by suppressing aggregation. Biochemistry 30: 1586–1591
6. Büeler H, Fischer M, Lang Y, Bluethmann H, Lipp H-P, DeArmond SJ, Prusiner SB, Aguet M, Weissmann C (1992) Normal development and behaviour of mice lacking the neuronal cell-surface PrP protein. Nature 356: 577–582
7. Bycroft M, Matouschek A, Kellis JT, Serrano L, Fersht AR (1990) Detection and characterization of a folding intermediate in barnase by RMN. Nature 346: 488–490
8. Carlson GA, Hsiao K, Oesch B, Westaway D, Prusiner SB (1991) Genetics of prion infections. Trends Genet 7: 61–65
9. Cheng MY, Hartl F-U, Horwich AI (1990) The mitochondrial chaperonin hsp60 is required for its own assembly. Nature 348: 455–458
10. Chou PY, Fasman G (1974) Conformational parameters for amino acids in helical, β-sheet, and random coil regions calculated from proteins. Biochemistry 13: 211–245
11. Degen E, Williams D (1991) Participation of a novel 88-kD protein in the biogenesis of murine Class I histocompatibility molecules. J Cell Biol 112: 1099–1115
12. Eisenberg D, Weiss RM, Terwilliger TC (1982) The helical hydrophobic moment a measure of the amphilicity of a helix. Nature 299: 371–374
13. Elis RJ (1987) Proteins as molecular chaperones. Nature 328: 378–379
14. Flaherty KM, DeLuca-Flaherty C, McKay DB (1990) Three-dimensional structure of the ATPase fragment of a 70 K heat-shock cognate protein. Nature 346: 623–628
15. Garnier J, Osguthorpe DJ, Robson B (1978) Analysis of the accuracy and implications of simple methods for predicting the secondary structure of globular proteins. J Mol Biol 120: 97–120
16. Gething M-J, Sambrook J (1992) Protein folding in the cell. Nature 355: 33–45
17. Georgopoulos C (1992) The emergence of the chaperone machines. Trends Biochem Sci 17: 295–299
18. Ghelis C, Yon J (1982) Protein folding. Academic Press, New York
19. Gibrat JF, Garnier J, Robson B (1987) Further develoments of protein secondary structure prediction using information theory. New parameters and consideration of residue pairs. J Mol Biol 198: 425–443
20. Grateau G (1992) Les amylose héréditaires. Medecine/Sciences 6: 524–531
21. Jou D, Llebot JE (1991) Introduction a la thermodynamique des processus biologiques. Techniques & Documentation-Lavoisier, Paris
22. Katchalsky A, Curran PF (1965) Nonequilibrium thermodynamics in biophysics. Harvard University Press, Cambridge Massachussets

23. Kim PS, Baldwin RL (1990) Intermediates in the folding reactions of small proteins. Annu Rev Biochem 59: 631–660
24. Landry SJ, Gierasch LM (1991) The chaperonin GroEL binds a polypeptide in an alpha-helical conformation. Biochemistry 307: 359–7362
25. Landry SJ, Jordan R, McMachen R, Gierash L (1992) Different conformation for the same polypeptide bound the chaperone DnaK and GroEL. Nature 355: 455–457
26. Liautard JP (1990) A thermo-kinetic model for protein folding. CR Acad Sci 311: 385–389
27. Liautard JP (1991) Are prions misfolded molecular chaperones? FEBS Lett 294: 155–157
28. Liautard JP (1993) A new theoritical model of protein folding could explain the etiology of degenerative encephalopathy. In: Siest G (ed) Biology prospective. John Libbey, London (in press)
29. Matthews RC (1991) The mechanism of protein folding. Curr Opin Struct Biol 1: 28–35
30. Maury CPJ (1990) β-microglobulin amyloidosis. Rheumatol Int 10: 1–8
31. McLachlan AD (1976) Periodic fearures in the amino acid sequence of nematode myosin rod. J Mol Biol 103: 271–298
32. Osterman J, Horwich AL, Neupert W, Hartl F-U (1989) Protein folding in mitochondria requires complex formation with hsp60 and ATP hydrolysis. Nature 341: 125–130
33. Prigogine I (1968) Introduction à la thermodynamique des processus irreversibles. Dunod, Paris
34. Prusiner SB, Groth DF, Bolton DC, Kent SB, Hood LE (1984) Purification and structural studies of a major scrapie prion protein. Cell 38: 127–134
35. Prusiner SB, Scott M, Foster D, Pan KM, Groth D, Mirenda C, Torchia M, Yang SL, Serban D, Carlson GA, Hoppe PC, Westaway D, DeArmond SJ (1990) Transgenic studies implicate interactions between homologous PrP isoforms in scrapie prion replication. Cell 63: 673–686
36. Ptitsyn OB, Pain RH, Semisotnov GV, Zerovnik E, Razguliaev OI (1990) Evidence for molten-globule state as a general intermediate in protein folding. FEBS Lett 262: 20–24
37. Rippmann F, Taylor WR, Rothbard JB, Green M (1991) A hypothetical model for the peptide binding domain of hsp70 based on the peptide binding domain of HLA. EMBO J 10: 1053–1059
38. Schmid FX (1991) Catalysis and assistance of protein folding. Curr Opin Struct Biol 1: 36–41
39. Stahl N, Borchelt DR, Hsiao K, Prusiner S (1987) Scrapie prion protein contain a phosphatidylinositol glycopipid. Cell 51: 229–240
40. Sugawara T, Kuwajima K, Sugai S (1991) Folding of staphylococcal nuclease A studied by equilibrium and kinetic circular dichroism spectra. Biochemistry 30: 2698–2706
41. Vitanen PV, Lubben TH, Goloubinoff P, O'Keefe PO, Lorimer GH (1990) Chaperonin-facilited refolding of ribulose-bisphosphate carboxylase and ATP hydrolysis by chaperonin 60 (GroEL) are K+-dependent. Biochemistry 29: 5665–5671
42. Weissmann (1991) A "unified theory" of prion propagation. Nature 352: 679–683
43. Wiech H, Buchner J, Zimmermann R, Jakob U (1992) Hsp90 chaperones protein folding *in vitro*. Nature 358: 169–170

44. Zimm BH, Bragg JK (1959) Theory of the phase transition between helix and random coil in polypeptide chains. J Chem Phys 31: 526–532
45. Zhu X, Ohta Y, Jordan F, Inouye M (1989) Pro-sequence of subtilisin can guide the refolding of denatured subtilisin in an intermolecular process. Nature 339: 483–484

Author's address: Dr. J.-P. Liautard, INSERM U-65, Departement Biologie Santé, Case Postale 100, Université de Montpellier II, Place E. Bataillon, 34095 Montpellier, France.

Arch Virol (1993) [Suppl] 7: 245–254

Epidemiology of bovine spongiform encephalopathy and related diseases

J.W. Wilesmith

Epidemiology Department, Central Veterinary Laboratory, New Haw, Addlestone, Surrey, U.K.

Summary. The occurrence of bovine spongiform encephalopathy in Great Britain, first detected in 1986, has necessarily stimulated a large research programme. Encompassed in this are studies of the epidemiology of the disease. These commenced in 1987 and are continuing. The ensuing results have been reported and reviewed during the course of the epidemic [13, 17, 20]. This paper provides a brief overview of the results of the epidemiological studies, but concentrates on the more recent features of the epidemic which are important in assessing the future course of the disease.

The beginnings of the epidemic

The initial cases of BSE were detected by histopathological examinations of brains submitted by practising veterinary surgeons in November 1986 and included in the routine surveillance of animal diseases in Great Britain. Information on the clinical signs associated with the disease was given to the veterinary profession to stimulate the voluntary notification of cases so that clinical observations of suspect cases could be made and collated together with histopathological examination of their brains following their slaughter. As a result a more detailed pathological study and a more complete description of the clinical course of the disease and the clinical signs was formulated. This indicated that histopathological examination of brain was necessary to confirm the diagnosis of BSE [19]. The identification of these cases also provided a basis for the initial epidemiological study to investigate potential aetiological hypotheses and possible sources and vehicles of a scrapie-like agent for cattle. This study included collection of detailed data to provide the descriptive epidemiological features of the disease and had the objective of determining whether the disease was novel and, if so, when the first cases occurred.

In the meantime, histopathological, transmission and molecular biological studies were started. These were designed to confirm or deny the histopathological findings, which were reminiscent of those observed in sheep scrapie. These multi-disciplinary studies provided the convincing evidence that BSE was due to a scrapie-like agent [3, 4, 11].

The initial epidemiological study ruled out other possible aetiologies and provided evidence that cattle had become exposed to a scrapie-like agent via the consumption of meat and bone meal contained in cattle feedstuffs [20]. It also indicated that the disease was new and the first cases, based on clinical histories obtained from herdsmen and their veterinary surgeons, had occurred in April 1985. Another interesting finding was that cases occurred contemporaneously throughout Great Britain [12], but the risk of herds becoming affected was considerably greater in the south of England.

Following these initial studies statutory action was taken to control the disease. On 21 June 1988 BSE became a statutorily notifiable disease, and on 18 July 1988 the inclusion of ruminant derived protein in ruminant rations was prohibited. These control measures were naturally of significance in the epidemiological studies of the disease. Most importantly, a fuller understanding of the true incidence was possible, because all, as opposed to some, cases were then made known to the authorities. Of equal importance, the statutory ban prevented further exposure to the most probable food borne source. This in turn provided the necessary intervention study involving the whole cattle population of Great Britain, enabling the most powerful epidemiological study of the hypothesised source of infection.

Investigations of the source of a scrapie-like agent and reasons for exposure of cattle

The initial epidemiological study provided considerable evidence indicating the start of this novel epidemic in early 1985. Further studies, principally by means of simulation modelling, indicated that exposure of the cattle population to a scrapie-like agent at a level sufficient to cause clinical expression of disease had commenced suddenly in 1981/82 [20]. This provided a starting point to determine the reasons for exposure of the cattle population in Great Britain to a scrapie-like agent at this time.

The inclusion of meat and bone meal in cattle rations was not novel, and attention was therefore turned to the processes used in the production of meat and bone meal together with the changes in these processes over time. Preliminary investigations had identified two major changes in the production process. One was a change from batch to continuous rendering. The second was a reduction in the use of hydrocarbon solvents to extract fat from the intermediate product, greaves, in the production of meat and bone meal.

These changes *inter alia* were investigated by a detailed survey of the rendering plants still in operation in the autumn of 1988. In summary, this survey revealed that the initial estimate of the time of exposure of the cattle population, sufficient to cause clinical disease, to a scrapie-like agent was associated with the reduction in the use of hydrocarbon solvent extraction of fat in the rendering of animal tissues in the production of meat and bone meal [12, 15].

The results of this survey therefore substantiated the original hypothesis with respect to the vehicle for a scrapie-like agent, in addition to supporting the initial estimate of the time of the start of effective exposure of the cattle population. However, the hypothesis that meat and bone meal were the source of infection was investigated formally by means of a case-control study. This essentially compared the exposure of BSE affected animals to meat and bone meal in their first year of life with that for animals which did not develop BSE and which were born in the same calving season, in 1983/84. The feeding of proprietary concentrates containing meat and bone meal was found to be a statistically significant risk factor for the occurrence of BSE [16]. This study therefore provided further supporting evidence that BSE occurred as a result of exposure to a scrapie-like agent via meat and bone meal.

The conclusive testing of the hypothesis will be provided by the detailed monitoring of the epidemic following statutory action to prevent further exposure of cattle to the food borne source from July 1988. The results of this monitoring with respect to the expected effects on the descriptive epidemiology are discussed below. However, monitoring provided evidence of a significant increase in the incidence of BSE, which began during the latter six months of 1989. This required further investigations and analyses to determine whether the increase was real, rather than an unexpected change in the case ascertainment rate. The investigations revealed that the increase in incidence was real and that it had occurred uniformly and simultaneously throughout Great Britain [13]. The increase, however, could not be attributed to the horizontal transmission of BSE between cattle as the within-herd incidence of BSE did not increase significantly after the increase in the national incidence of cases [17]. This significant increase in incidence was not entirely unexpected, as the probable epidemiological consequences arising from the inclusion of tissues from BSE infected cattle in meat and bone meal had been identified early in the course of the epidemic [20]. Further investigations and studies have revealed that recycled infection from cattle was responsible for the increase in incidence, and that the risk of exposure for cattle from infected cattle tissues commenced in 1984 and continued until the introduction of the legislation in July 1988 [19].

In summary, the epidemic has been entirely from the food borne source, meat and bone meal. The initial epidemic is most likely to have been from exposure to remnant tissues of pre-clinical, scrapie-infected sheep carcases following normal slaughtering and butchery, which were rendered and included in meat and bone meal. This primary exposure was then enhanced, by infected material, from pre-clinical, BSE-infected cattle in the same vehicle. In other words an epidemic from the cattle source has been superimposed on the epidemic from the original sheep source.

Clinical and descriptive epidemiological features of BSE

The clinical features of the disease as discerned by herdsmen and their veterinary surgeons have been described in detail previously [20].

Attention has also been paid to possible changes in the clinical features as a result of the exposure of cattle to BSE-infected cattle tissues, which is akin to experimental passage of the scrapie-like agent. The results of analyses so far indicate that no major changes in the clinical picture have occurred over time [18].

The initial signs are usually non-specific, but the eventual predominant neurological signs remain as apprehensive behaviour, hyperaesthesia and ataxia of gait. These signs are accompanied in the majority of cases by loss of general body condition and reduced liveweight, and diminished milk yield. Pruritus does not occur as commonly as in sheep scrapie. This constellation of clinical signs is present in the majority of cases 6–8 weeks after onset. The original series of cases indicated that the clinical course ranged from 7 days to 14 months with the majority of animals dying or having to be slaughtered within 4 months. The disease is confined to adult animals with a modal age at onset between 4 and 5 years.

Turning to the basic descriptive epidemiological features of the disease, there is no evidence of any breed or sex predisposition. There are, however, considerable differences in the risk of infection for the two herd production types, dairy and beef suckler herds, and for all herds dependent on their geographical location.

Taking the former difference first, the risk of infection for dairy herds is considerably greater than that for beef suckler herds [17]. This is simply explained by the different feeding practices in these two types of herds. Dairy herds are more reliant on commercial concentrate rations which could potentially contain meat and bone meal. The second element of variation in risk, geographical location, can be explained by three major factors. The first is the use of hydrocarbon solvent extraction in the production of meat and bone meal. This has only continued in two plants in Scotland which have produced the majority of meat and bone meal in this part of Great Britain since the early 1980's, and therefore provides an explanation for the lower risk in Scotland and northern England. The second is that individual manufacturers of concentrate cattle feedstuffs differ in their use and inclusion rates of meat and bone meal and in their share of the market geographically. The effect of this difference is most evident in the comparison of the incidence of BSE on the two principal Channel Islands of the British Isles, Guernsey and Jersey [12]. Thirdly, the survey of rendering plants in Great Britain revealed a geographical variation in the proportion of meat and bone meal produced as a result of reprocessing greaves, requiring a second heat treatment, which was co-incident with the geographical variation in the incidence of BSE [15].

The occurrence of spongiform encephalopathies in other species

Cases of spongiform encephalopathy have now been identified in five other species of the family Bovidae in captive groups of animals in British zoological collections. The first of these cases, in a nyala (*Tragelaphus angasi*), occurred in 1986 [5]. This was initially diagnosed by histopathological examination of the brain and later confirmed as a transmissible spongiform encephalopathy by experimental transmission to mice from fixed brain tissue. This case preceded the recognition of BSE and, with the benefit of hindsight and further epidemiological investigation, was the herald of the epidemic in cattle.

Enquiry revealed that this animal had received meat and bone meal as a constituent of the proprietary concentrate ration fed and therefore appeared to have been exposed from the same source as cattle [20]. In the following year, 1987, spongiform encephalopathy was confirmed in a gemsbok (*Oryx gazella*) in the same wildlife collection as the nyala [5]. This animal also received the same proprietary feedstuff as the nyala [20]. Subsequently, cases of spongiform encephalopathy have occurred in one Arabian oryx (*Oryx leucoryx*) [6], in four eland (*Taurotragus oryx*) [1, 2] and in three greater kudu (*Tragelaphus strepsiceros*) [1, 6, 7].

All of these animals with the exception of one of the greater kudu were fed proprietary cattle feedstuffs containing meat and bone meal and therefore were also exposed to the same source of infection as cattle. There is, however, evidence from one of the cases in a greater kudu, which was the offspring of an affected dam, that, in this species the transmission of infection can occur by means other than the food-borne meal [7]. Further investigations on this affected group, all maintained in one zoological collection, have provided further indications that natural transmission may be possible in this species, at least under the husbandry conditions under which these animals have been maintained. In contrast, no further cases have occurred in the collections of nyala and gemsbok suggesting that greater kudu may be different to other members of the family Bovidae in the epidemiology of spongiform encephalopathy.

In addition to this apparent specific difference, these exotic ungulates have exhibited a number of differences compared to the disease in domestic cattle. First, although there are obvious differences between the management of these animals and of cattle, particularly dairy cows, which could potentially result in an observation bias with respect to the recognition of the onset of clinical signs, the evidence suggests that the clinical duration was considerably shorter than that in cattle, being sometimes days rather than months. Secondly, the incidence of disease was apparently greater in the captive ungulates. Thirdly, the incubation period distribution has a considerably lesser mean and mode, as adjudged by the age at onset of clinical signs. The mean age at onset in cattle has been 60 to 62 months, whereas in the captive ungulates the range at clinical onset was 30 to 38 months.

In addition to these cases in species of the family Bovidae, natural cases of spongiform encephalopathy have also occurred in felidae with the majority occurring in domestic cats [8, 21–23]. The initial case in a domestic cat was confirmed in 1990, in which the first clinical signs had been observed in November 1989. Since then a further 29 cases have been confirmed in domestic cats in the United Kingdom, in which the total population is approximately 7 million. Experimental transmission to mice has been demonstrated by the injection of brain from one of these cats. The other two species of the felidae which have been affected are the puma (*Felis concolor*) [2] and a cheetah (*Acinoryx jubatas*) [10], with one case in each. The former occurred in 1991 in an animal kept in a zoological collection in Great Britain. The latter case occurred in Australia in an imported animal that was born in a British wildlife park. An insufficient number of cases in domestic cats have occurred so far to enable a formal epidemiological study to investigate the likely source. However, case studies of both the domestic cats and the two captive-held species indicate that a food borne source cannot be ruled out nor can the possibility that these cases in felidae were a result of exposure to BSE infected bovine tissue in food. Fortunately the incidence in domestic cats has been very low and the potential risk of infection from commercial petfoods in the United Kingdom was eliminated by a voluntary ban on the inclusion of the specified bovine offals in 1989 which became mandatory in 1990 [9].

Prospects for the future incidence of BSE in Great Britain

The most important factor in determining the prospects of the future incidence of BSE is whether cattle are a dead-end host for the causal scrapie-like agent or whether the disease is capable of being maintained in the cattle population as has occurred in scrapie of sheep. In the case of the latter species, a component of vertical or pseudo-vertical transmission has been postulated. The potential epidemiological role of vertical/pseudo-vertical (or maternal) transmission (that is transmission only from affected dam to its offspring) for BSE is being studied in detail. The results indicate that given the methods of cattle production such maternal transmission could not, by itself, maintain the disease in the cattle population. Essentially, the dairy cattle production system in Great Britain does not result in the necessary contact rate of 1:1 [19, Wilesmith and Ryan, unpubl.]. Maternal transmission alone is probably incapable of maintaining scrapie in sheep either.

Despite this finding, detecting evidence of maternal transmission has not been ignored in the study of the possible dynamics of BSE infection. This is because the occurrence of a degree of maternal transmission would be the component of horizontal transmission which would be discernible first. In other words, the means of transmission could be the same as for sheep scrapie, where the placenta is believed to be a source of infection for other animals in the flock as well as for the ewe's own lamb. This results not only in an apparently elevated incidence in the offspring of scrapie-affected dams, but also a source of infection for unrelated, in-contact sheep.

Two essential studies are therefore in progress on this aspect. One concerns the analyses of the data arising from the monitoring of the epidemic. The other is rather more specific being a cohort study of the incidence of BSE in the offspring of dams which have developed BSE and those of BSE-unaffected animals. The former analyses involve a comparison of the observed incidence of BSE in offspring of confirmed cases of BSE with (the expected incidence from the food borne source alone). The results of these analyses, as at mid-June 1992, do not provide any evidence of a greater risk for offspring. The cohort study comprises some 316 pairs of offspring, each pair consisting of an offspring of BSE-affected dam and an offspring of BSE-unaffected dam. The members of each pair were born in the same calving season and herd. The age of the animals in the study is currently 3 to 5 years. The study is necessarily long-term because of the incubation period distribution associated with the disease. At the time of writing only 3 of the 632 trial animals had developed BSE and, whilst it was not known whether these were offspring of confirmed cases or not (because the trial is blind), all were born before 18 July 1988 and so potential recipients of ruminant protein.

These and other analyses, concerned with identifying the initial signs of cattle to cattle transmission are obviously of fundamental importance. However, at present, the results have failed to reveal any evidence of cattle to cattle transmission, let alone at a level sufficient to maintain the disease in the cattle population of Great Britain.

It has been necessary obviously also to determine the effects of the steps taken to prevent exposure of cattle to the food borne source after 18 July 1988. The most instructive analyses in this respect are the age specific incidences. Currently the incidence in the two year old age class is of interest because during 1991 and subsequently there will be a diminishing number of two year old animals exposed from the common food source. The effect of the intervention should therefore be observed first in this sector of the population. Analyses of the age specific incidences in herds in which cases occurred in homebred animals from January to June 1991 and in the same period in the preceding two years has revealed a reduction in the incidence in the two year old age class [14]. This analysis has been updated for the complete calendar years with the same findings. During 1992 a further reduction in the two year old class is therefore expected together with a reduction in the incidence in the three year old age class. These expected effects are the subject of current analyses.

In addition, cases of BSE which were born after 18 July 1988 are being subject to detailed investigations. These have the objective of assessing the risks of infection from all potential sources. By 1 October 1992, 291 such cases of BSE had been confirmed. All of these animals had a high risk of exposure to meat and bone meal containing ruminant protein as a result of using supplies of feedstuff manufactured before 18 July, but still on farms or in the supply chain. If a ban on feeding meat and bone meal to ruminants had not been introduced on 18 July 1988, then the incidence of BSE in this age group would have been at least 6 times that observed.

The current evidence, at June 1992, is therefore that the expected initial effects of the intervention on the descriptive epidemiological features are apparent. Perhaps more importantly the relevant epidemiological studies have not revealed any evidence of cattle to cattle transmission let alone at a level sufficient to maintain the disease in the cattle population. However, these studies are being subjected to constant review and analysis. The epidemiological findings during 1992 are of considerable interest as the initial effect of the intervention on the national incidence can only be expected to be demonstrable in the latter months of the year. It will therefore be some time before any conclusive remarks on the future incidence can be made.

References

1. Bradley R, Matthews D (1992) Sub-acute, transmissible spongiform encephalopathies: current concepts and future needs. Rev Sci Tech Off Int Epiz 11: 605–634
2. Fleetwood AJ, Furley CW (1990) Spongiform encephalopathy in an eland. Vet Rec 126: 408–409
3. Fraser H, McConnell I, Wells GAH, Dawson M (1988) Transmission of bovine spongiform encephalopathy to mice. Vet Rec 123: 472
4. Hope J, Reekie LJD, Hunter N, Multhaup G, Beyreuther K, White H, Scott AC, Stack MJ, Dawson M, Wells GAH (1988) Brain fibrils of novel British cattle disease contain scrapie associated protein. Nature 336: 390–392
5. Jeffrey M, Wells GAH (1988) Spongiform encephalopathy in a nyala (*Tragelaphus angasi*). Vet Pathol 25: 398–399
6. Kirkwood JK, Wells GAH, Wilesmith JW, Cunningham AA, Jackson SI (1990) Spongiform encephalopathy in an Arabian oryx (*Oryx leucoryx*) and a greater kudu (*Tragelaphus strepsiceros*). Vet Rec 127–420
7. Kirkwood JK, Wells GAH, Cunningham AA, Jackson SI, Scott AC, Dawson M, Wilesmith JW (1992) Scrapie-like encephalopathy in greater kudu (*Tragelaphus strepsiceros*): morbidity in an index case offspring without dietary exposure to ruminant derived protein. Vet Rec 130: 365–367
8. Leggett MM, Dukes J, Pirie HM (1990) A spongiform encephalopathy in a cat. Vet Rec 127: 586–588
9. Order (1990) The bovine spongiform encephalopathy (No 2) (Amendment) Order 1990. Statutory Instrument 1990, No 1930. HMSO, London
10. Peet RL, Curran JM (1992) Spongiform encephalopathy in an imported cheetah (*Acinoryx jubatus*). Aust Vet J 69: 171
11. Wells GAH, Scott AC, Johnson CT, Gunning RF, Hancock RD, Jeffrey M, Dawson M, Bradley R (1987) A novel progressive spongiform encephalopathy in cattle. Vet Rec 121: 419–420
12. Wilesmith JW (1991) Bovine spongiform encephalopathy: epidemiological approaches, trials and tribulations. In: Proceedings of the sixth international symposium on veterinary epidemiology and economics, pp 32–43
13. Wilesmith JW (1992) Epidemiology of bovine spongiform encephalopathy. Semin Virol 2: 239–245
14. Wilesmith JW, Ryan JBM (1992) Bovine spongiform encephalopathy: recent observations on the age-specific incidences. Vet Rec 130: 491–492
15. Wilesmith JW, Ryan JBM, Atkinson MJ (1991) Bovine spongiform encephalopathy: epidemiological studies on the origin. Vet Rec 128: 199–203
16. Wilesmith JW, Ryan JBM, Hueston WD (1992) Bovine spongiform encephalopathy: case-control studies of calf feeding practices and meat and bone meal inclusion in proprietary concentrates. Res Vet Sci 52: 325–331

17. Wilesmith JW, Ryan JBM, Hueston WD, Hoinville LJ (1992) Bovine spongiform encephalopathy: descriptive epidemiological features 1985–1990. Vet Rec 130: 90–94
18. Wilesmith JW, Hoinville LJ, Ryan JBM, Sayers AR (1992) Bovine spongiform encephalopathy: aspects of the clinical picture and analyses of possible changes 1986–1990. Vet Rec 130: 197–201
19. Wilesmith JW, Wells GAH (1991) Bovine spongiform encephalopathy. Curr Top Microbiol Immunol 172: 21–38
20. Wilesmith JW, Wells GAH, Cranwell MP, Ryan JBM (1988) Bovine spongiform encephalopathy: epidemiological studies. Vet Rec 123: 638–644
21. Willoughby K, Kelly DF, Lyon DG, Wells GAH (1992) Spongiform encephalopathy in a captive puma (*Felis concolor*). Vet Rec 131: 431–434
22. Wyatt JM, Pearson GR, Smerdon TN, Gruffyd-Jones TJ, Wells GAH (1990) Spongiform encephalopathy in a cat. Vet Rec 126: 513
23. Wyatt JM, Pearson GR, Smerdon TN, Gruffyd-Jones TJ, Wells GAH, Wilesmith JW (1991) Naturally occurring scrapie-like spongiform encephalopathy in five domestic cats. Vet Rec 129: 233–236

Author's address: Dr. J.W. Wilesmith, Epidemiology Department, Central Veterinary Laboratory, New Haw, Addlestone, Surrey, KT15 3NB, U.K.

Arch Virol (1993) [Suppl] 7: 255–259

Bovine spongiform encephalopathy: a new disease of cattle?

R.F. Marsh

Department of Animal Health and Biomedical Sciences,
University of Wisconsin-Madison, Madison, Wisconsin, U.S.A.

Summary. Bovine spongiform encephalopathy (BSE) was first recognized in Great Britain in 1985. Most believe that the disease is of recent origin initiated by feeding rendered animal protein from scrapie-infected sheep to cattle, then perpetuated by feeding rendered infected cattle to other cattle. This paper explores an alternative hypothesis that BSE existed in cattle populations in an unrecognized form for a much longer time until amplified by changes in the rendering process that allowed cattle to cattle transmission to occur. This viewpoint is supported by observations that transmissible mink encephalopathy, a disease that first occurred 45 years ago, is likely caused by feeding downer cows to mink, and that the sporadic form of Creutzfeldt-Jakob disease occurs spontaneously with no evidence of natural transmission. This epidemiologic scenario on the origin of BSE has important implications for prevention of the disease in BSE-free countries. Mainly, emphasis needs to put on practices of feeding animal protein to cattle rather than in reducing the prevalence of sheep scrapie. If BSE is already present in the cattle population, the major threat becomes feeding cows to cows.

Introduction

"Bovine spongiform encephalopathy (BSE) is a new disease of cattle" [1]. So begins an excellent recent review of BSE written for the International Office of Epizootics. This article thoroughly describes the epidemiology of the disease and the economic impact it has had on Great Britain. But is BSE truly a "new disease" or does it represent an outbreak of a previously unrecognized disease caused by changes in the feeding of animal protein that allowed cattle to cattle transmission to occur? This chapter will review this possibility emphasizing its implication for the prevention of other BSE outbreaks in countries presently thought to be free of the disease.

Origin of the BSE outbreak

Most scientists believe that BSE was initiated by feeding cattle rendered animal protein from scrapie-infected sheep. This is certainly possible, especially in light of recent studies in the United States showing that cattle are quite susceptible to American sources of sheep scrapie (Cutlip et al., in prep.). A "second possibility is that BSE resulted from a similar increase in exposure, but to a strain of a scrapie-like agent which had been previously selected by cattle and passaged in the species. If this hypothesis is correct, there is also the possibility that such subclinical infection of cattle with a scrapie-like agent or, indeed, a very low incidence of clinical BSE is similarly present in other countries" [2].

This statement by Wilesmith and Wells may prove to be prothetic, but with some modification. A scrapie-like disease of cattle need not be "passaged in the species" in order to maintain an endemic state. There is no evidence that the sporadic form of Creutzfeldt-Jakob disease (CJD) is naturally transmissible, yet the disease maintains itself in the human population at a constant incidence of one case per one million population. These affected individuals have no mutation of their prion protein gene as seen in familial cases of CJD [3], but their disease is transmissible to subhuman primates by experimental inoculation. Thus it would appear that sporadic CJD occurs spontaneously, perhaps due to a mutation of a regulatory gene not yet identified, or to a rare event influencing the conformation of the prion protein.

Fortunately, these spontaneous cases of scrapie-like disease are not transmitted to cohorts or offspring. Scrapie in sheep is the only transmissible spongiform encephalopathy known to be naturally transmitted by horizontal means at a relatively high rate. Therefore, these spontaneous cases are insignificant unless an affected human brain is ingested by the unnatural act of cannibalism, or a bovine brain ingested by other cattle or by a mink.

Transmissible mink encephalopathy

Transmissible mink encephalopathy (TME) is a rare disease of ranch-raised mink that has been observed in the United States, Canada, Finland, Russia, and Germany [4]. Because of the similarity between TME and scrapie, it was first thought that TME was caused by feeding mink scrapie-infected sheep. However, testing of oral susceptibility of mink to several American sources of sheep scrapie have been unsuccessful. Furthermore, epidemiologic investigations of TME have not revealed consistent evidence that sheep tissues were fed to the mink.

There are two incidents of TME, one in Canada in 1963 [5] and one in the United States in 1985 [6], where the ranch operators were certain that sheep had not been fed. The latter incident is especially interesting because this rancher was a "dead stock" feeder using mostly downer dairy cows for the meat portion of the diet.

Cattle tissues are commonly fed to mink on most ranches worldwide. Other diseases mink have contracted from eating contaminated cattle include anthrax, botulism, black leg, brucellosis, tuberculosis, and urinary lithiasis (diethylstilbestrol).

Risk assessment and prevention of BSE

Some previous risk assessments for the possible occurrence of BSE in the United States have only considered feeding rendered animal protein from *sheep* to cattle [7]. This risk factor is low in the United States mainly because of action taken by the Animal Protein Producers Industry (APPI) of the National Renderers Association, Inc. The APPI has made the following recommendations concerning sheep carcasses [8]:

1. Do not pick up diseased, dying, disabled and dead sheep until further notice.
2. Lambs may be processed from a certified scrapie-free flock, but animals more than 1 year of age should be avoided since they represent a greater risk.
3. Rendered sheep offal should be diverted to feeds other than dairy or beef cattle feed.

These are sound recommendations that should be implemented in all countries having scrapie-infected sheep. But what of the risk of feeding rendered *cattle* protein back to other cattle, a practice that varies from country to country and even has significant regional differences within a country? Rendered products tend to be used locally because of the high cost of transportation. Therefore, poultry are more likely to be fed back to other poultry in the southeastern United States, swine to swine in parts of the midwest, and cows to cows in Wisconsin [9].

Every country needs to evaluate their own use of animal protein. Even countries free of scrapie, such as Australia and New Zealand, remain at risk because of the possibility of spontaneous cases of BSE-like disease.

The evidence is convincing that, whatever the source of BSE, the major reason for the outbreak was changes in the rendering process beginning in the later part of the 1970s [2]. The change from batch to continuous processes and the reduction in the use of hydrocarbon solvent exaction are thought to have allowed the increased survival of a scrapie-like agent capable of infecting cattle. Studies are presently underway to investigate this possibility and to determine the most optimal rendering conditions to assure the greatest degree of BSE and scrapie agent inactivation while preserving nutritional value.

The only way to absolutely prevent cattle to cattle transmission of these unconventional slow viral infections is to ban the feeding of ruminant animal protein to other ruminants. A variation (specific offal ban) of this has been implemented in Great Britain and the gradual increase in age incidence of BSE indicates that it will be effective. Other countries with BSE, the Republic of Ireland, France, and Switzerland, have also implemented ruminant protein feed bans of one type or another.

There has yet been no country free of BSE that has taken official preventive measures regarding the feeding of animal protein from cattle. Only time will tell if this response is correct. It is unlikely that surveillance programs will provide adequate warning since it will be almost impossible to detect the disease in the first few animals affected, and an incubation period of 3–8 years may allow a decade of exposure before the disease is recognized.

References

1. Kimberlin RH (1992) Bovine spongiform encephalopathy. In: Transmissible spongiform encephalopathies of animals. Scientific and Technical Review, vol 11(2). Office International Des Epizooties, Paris, pp 347–390
2. Wilesmith JW, Wells GAH (1991) Bovine spongiform encephalopathy. Curr Top Microbiol Immunol 172: 21–38
3. Goldgaber D, Goldfarb LG, Brown P, Asher DM, Brown WT, Lin S, Teener JW, Feinstone SM, Rubenstein R, Kascsak RJ, Boellaard JW, Gajdusek DC (1989) Mutations in familial Creutzfeldt-Jakob disease and Gerstmann-Sträussler-Scheinker's syndrome. J Exp Neurol 106: 204–212
4. Marsh RF, Hadlow WJ (1992) Transmissible mink encephalopathy. In: Transmissible spongiform encephalopathies of animals. Scientific and Technical Review, vol II(2). Office International Des Epizooties, Paris, pp 539–550
5. Hadlow WJ, Karstad L (1968) Transmissible encephalopathy of mink in Ontario. Can Vet J 9: 193–195
6. Marsh RF, Bessen RA, Lehmann S, Hartsough GR (1991) Epidemiological and experimental studies on a new incident of transmissible mink encephalopathy. J Gen Virol 72: 589–594
7. Anonymous (1991) Qualitative analysis of BSE risk factors in the United States, and Quantitative risk assessment of BSE in the United States. USDA: APHIS: VS, Animal Health Information (AH11.01/91), Fort Collins, Colorado
8. Franco, DA (1993) Bovine spongiform encephalopathy: The role of the rendering industry in prevention and control. Vet Med (in press)

9. Marsh RF, Bessen RA (1993) Epidemiologic and experimental studies on transmissible mink encephalopathy. In: Brown F, Schwerdtfeger W (eds) Transmissible spongiform encephalopathies: Impact on animal and human health. Developments in biological standardization, Karger, Basel, pp 105–112

Author's address: Dr. R.F. Marsh, Department of Animal Health and Biomedical Sciences, University of Wisconsin-Madison, Madison, WI 53706, U.S.A.

Arch Virol (1993) [Suppl] 7: 261–293

Human prion diseases (spongiform encephalopathies)

H.A. Kretzschmar

Department of Neuropathology, University of Göttingen, Göttingen,
Federal Republic of Germany

Summary. Prion diseases (spongiform encephalopathies) in humans are Creutzfeldt-Jakob disease (CJD), Gerstmann-Sträussler-Scheinker syndrome (GSS), and kuru. Clinically, they are characterized by an inexorably progressing neurological illness with dementia and ataxia as the most prominent signs. The classical neuropathological changes are limited to the central nervous system and consist of spongiform degeneration, amyloid plaques, astrocytic gliosis, and nerve cell loss. The human spongiform encephalopathies, which for many years were considered neurodegenerative disorders of unknown etiology, were finally recognized as transmissible diseases similar to scrapie in sheep in the late 1960's. The infectious agent appears to consist of protein devoid of functional nucleic acid and has been termed prion to distinguish it from viruses. The prion hypothesis has gained wide acceptance through the finding that mutations of the prion protein gene are associated with heritable human prion disease. Different mutations appear to cause prion disease with a distinct pattern of clinical and pathological features in a great number of families. Certain mutations of the PrP gene have been shown to be associated with clinical and neuropathological changes not typical of any variant of human prion disease known to date. A new classification of prion diseases based on the molecular biology and biochemistry of the prion protein is likely to emerge.

Introduction

Prion diseases (spongiform encephalopathies), which in humans comprise kuru, Creutzfeldt-Jakob disease (CJD), and Gerstmann-Sträussler-Scheinker syndrome (GSS), are a group of transmissible neurodegenerative disorders with long incubation periods and invariably fatal outcome. Related diseases in animals are scrapie in sheep and goats, bovine spongiform encephalopathy (BSE) in cattle, transmissible mink encephalopathy, and chronic wasting disease of captive mule deer and elk.

Although the transmissible nature of scrapie in sheep has been known since the 1930's [39, 67], it has taken a long time for the human spongiform encephalopathies to be recognized as members of the same family of diseases. Because of the lack of an inflammatory response, human prion diseases were thought to be genetic, toxic, or neurodegenerative disorders comparable to Parkinson's disease.

The recognition of kuru and CJD as transmissible diseases similar to scrapie instigated the search for a postulated slow virus. After years of futile effort to identify a causative virus, a host of experimental findings was accumulated testifying to the unique biochemical properties of the transmissible agent, and a completely different concept, the prion hypothesis, was put forth [145]. According to this hypothesis, the infectious agent, termed prion, consists mainly – and perhaps exclusively – of a protein, the prion protein. Attempts to identify a nucleic acid associated with infectivity have failed thus far; it now seems very unlikely that a nucleic acid >50 nucleotides is essential for infectivity [1, 44, 47, 48, 128, 175].

The prion protein exists in two isoforms: PrP^C, the normal cellular isoform, and a pathological isoform, which is part of the infectious particle and has been termed PrP^{Sc} (scrapie), PrP^P (pathological), PrP^{CJD} (Creutzfeldt-Jakob disease), etc. PrP^C is a normal host protein [98, 138] of unknown function whose amino acid sequence is encoded on a single exon by a single-copy gene [6, 149], and is found predominantly on the surface of neurons [97, 112, 161]. In contrast to PrP^C, PrP^{Sc} is relatively resistant to digestion with proteases and accumulates in cytoplasmic vacuoles [125, 165]. It has been proposed that upon entering a host cell, PrP^{Sc} causes the conversion of PrP^C to PrP^{Sc} by an unknown process [146], thereby initiating a self-amplifying process of PrP^{Sc} which recruits the available PrP^C, and thus propagates prion disease.

Mutations and insertions in the human PrP gene have been linked to inheritable human prion diseases (for reviews cf. Hsiao et al. [78], Brown [15]). Transgenic animals expressing one of these mutations spontaneously develop neurodegeneration and subsequently die of a disease very similar to scrapie [79]. Although the detailed study of heritable human prion diseases and consequent experiments in animals are convincing in terms of the importance of the prion protein for neurodegeneration and transmission of spongiform encephalopathies, a number of paramount questions remain unanswered. Among these are the specificity of scrapie-strains in the absence of nucleic acid [24, 25, 72, 84, 174], molecular mechanisms influencing the posttranslational modification or conformation of PrP^C and PrP^{Sc} [146, 148], and the transmission or occurrence of spontaneous human prion diseases.

The human spongiform encephalopathies

Pathologically, prion diseases are characterized by spongiform degeneration (Figs. 1 and 2), atrophy and loss of nerve cells, astrocytic gliosis, and PrP-containing amyloid plaques. Prion diseases are unique diseases which, in the face of their transmissible nature, do not show overt inflammatory response. The only cells that seem to react to the disease are astrocytes (Fig. 3). Astrocytic glial reaction has been described as excessive and out of proportion to the degree of nerve cell loss and injury in scrapie [109]. In comparison to astrocytic gliosis, microglial reaction and the appearance of macrophages seem minimal.

The molecular mechanism leading to astrocytic proliferation in prion diseases remains to be elucidated. Prions may possess some glial growth or maturation activity; alternatively, changes in neuronal function or simply nerve cell loss may provoke astrocytic gliosis.

Nerve cell loss does not seem to be a result of the function of PrP^C, since mice that are homozygous for a deletion of the PrP gene ("knock-

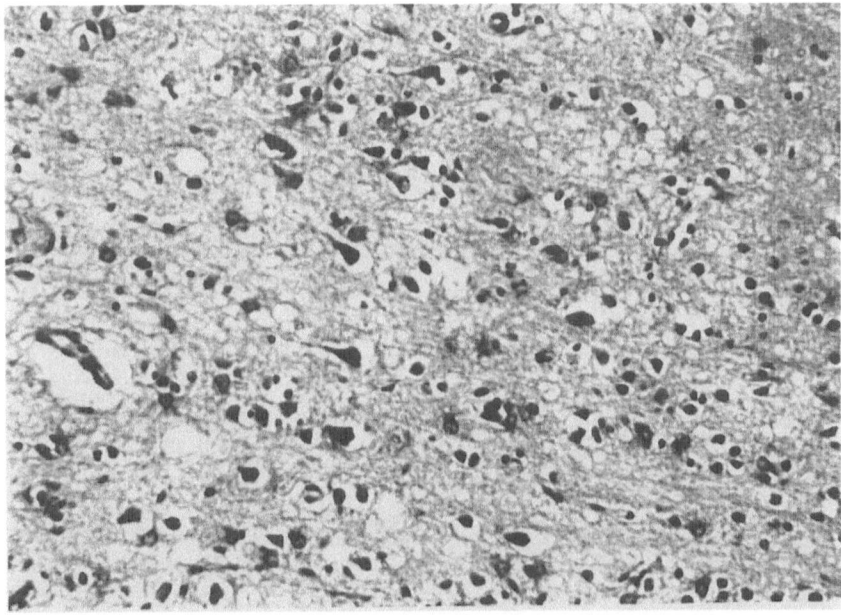

Fig. 1. Spongiform changes. This section from the cerebral cortex of a CJD patient shows mild to moderate spongiform changes with vacuoles of varying size in the neuropil. Note that there are numerous nerve cells visible in this area. H&E, ×100

out mice") develop normally and show no morphological or behavioral peculiarities [28]. Rather, there are indications that pathology and loss of nerve cells are related to accumulation of PrP^{Sc} in lysosomes [103, 125]. Since in infected nerve cell cultures PrP^{Sc} accumulations are found in cells that do not undergo degeneration [165], it remains to be determined whether this accumulation causes nerve cell degeneration or merely represents an epiphenomenon of prion disease. The two most distinctive neuropathologic changes in prion diseases, spongiform degeneration and amyloid plaques, call for special consideration.

Spongiform degeneration

Spongiform change describes small vacuoles in the neuropil that are usually round to ovoid and occasionally coalesce (Figs. 1 and 2). Electron microscopy has shown that this is associated with cystic dilation of neurons and focal necrosis of cellular membranes [30, 68]. Most observers agree that the vacuolation of the neuropil is the result of membranous dilations occurring mainly in the dendrites of neurons. Vacuolation has also been described in the perikaryon of neurons, in astrocytes, and within the myelin sheaths of oligodendroglia, the latter being apparently of importance to the white matter lesions seen occa-

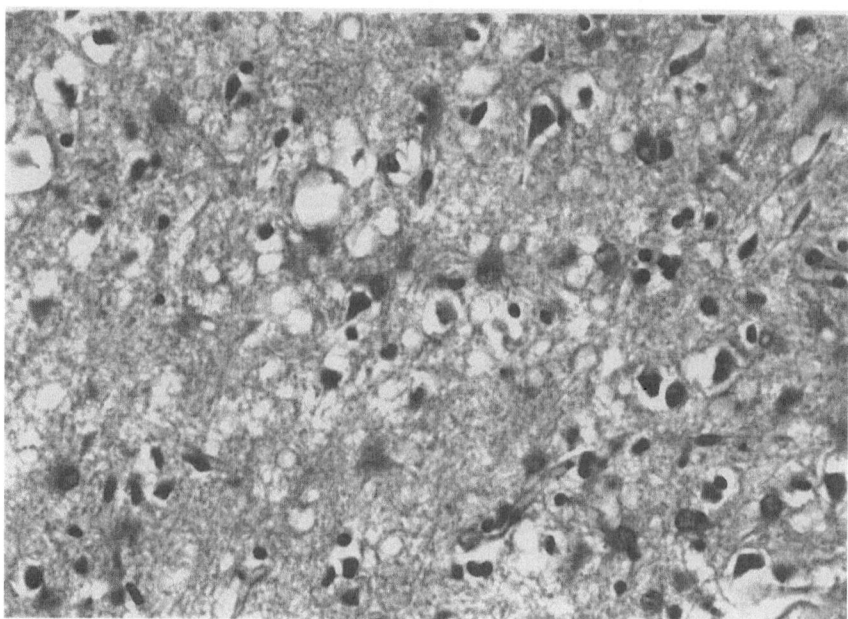

Fig. 2. Spongiform changes. Larger vacuoles are seen in other areas of the cerebral cortex of the same patient. Note the astrocytic gliosis and severe loss of nerve cells. H&E, ×200

sionally in both natural and experimental infection. Vacuolation of the perikaryon of neurons seems to be quite characteristic of prion diseases in animals; with the exception of kuru, it is not very striking in human prion diseases.

The anatomical distribution of spongiform change in the human brain is rather characteristic. The lesions are usually widespread in the cerebral cortex, striatum, thalamus, upper brain stem, and the molecular layer of the cerebellum. Within the cerebral cortex, there often appears to be a laminar distribution of the spongiform change; however, spongiform change in CJD is never confined to the more superficial layers (layers I and II). This may be an important point for differential diagnosis, since many unrelated anoxic and metabolic disturbances cause superficial vacuolation of the cerebral cortex. Another characteristic feature is the peculiar distribution of spongiform change within the hippocampal region, where it is often absent from the subiculum and parahippocampal gyrus [116].

It is of great practical importance to distinguish between spongiform change and status spongiosus as defined by Masters and Richardson in 1978 [121]. Status spongiosus describes cavities within the neuropil in the presence of hypertrophied glial processes. These cavities are larger than the vacuoles of spongiform change and are of irregular size and shape. The term status spongiosus carries no connotation of specificity and

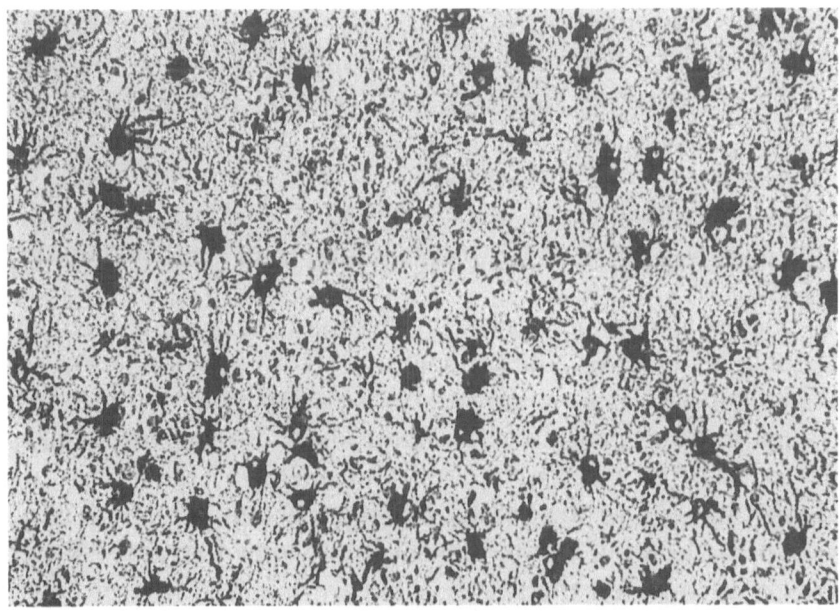

Fig. 3. Astrocytic gliosis. Pronounced astrocytic gliosis is often seen in human CJD cases. Immunohistochemistry with antibodies against GFAP (glial fibrillary acidic protein). ×200

may be used for a variety of conditions in which there is disruption and rarefaction of the neuropil. In the context of prion diseases status spongiosus describes the end-stage gliotic cerebrum. Other investigators do not make this distinction and use the term "status spongiosus" for the spongy lesions as introduced by Spielmeyer [7, 159].

Accepting the premise of spongiform changes as the pathologic hallmark, Daniel [40] used the distribution of the lesions as a basis for distinguishing four major clinical syndromes:

1. The Jakob type or cortico-striato-spinal variant, clinically characterized by mental disturbance and usually ending with severe dementia, curious painful sensations and spasticity of the limbs;
2. The Heidenhain type with lesions predominating in the occipital cortex, and visual disturbances constituting the presenting symptoms;
3. The diffuse type in which lesions are concentrated in the basal ganglia and thalamus, the prominent symptoms being dementia and extra-pyramidal signs;
4. The ataxic type in which the cerebellum is most severely affected, with ataxia as the most prominent feature.

Another variant of CJD, the panencephalopathic form, was first reported in Japan and later also in other locations [99, 131]. The involvement of white matter in these cases seems to be out of proportion to the degeneration of grey matter.

In the light of recent findings in experimental scrapie [72], the region-specific distribution of spongiform changes in subtypes of human CJD may correspond to region-specific pathology and accumulation of PrP in various scrapie strains. Targeting of PrP^{Sc} to certain brain areas in experimental scrapie may be mediated by differing conformations of PrP^{Sc}, post-translational modifications of PrP^{Sc} such as N-glycosylation, or an as yet undetermined molecule linked to PrP^{Sc} in the prion. These questions have not been dealt with adequately in human pathology and epidemiology.

It is evident that at one point the finding of spongiform changes was instrumental in defining a disease entity. However, the term "spongiform encephalopathy" now appears to be a misnomer; the term "prion diseases" is more appropriate. In naturally occurring prion diseases in animals as well as in experimental transmission, there are often no spongiform changes in the brains of affected animals [114]. Dickinson argued that vacuolation can be absent in advanced scrapie "because we know when we injected the sheep, what we injected it with and we can predict the course of the disease. If such a sheep had presented itself as a natural case, we would have no way of knowing why the sheep died without doing a transmission from it" [116]. In addition, we now know there are cases of heritable human CJD with mutations in the PrP gene and clinical disease but no spongiform changes [34, 127]. Thus, by solely relying on spongiform changes as a diagnostic criterion, an unknown percentage of cases, perhaps even of sporadic human CJD, may be overlooked.

Amyloid plaques

Klatzo [91] was the first to note the presence of amyloid plaques in kuru. These unicentric plaques, whether encountered in kuru, CJD, or GSS, have since been called kuru plaques (Fig. 4). They are PAS-positive, moderately argentophilic, and surrounded by a halo of delicate, radially arranged filaments. They are congophilic and show apple-green birefringence under polarized light; at the electron microscopic level, they are composed of fibrils of 10 nm width and thus fulfil the criteria of amyloid. These plaques have also been shown to stain with antibodies raised against the prion protein [8, 90]. In GSS, both kuru plaques and multicentric plaques are found, the latter being pathognomonic of the disease. Multicentric plaques have been termed cockade plaques, satellite plaques or multilobular plaques (Fig. 5).

While prion rods can be isolated from all prion diseases, PrP plaques are not an absolutely regular finding. By definition they are always present in GSS. According to data by Masters et al. [119] they are found in 50 to 70% of human kuru and 5 to 10% of human CJD cases. These data are derived from classical light microscopy with H&E and Congored stained sections. Immunohistochemistry with antibodies against PrP reveals a much higher percentage of plaques. Kitamoto and Tateishi [89] claim that by using immunohistochemistry PrP-positive plaques can be found in 95% of longterm survivors [i.e. patients with clinical disease exceeding one year]. There seemed to be a general drift for positivity of plaques in longstanding clinical disease [89]. The percentage of CJD cases with clinical disease exceeding 13 months appears to be highly dependent on the geographical location. Thus, it is very high in Japan [43 to 61%] [132, 172] and much lower in England [20%] [123] and France [11%] [16].

Liberski et al. [106] recently surveyed 31 CJD cases from Europe and found plaques in only one case from Poland with a very short clinical course [8 weeks]. They conclude that the tendency towards plaque formation seems to be more dependent on certain geographical locations than on the duration of clinical disease.

PrP amyloid plaques clearly differ from amyloid plaques in Alzheimer's disease. The latter are called "senile" or "neuritic" plaques; they consist largely of the ßA4 protein and do not contain PrP.

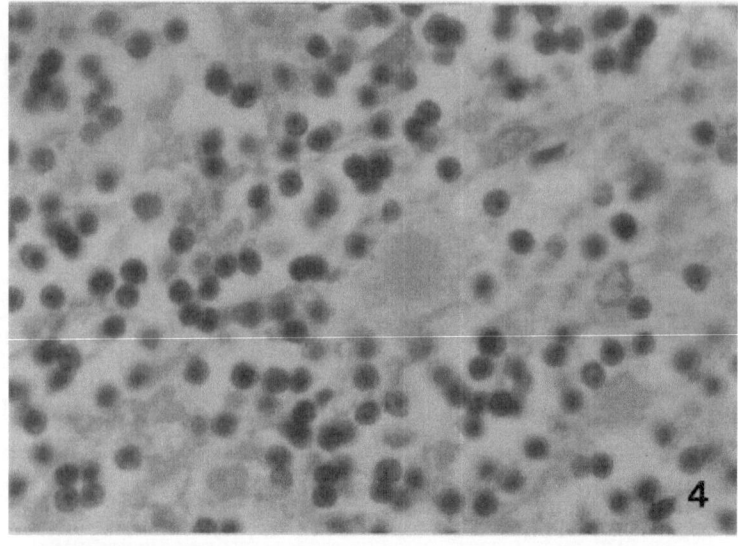

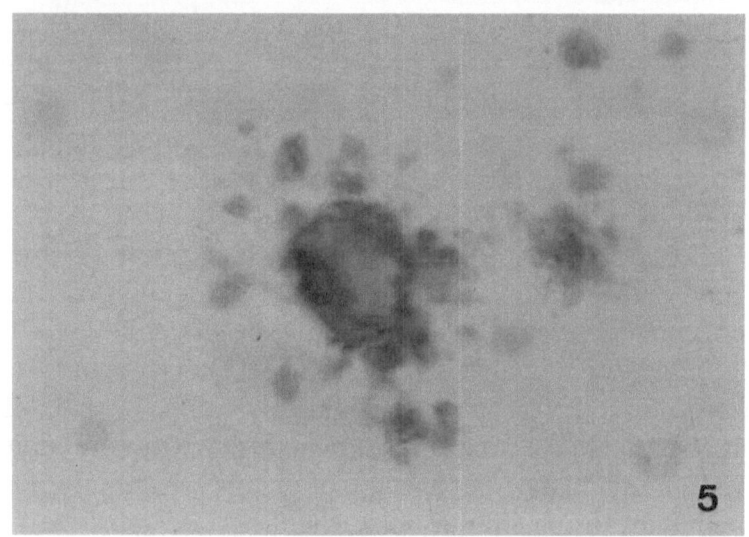

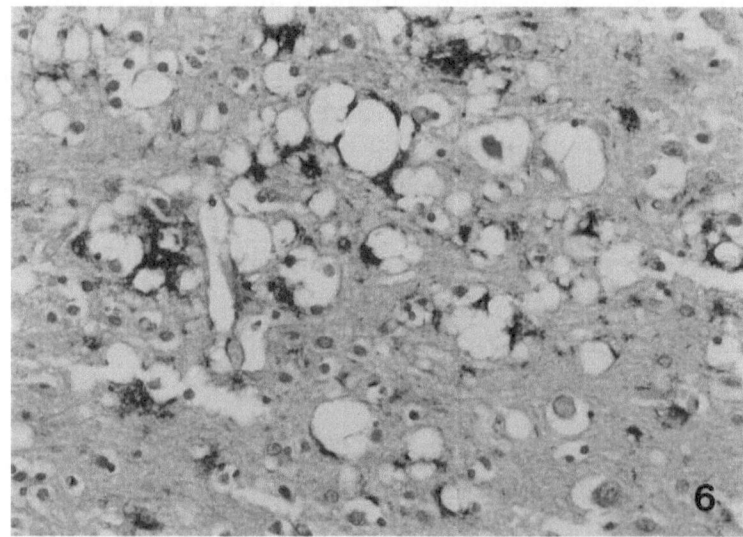

However, Miyazono et al. [130] recently reported the colocalization of prion protein and ßA4 protein in amyloid plaques in GSS patients older than 60 years. There was no ßA4 protein deposition in plaques in a GSS patient younger than 60 years.

Kuru

Kuru played a crucial role in establishing the concept of transmissible encephalopathies in humans. It was first reported by Gajdusek and Zigas [53] as a syndrome similar to paralysis agitans affecting children and adults, predominantly females, in the Eastern Highlands of New Guinea, and was thought to be most closely related to "the ill-defined group of heredofamilial neurologic degenerative disorders of the central nervous system, particularly the cerebellar and spinal ataxias". It was William Hadlow [69] who noticed a number of similarities between scrapie and kuru, which led him to hypothesize that kuru in man might be analogous to scrapie in sheep, and that nonhuman primates should be inoculated with specimens from patients dying of kuru. Indeed, a few years later Gajdusek, Gibbs, and Alpers were able to show that kuru is transmissible to chimpanzees [52].

Kuru was transmitted by ritualistic cannibalism in the Fore population; it is rapidly disappearing since endocannibalism has ceased in New Guinea. The patients now developing kuru were exposed to the kuru agent presumably more than two decades ago. Incubation periods in some monkeys inoculated with the kuru agent exceeded 7 years [57].

Typically, kuru begins with a prodromal phase with headaches and joint pains. Within a few months the victims develop a fine tremor, ataxia, and difficulty walking. Both ataxia and tremor become more pronounced as the disease progresses and are joined by behavioral abnormalities such as hilarity and inappropriate mood changes. Intellectual deterioration is usually minimal and supervenes only in the late

---

Fig. 4. Kuru plaque. A typical solitary kuru plaque in the cerebellar granule cell layer in a CJD patient. Kuru plaques appear pink in sections stained with hematoxylin and eosin. Note there is no visible relationship to other tissue elements. H&E, ×400. **Fig. 5.** GSS-Plaque. Typical multicentric amyloid plaque in the cerebral cortex of a case of Gerstmann-Sträussler-Scheinker disease. There is a dense central plaque surrounded by "satellites". Immunohistochemistry with an antiserum against the prion protein (gift from J. Tateishi). ×400. **Fig. 6.** PrP. Hydrolytic autoclaving. The amount of immunohistochemically labeled PrP is increased considerably by hydrolytic autoclaving of tissue sections prior to immunohistochemistry [86]. PrP is seen in vesicle membranes and in the neuropil as well in a section from the cerebral cortex of a CJD patient. ×400

stages of the disease. The morphology of kuru is characterized by wide-spread vacuolation of neurons and proliferation of glial cells, the severest degeneration always being found within the cerebellum and its connections [7]. Most but not all of the cases examined have shown the presence of kuru plaques.

Creutzfeldt-Jakob disease (CJD)

CJD was officially recognized as a member of the transmissible encephalopathies caused by unconventional infectious agents after its experimental transmission to chimpanzees in 1968 [58]. Its transmissible nature, however, was suspected many years before with the observation of morphologic similarities between kuru and scrapie by Klatzo et al. [91] and Hadlow [69], and even earlier by Alfons Jakob [81].

Walter Spielmeyer in 1922 [160] proposed the name Creutzfeldt-Jakob disease to cover a series of degenerative disorders of the CNS characterized by loss of nerve cells and gliosis, with preferential involvement of cortical, pyramidal and extrapyramidal systems. Our notion of the disease bearing Hans Gerhard Creutzfeldt's and Alfons Jakob's names has undergone considerable change since their first description in the early 20's [37, 80]. Most authors agree that at least three of their six original cases would not meet our present criteria for CJD.

One of the reasons for this is the fact that spongiform changes were not considered a hallmark of the disease in the early days. In fact, the term subacute spongiform encephalopathy was first used by Nevin et al. in 1960 [134] who reported eight cases with the characteristic pathological triad of neuronal loss, astrocytic gliosis, and spongiform changes.

In this context one should bear in mind that in the early days of neuropathology, the Nissl cresyl violet stain was most widely used; spongiform changes are hardly visible in this stain. In a large series collected in 1961, spongiform change was absent in more than half the cases [85]. This changed dramatically when the hematoxylin and eosin stain became standard in the 1960's. Masters and Gajdusek [117] had the opportunity to restain the slides of Jakob's first five cases; they found that at least two of them demonstrated spongiform changes. Two of the remaining cases were more consistent with a toxic-metabolic encephalopathy, while Jakob's first case was a motor-neuron disease with dementia. Creutzfeldt's case now is considered unclassifiable.

Pathologically, the gross appearance of CJD brains at autopsy varies from almost normal to severely atrophic with narrow gyri, gaping sulci and weights as low as 800 grams [40, 85]. Histologically, the preeminent feature is spongiform change of the grey matter, which may vary from slight to severe with cleft formation and a complete collapse of the cortical ribbon accompanied by proliferation and hypertrophy of astrocytes (Figs. 1, 2 and 3); gemistocytes are often numerous. In cortical variants of the disease, spongiform change may be patchy and sometimes restricted to certain areas [7]. In the cerebellum, there may be variable loss of Purkinje and granule cells, and "torpedoes" in the axons of Purkinje cells may be found. A variety of pathological changes may be seen in those areas, including chromatolysis of Betz cells in the motor cortex as well as satellitosis and neuronophagia (Jakob's "glial rosettes") [80, 81]. Other neuronal abnormalities include swelling and loss of dendrites and axons [68, 82], and accumulation of abnormal cytoskeletal protein [133]. There is astrocytic gliosis predominantly in the gray matter, but the white matter may also be heavily affected [27]. Kuru plaques in CJD are most often seen in the cerebellum (Fig. 4).

Clinically, nonspecific prodromal symptoms are encountered in about one-third of CJD patients, and may include fatigue, sleep disturbance, headache, general malaise, and ill-defined pain [29]. About two-thirds of the patients present with mental deterioration such as dementia, behavioral abnormalities or higher cortical dysfunction (Table 2); in about half of the patients mental deterioration is the only beginning sign. In about one-third of the cases cerebellar and/or visual disturbances, without mental deterioration, are the presenting signs [17, 29, 85]. Those with cerebellar problems generally exhibit difficulty walking as well as incoordination. Frequently, the cerebellar deficits are rapidly followed by the onset of a progressive dementia. Visual problems often begin with blurred vision and diminished acuity. As with cerebellar signs, the appearance of visual symptoms is rapidly followed by mental deterioration, which virtually always progresses to a state of profound dementia. Other signs and symptoms usually occurring later in the disease include extrapyramidal dysfunction with rigidity, mask-like facies, or choreo-athetoid movements and pyramidal signs.

Myoclonic movements are the most constant physical sign in CJD and are exhibited by nearly 90% of the patients. These are quick jerky movements of the extremities or the head that usually are not present throughout the evolution of the disease, and in the majority of cases become manifest only in its later stages [29, 151, 177]. It is important to note that myoclonus is neither specific nor confined to CJD. It has also been described in Alzheimer's disease [118, 173], cryptococcal encephalitis [162], and toxic and metabolic encephalopathies [29].

The electroencephalogram (EEG) may be a useful tool in the diagnosis of CJD. It is diffusely abnormal in all cases and is considered "evocative" in 75% [29]. Typically, high-voltage, triphasic and polyphasic sharp discharges are seen. The diagnosis of CJD is very likely

Table 1. Mutations of the PrP gene and related morphologic changes

Mutation	Spongiform changes	PrP plaques	NFTs
Inserts	+/−	−/+	−
178 Asn	+/−	−	−
200 Lys	+	−	−
102 Leu	+	+	−
117 Val	+	+	−
198 Ser	+	+	+
217 Arg	+	+	+

Table 2. Comparison of important clinical and pathological data of Creutzfeldt-Jakob disease and Gerstmann-Sträussler-Scheinker syndrome

	Creutzfeldt-Jakob disease(CJD)	Gerstmann-Sträussler-Scheinker syndrome GSS)
Clinical appearance	dementia, ataxia and visual deficits	spinocerebellar ataxia
	later: myoclonus, periodic EEG activity, (extra-) pyramidal signs	later: dementia, pyramidal signs, amyotrophy
Age at onset	6th decade (50–75 years)	5th decade (35–50 years)
Duration	months	2–8 years
Morphology		
– spongiform change	0^a/+++	+
– plaques	0/+ (kuru-type)	+++ (multicentric and kuru-type)
– tract degeneration	0	+
Incidence	1/1,000,000/year	1–10/100,000,000/year
Epidemiology	sporadic 15% autosomal dominant	autosomal dominant

[a] In some familial cases

when these stereotyped periodic bursts of ~200 ms duration occurring every 1 to 2 seconds are present [100, 134, 171].

CJD is a rare disorder which is found throughout the world with an incidence of approximately one case per million per year [121]. The mode of transmission of most cases of sporadic CJD is not known. Somatic mutation of the PrP gene or spontaneous conversion of PrP^C to

PrP^{Sc} may trigger sporadic CJD [146]. Many geographic clusters have been reported; it now appears that these clusters actually represent familial cases of CJD (v.i.) rather than transmission through local common exposure to some etiologic agent.

For many years, investigators have attempted to implicate the ingestion of scrapie-infected sheep meat in the pathogenesis of CJD. Numerous epidemiologic studies have failed to convincingly establish this relationship, but speculation continues [2, 10, 12, 108, 121]. Recent spread of the disease to cattle and the experimental transmission of BSE to pigs, a species not previously known to be affected, has rekindled old concerns. Since passage of the disease through different hosts is known to be accompanied by changed characteristics of the infectious agent [83], BSE presents an unknown risk to humans.

It has been known for some time that 10 to 15% of CJD cases are familial [118]. Familial CJD has been associated with mutations of the PrP gene, specifically mutations at codon 178 and codon 200 as well as octapeptide repeat insertions near the N-terminus of the protein (v.i., Table 1, Fig. 6).

Iatrogenic transmission of CJD has been reported in a number of cases. It appears to have occurred with corneal transplantation [46], contaminated EEG electrode implantation [9], and surgical operations using contaminated instruments [41, 93, 120, 176], or dura mater implants [122, 129, 136, 143, 168, 178]. Therapy with human pituitary growth hormone (hGH) [26, 38, 43, 51, 59, 92, 113, 115, 135, 144, 150, 169] and gonadotrophin [32] have been implicated in a great number of cases. However, this outbreak has not attained the major proportions that were foretold when the problem first became known [18]. Incubation periods after iatrogenic transmission range from 18 months to 21 years [14].

Direct iatrogenic inoculation of the infectious agent primarily leads to a disease characterized by a dementia similar to most cases of sporadically occurring CJD, whereas peripheral inoculation results in a predominantly cerebellar syndrome similar to kuru which is acquired through a peripheral infection. This peculiarity has been used as an additional argument for the hypothesis that in sporadic cases the infection arises *de novo* – possibly triggered by a somatic mutation or by spontaneous conversion of PrP^{C} to PrP^{Sc} – rather than by exogenous infection [14].

Gerstmann-Sträussler-Scheinker syndrome (GSS)

GSS was first described by J. Gerstmann in 1928 [54, 55]. It is a familial disease with autosomal dominant inheritance. As with CJD, many decades passed before the transmissibility of GSS was demonstrated [119].

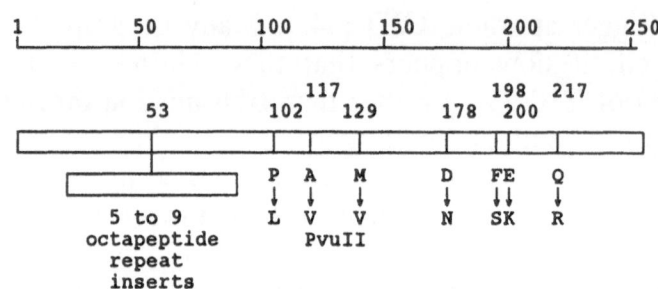

Fig. 7. Mutations and polymorphisms of the human PrP gene

Affected patients typically first complain of difficulty walking and unsteadiness (Table 2); however, a high degree of diversity of clinical manifestations in GSS, even within one family, is well documented. On physical examination, cerebellar ataxia, dysarthria, ocular dysmetria, hyporeflexia or areflexia in the lower extremities may be found. Later in the course, mental deterioration occurs. It may be mild and difficult to diagnose because of severe dysarthria. In the later stages, there is often dysphagia. Extrapyramidal rigidity and amyotrophy have been reported in a number of cases. Myoclonus seldom occurs and may be confined to the lower extremities. EEG is normal or shows nonspecific diffuse changes.

The majority of patients present first symptoms in the fifth decade, with an average of 43 years. The age of onset may vary considerably within a single pedigree. Since symptoms may initially be relapsing, GSS can be misinterpreted as multiple sclerosis. The mean duration of illness is five years.

Five different mutations (Table 1, Fig. 7) of the prion protein gene on chromosome 20 of the human genome [158] have been reported in GSS families (v.i.). Attempts have been made to assign certain mutations of the PrP gene to defined clinical subtypes [76]. On the other hand, a great diversity of clinical symptoms, age of onset and duration of clinical disease has been documented even within one and the same family.

Neuropathologically, the distinctive and pathognomonic feature is the presence of multicentric amyloid plaques in the cerebellum and the cerebral cortex (Fig. 5) associated with variable degrees of spongiform change and gliosis as well as degeneration of white matter tracts [153–155]. The distribution and extent of these neuropathologic changes differs widely between patients, even within one family.

The pattern of degeneration of white matter tracts is comparable to that of other system degenerations such as Friedreich's ataxia or cerebellar degeneration (Marie). Neuronal loss occurs in scattered areas throughout the CNS and may affect anterior horn cells, dentate nuclei,

Purkinje and granule cells, inferior olive, pontine nuclei, basal ganglia, thalamus and all layers of the cerebral cortex. Neurofibrillary tangles have been described in families with a mutation of the PrP gene at codons 198 and 217 (v.i., Table 1).

Genetic variants and morphologic correlates

Familial cases of prion diseases were shown to be transmissible to experimental animals almost 20 years ago [49]. In the meantime, a number of mutations of the PrP gene have been associated with clinical variants of human prion diseases. Transgenic mice expressing one of these mutations (102 Leu) spontaneously develop a fatal spongiform encephalopathy similar to scrapie in mice [79].

Attempts have been made to assign clinical and pathological characteristics to certain mutations of the PrP gene. Although there is some degree of correlation, one should bear in mind that for each described mutation there are patients whose clinical and pathological features are significantly different from others having the same mutation.

The **102 proline → leucine** mutation was first reported in inherited human prion disease (GSS) in English and American families by Hsiao et al. [74]. This seems to be the most common mutation associated with GSS which has subsequently been identified in German, Austrian, Italian, and Japanese families [45, 66, 94, 96]. The original family published by Gerstmann in 1928 and 1936 [54, 55] has recently been shown to carry a 102^{Leu} mutation [94].

Clinically, the form of the disease originally described produced primarily ataxia and other cerebellar signs, including gaze abnormalities. Pyramidal signs such as spasticity, weakness and Babinski responses were also present, as was mild dementia. In other families inconstant signs such as seizures, areflexia, and posterior column signs have also been reported. The 102^{Leu} mutation was identified in an Italian family, who in addition to ataxia and dementia showed strong features of amyotrophy with muscle fasicullations, fibrillations, and atrophy as the prominent signs in some but not all family members [96].

PrP plaques high in number are seen in the cerebral and cerebellar cortex. These are kuru plaques, multilobular plaques and plaques with satellites (Fig. 5). There are only slight spongiform changes of the cerebral neuropil. Systemic atrophies affecting different anatomic systems in varying degrees are noted. These include the spinocerebellar tracts, posterior columns, and pyramidal tracts. There is also diffuse atrophy of the cerebral and cerebellar cortex with corresponding glial reaction. In addition, varying degrees of atrophy of brain stem and subcortical nuclei and fiber systems may be found.

A **117 alanine → valine** mutation thus far has only been reported in an Alsatian family [45, 170] and an American family of German descent [77, 137]. The clinical picture is variable and has become more complex over the generations: isolated dementia in the first generations, then in the following generations, a triad of pyramidal and pseudobulbar syndrome and dementia associated with symptoms indicating spread of damage to the spinal cord and cerebellum. This variant of GSS has also been called dementing GSS, in contrast to the 102^{Leu} mutation which was designated as ataxic GSS [76]. Pathologically, numerous PrP-positive unicentric or multicentric amyloid plaques, neuronal degeneration and moderate spongiform changes are found.

A **198 phenylalanine → serine** mutation was reported in a large kindred from Indiana [56, 75]. The age at onset ranges from the 4th to the 7th decade of life. Death occurs within 2–3 years. Clinically, this familial disease is characterized by ataxia, parkinsonism, and dementia. Ocular movement abnormalities due to cerebellar dysfunction are detected in the early course. PrP-containing amyloid plaques are widespread throughout the cerebrum and cerebellum. Neurofibrillary tangles (NFTs) are numerous in the cerebral cortex, the hippocampus, and the substantia innominata; these are composed of paired helical filaments and are morphologically indistinguishable from those seen in Alzheimer's disease. Spongiform changes were occasionally observed and were mild.

The mutant serine at PrP residue 198 does not appear to be present in the 11-kd PrP fragment, spanning residues 58 to about 150 which was isolated from PrP amyloid plaques from the Indiana kindred [163]. This does not exclude its potential importance in the pathogenesis of disease, since PrP plaques presumably represent an endstage product of degradation.

A **217 glutamine → arginine** substitution was identified in a Swedish family with GSS and neocortical NFTs. In two patients from this family dementia was noted four years prior to the development of gait ataxia, dysphagia, and confusion occurring in the months preceding their deaths at the ages of 67 and 71 [75].

Familial CJD and GSS are dominantly inherited diseases; nevertheless, brain extracts from affected patients can sometimes induce neurodegeneration in primates and rodents that are inoculated intracerebrally [74, 119]. At 450 days after inoculation with brain homogenates from patients with the 198^{Ser} and the 217^{Arg} mutations, no such transmission was observed in more than a dozen animals [75].

A **178 aspartic acid → asparagine** mutation was first identified in a large kindred from Finland [70] and was later reported in a number of

European families and families of European origin. Affected members of seven families were studied clinically (43 patients) and neuropathologically (25 patients) [19, 61]. Clinically, there was a notably early age at onset (i.e. 46 years versus 62 years in sporadic CJD cases) and a relatively long duration (23 months versus an average of 6 months in sporadic cases). The disease almost always presented with insidious memory loss. Sleep disturbances were not noted in these families. The high frequency of myoclonus in the absence of EEG periodicity in almost all patients is of particular interest.

Pathologically, there was considerable diversity in both the intensity and the topography of the characteristic triad of spongiform change, gliosis and neuronal loss. The cerebral cortex and basal ganglia were usually most severely involved, whereas the cerebellum rarely showed more than minimal abnormalities, regardless of the clinical picture. There was no prominent thalamic pathology. In general, spongiform change and gliosis were more prominent than neuronal loss; necrotic white matter demyelination was found in two members of one family, but minimal or absent neuropathology in all four autopsied members of another family. Plaques were not observed.

It is not yet clear whether the families with a 178^{Asn} mutation reported by Medori et al. [126, 127] represent a significantly different clinical subgroup. The authors describe a rapidly progressive familial disease with a 7-to 36-month duration, characterized by the progressive loss of the ability to sleep ("fatal familial insomnia"), dysautonomia, and endocrine disturbances. In the later course of the disease more typical signs such as dementia, cerebellar ataxia, and myoclonus appear. There is marked atrophy of the anterior ventral and mediodorsal thalamic nuclei and varying degrees of cerebral and cerebellar cortical gliosis, as well as olivary atrophy. Spongiosis of the cerebral cortex is present only in some of the cases. There are no plaques.

200 glutamic acid → lysine. This mutation has been reported in a great number of families in Slovakia [65], Chile [3, 15], in Sephardic Jews [64] and in Japan [166]. The clinical picture is very similar to sporadic CJD, and the age at onset is a few years earlier (56 years) than in patients with sporadic CJD (62 years). Pathological features are spongiform degeneration with gliosis and no amyloid plaques, which makes this variant practically indistinguishable from sporadic CJD. The penetrance of this mutation is around 56%, which is the likely explanation for "sporadic cases" and skipped generations [15].

Octapeptide repeat inserts. An insertion in the prion protein gene was first reported in an English kindred by Owen et al. [140]. This was a 144 bp insert coding for six extra uninterrupted octapeptide repeats in the N-terminal region of the protein. Since then, families in England,

France, Japan, and the US with a total of 10, 11, 12, 13, and 14 repeats and varying clinical and pathological features have been recorded [20, 33, 34, 62, 63, 139, 166].

Clinically, the first patient to be reported with this mutation began to exhibit antisocial traits in his late teens; he was noted to have an intermittently unsteady gate and became dysarthric with jerky abrupt speech in his early 20's. At age 27 he demonstrated severe cognitive impairment with profound dyspraxia which resulted in complete dependency. In the later course he developed seizures and died at age 36. At autopsy, the brain showed no evidence of CJD or GSS. There was no cortical neuronal loss, gliosis, or spongiform change. "Torpedo" swellings and swelling of dendrites were noted in the cerebellar cortex. There were no amyloid plaques. Immunohistochemistry, however, revealed "definite immunoreactivity in the Purkinje cells of the cerebellum and in some neurons of the superficial laminae of the cortex, indicating the aberrant isoform of the prion protein. This immunoreactivity was detectable despite the absence of overt lesions" [31]. This by itself is remarkable since these are not the usual immunohistochemical findings in CJD.

It now seems that each of the affected families have their individual number of additional octapeptide inserts. Therefore, generalizations on the group as a whole must be considered with caution. Nevertheless, this group seems to have the longest duration of illness (average 7 years) and the earliest age at onset (average 34 years) of all mutations. The clinical features in general are similar to sporadic CJD except for lower frequencies of myoclonus and periodic EEG activity. Neuropathologic changes seem to correlate to a certain degree with the number of octapeptide repeats; families with 10 and 11 repeats have almost normal to minimally altered morphology, while those with 12 repeats are similar to sporadic CJD cases, and those with 13 repeats show changes typical of GSS.

Other mutations and polymorphisms. Deletions of octapeptide coding sequences have been reported in the same areas in which insertions are found [102, 149]. However, their relation to prion diseases has not been established.

There is a methionine-valine polymorphism at codon 129 with an approximate distribution of 40% Met-Met, 50% Met-Val, and 10% Val-Val [35, 45]. The homozygous Val-Val state has been suggested as a predisposing factor for developing sporadic CJD, since it was found in an unexpectedly high number of iatrogenic (4/7) [35, 60] and sporadic CJD cases (5/22) [35, 141]. Pending the investigation of larger numbers of affected persons, this question remains unresolved.

Recent developments and morphological findings

The majority of recent morphological findings were only possible after antibodies against the prion protein had become available. Using these antibodies it became clear that the prion protein is not only part of the infectious agent, but is also deposited in diseased tissue, sometimes in the form of amyloid plaques visible at the light microscopic level [8, 90]. More recently developed histochemical methods using antibodies against PrP have made possible the visualization of a far greater proportion of PrP in tissue sections.

Kitamoto et al. [87] introduced formic acid pretreatment of tissue sections to enhance the immunostaining of cerebral amyloids. This is a very simple and effective method which only requires the immersion of sections in 100% formic acid for 5 minutes. Using this method, immuno-histochemical staining of PrP plaques can be increased considerably.

The mechanism by which formic acid pretreatment enhances amyloid antigenicity and at the same time results in the loss of Congophilia is unknown; possible mechanisms include limited proteolysis as well as denaturation of amyloid protein polymers, which may reveal buried epitopes of amyloid deposits in tissue sections.

Intracellular location of PrP has been shown in experimental animals using special embedding techniques and immunohistochemistry [42]. Using in situ hybridization with a hamster PrP-cDNA, PrP has been shown to be mainly synthesized in neurons in the brain [97]. Northern blot analysis has shown lower levels of PrP mRNA to be present in the heart, lung and spleen [138]. By in situ hybridization in developing mice, PrP mRNA was detectable in the brain, the peripheral nervous system, tooth buds, kidneys, and extra-embryonic tissue [112].

In scrapie-infected mouse neuroblastoma cells, PrP^{Sc} accumulates primarily within the cell cytoplasm whereas cellular PrP is attached to the cell surface by a GPI anchor. McKinley et al. [124] recently were able to show that in scrapie-infected cell cultures PrP^{Sc} is localized in secondary lysosomes, in contrast to earlier findings which had lead to the assumption that PrP^{Sc} is present in the Golgi stack. The role played by secondary lysosomes in the post-translational formation of PrP^{Sc} has not been clarified.

There are a few reports on neuronal staining in humans, including a 5-year old boy with idiopathic chronic encephalitis [142], an atypical familial CJD case with octapeptide repeat inserts in the prion protein who presented with staining in some Purkinje cells [31], and another case of CJD with diffuse deposition in the granular layer of the cerebellum [95].

A new technique to visualize PrP^{CJD} in tissue sections recently was developed by Kitamoto et al. [86] and was called "hydrolytic autoclaving". Briefly, their protocol consists of autoclaving tissue sections at 121°C for 10 min in 3 or 9 mM HCl (hydrochloric acid) followed by the usual immunohistochemical procedures. Using this technique they were able to detect very delicate deposits of PrP in the grey matter in all CJD and GSS cases, even in cases with a clinical course shorter than 11 months. They described punctate staining for PrP^{CJD} around neuronal cell bodies and dendrites in CJD brains. This staining was found to be reminiscent of that of synaptophysin and suggested accumulation of PrP^{CJD} in synaptic structures [88]. This technique can be used with formalin-fixed and paraffin-embedded tissue. PrP^C is not labeled with this method. In our own experience immunohistochemical staining of PrP^{CJD} is considerably enhanced; however, diffuse grey matter staining is not seen (cf. Fig. 6).

Using traditional immunohistochemical methods, the differential localization of the PrP isoforms in CJD or scrapie brains cannot be achieved with certainty; the main reason for this is a lack of antibodies specific for PrP^{CJD}. In particular, while PrP^{CJD}-containing extracellular amyloid plaques are easily distinguished, cellular PrP^{CJD} immunostaining cannot be unequivocally discriminated from PrP^C. Earlier studies with Western blots had shown that the detection of proteinase-resistant PrP isoforms with currently available antibodies is enhanced by protein denaturants such as guanidine thiocyanate (GdnSCN). In contrast, immunoreactivity of PrP^C is not enhanced by this pretreatment. Specific detection of PrP^{CJD} is possible after limited proteolysis of PrP^C with proteinase K followed by incubation with GdnSCN. However, this method cannot be used with standard histological sections on glass slides, since the proteolysis needed to digest PrP^C largely destroys the tissue, even following fixation with formaldehyde.

In order to avoid tissue disintegration on glass slides, cryostat sections of unfixed brains can be transferred to nitrocellulose membranes and lysed in situ. To detect PrP^C, the blots are subjected to proteolysis by proteinase K followed by denaturation with GdnSCN, and labeling with PrP antibodies. This treatment digests PrP^C and exposes hidden epitopes of PrP^{CJD}. For detection of PrP^C in uninfected brains, the immunoblots are immunoprobed without proteolysis.

This method recently was developed by Taraboulos et al. [164] who were able to show in two cases of human CJD that PrP^{CJD} is confined to the grey matter, which correlates with the distribution of spongiform degeneration and reactive gliosis. Specifically, PrP^{CJD} was found in the cerebral cortex, basal ganglia, and thalamus as well as in other regions of grey matter. The cerebral cortical pattern of PrP^{CJD} accumulation varies from an intense band in the deep cortical layers to relatively uniform, diffuse distribution to all layers of the cortex. No PrP^{CJD} staining in the white matter was found in their CJD cases.

So-called tubulovesicular structures have been noted for a long time by Liberski et al. [107] in animals with transmissible encephalopathies such as natural and experimental scrapie, BSE and experimental Creutzfeldt-Jakob disease. Tubulovesicular structures were recently also seen in three cases of spontaneous human CJD [104, 105]. According to their description these are rare structures observed in distended pre- and postsynaptic terminals and measuring approximately 35 nm in diameter. They are smaller and of higher electron density than synaptic vesicles. The biochemical nature of tubulovesicular structures is yet unknown. According to Liberski's reports they are a regular marker of transmissible encephalopathies.

The number of techniques used in the diagnosis of prion diseases has been increasing in recent years. Among these are Western blots using homogenized CNS material and antibodies against PrP as well as the preparation of prion rods [11]. Other attempts to enlarge our diagnostic arsenal such as the identification of CJD-specific proteins in the CSF [71] or a proposed CJD-specific increase of ubiquitin in the CSF [110] did not turn out to be diagnostically relevant or are in need of further refinement and independent confirmation. Improved immunological techniques may enable us to develop tests for the detection of PrP in blood lymphocytes or serum [156].

Spongiform encephalopathies and other neurodegenerative diseases

In cases of rapidly progressive dementia with cerebellar and/or visual disturbances, myoclonus and EEG changes, the diagnosis of CJD is usually reliable and accurate. The absence of myoclonus and EEG changes almost excludes CJD [16].

This diagnosis of GSS has rarely been made prior to autopsy examination of an affected patient or family member. With our recent knowledge about mutations of the PrP gene in GSS, a correct diagnosis can be made from DNA extracted from peripheral blood lymphocytes.

The major differential diagnosis for both CJD and GSS is Alzheimer's disease, which is usually distinguished by its protracted course and lack of motor and visual dysfunction. However, one should bear in mind that many familial Alzheimer cases are believed to be clinically indistinguishable from CJD, yet show unequivocal Alzheimer pathology and are not transmissible to animals [13]. Moreover, there is also some overlap of pathological findings. Spongiform changes can be observed in Alzheimer's disease [5, 50, 111, 157], and CJD brains may contain Alzheimer type senile plaques [13, 73], neurofibrillary tangles [21], or show cell loss in the basal nucleus of Meynert [4].

Intracranial vasculitides may mimic most signs and symptoms that are typical of CJD; abnormal cerebrospinal fluid, focal CT or NMR abnormalities all are in favor of vasculitis.

The relation of prion diseases and amyotrophic lateral sclerosis (ALS) has long been a moot question. It is now believed that Jakob's first case was actually ALS. A number of otherwise typical CJD or GSS patients show some degree of amyotrophy. Some of these cases have also been successfully transmitted to laboratory animals [36]. However, most cases with the so-called amyotrophic variant of CJD are not transmissible to laboratory animals, suggesting that they may in fact be ALS patients demonstrating some features reminiscent of CJD [152]. The panoply of morphological, biochemical, and genetic methods in the diagnosis of prion diseases will help clarify these issues.

Safety precautions

There is a great deal of confusion and uncertainty concerning safety in handling CJD-infected tissue and performing autopsies. This apprehension is mainly caused by the fact that most chemical and physical procedures such as fixation in formalin or alcohol, UV-light etc., which are used to sterilize virus-infected material, are ineffective against prions.

Special procedures to inactivate prions have been tested, and guidelines for surgical and pathological practices have been established.

1. The brain and other parts of the nervous system are most infectious, while other organs are less likely to transmit the disease.
2. Contaminated skin should be disinfected by 1 N NaOH (sodium hydroxide) for 5 to 10 min followed by copious washing with water [22].
3. Fixation of 4 to 5 mm thick tissue specimens with formalin for 48 hours, transfer to formic acid ($\geq 96\%$) for 1 h, then transfer to fresh formalin almost completely eliminates prions without compromising tissue preservation for histological examination [23].
4. The use of a 1% dilution of hypochlorite containing 10,000 ppm chlorine is considered satisfactory for disinfecting instruments and exposed surfaces [101].
5. Prusiner et al. [147] recommend autoclaving for 4.5 h (121°C, 15 psi), or immersion in 2 M NaOH (sodium hydroxide) followed by autoclaving for 1.5 h or immersion in 2 M NaOH for 30 min at 25°C, which should be repeated three times.

6. Tateishi et al. [167] reported that boiling in SDS (sodium dodecyl sulfate) blocks transmission completely. Since SDS is not corrosive, for hospital and laboratory practice they recommend boiling for more than 3 min in 3% SDS, alone or in combination with other methods.

Conclusions

The prion hypothesis is largely supported by the available evidence, but it is not proven beyond doubt. Several important problems remain unsolved. What are the details of the molecular structure of the prion? How can we explain the existence of different disease variants in one and the same species ("scrapie strains")? What mechanisms are involved in the conversion of PrP^C to PrP^{Sc}, and what is the function of PrP^C?

The central role played by the prion protein in all known transmissible encephalopathies has provided us with new diagnostic tools; antibodies against PrP are now being used for immunochemical and immunohisto-chemical diagnosis of prion disease in humans and animals, PrP gene analysis is utilized for the analysis of heritable human prion disease. Improvements in immunological techniques may allow us to develop tests for detecting PrP in serum or in blood lymphocytes [156]. Furthermore, improved diagnostic techniques have helped demonstrate that prion diseases, although still considered very rare diseases, are much more common than previously thought.

New variants of human prion disease are being recognized with the refinement of diagnostic techniques. In some instances, a diagnosis of prion disease can now be made even in the absence of typical patho-logical changes [31, 34]. A new classification of prion diseases based on the molecular biology and biochemistry of PrP is likely to develop. Research on PrP and prions may provide new insights into normal brain function and mechanisms involved in neurodegenerative diseases.

Acknowledgement

This study was supported by a research grant from the Wilhelm-Sander-Stiftung (89.036.1).

References

1. Aiken JM, Marsh RF (1990) The search for scrapie agent nucleic acid. Microbiol Rev 54: 242–246

2. Alter M, Kahana E (1976) Creutzfeldt-Jakob disease among Libyan Jews in Israel. Science 192: 428

3. Araya G, Gálvez S, Cartier L, Gajdusek DC (1983) A spatiotemporal clustering of Creutzfeldt-Jakob disease in Chile. Rev Chil Neuropsiquiat 21: 291–295

4. Arendt T, Bigl V, Arendt A (1984) Neurone loss in the nucleus basalis of Meynert in Creutzfeldt-Jakob disease. Acta Neuropathol 65: 85–88

5. Ball MJ (1980) Features of Creutzfeldt-Jakob disease in brains of patients with familial dementia of Alzheimer type. Can J Neurol Sci 7: 51–57

6. Basler K, Oesch B, Scott M, Westaway D, Wälchli M, Groth DF, McKinley MP, Prusiner SB, Weissmann C (1986) Scrapie and cellular PrP isoforms are encoded by the same chromosomal gene. Cell 46: 417–428

7. Beck E, Daniel PM (1987) Neuropathology of transmissible spongiform encephalopathy. In: Pruisner SB, McKinley MP (eds) Prions: novel infectious pathogens causing scrapie and Creutzfeldt-Jakob disease. Academic Press, New York, pp 331–385

8. Bendheim PE, Barry RA, DeArmond SJ, Stites DP, Prusiner SB (1984) Antibodies to a scrapie prion protein. Nature 310: 418–421

9. Bernoulli C, Siegfried J, Baumgartner G, Regli F, Rabinowicz T, Gajdusek DC, Gibbs Jr CJ (1977) Danger of accidental person-to-person transmission of Creutzfeldt-Jakob disease by surgery. Lancet 1: 478–479

10. Bobowick AR, Brody JA, Matthews MR, Roos R, Gajdusek DC (1973) Creutzfeldt-Jakob disease: a case-control study. Am J Epidemiol 98: 381–394

11. Bockman JM, Kingsbury DT, McKinley MP, Bendheim PE, Prusiner SB (1985) Creutzfeldt-Jakob disease prion proteins in human brains. N Engl J Med 312: 73–78

12. Brown P (1980) An epidemiologic critique of Creutzfeldt-Jakob disease. Epidemiol Rev 2: 113–135

13. Brown P (1989) Central nervous system amyloidoses: a comparison of Alzheimer's disease and Creutzfeldt-Jakob disease. Neurology 39: 1103–1105

14. Brown P (1990) Iatrogenic Creutzfeldt-Jakob disease. Aust NZ J Med 20: 633–635

15. Brown P (1992) The phenotypic expression of different mutations in transmissible human spongiform encephalopathy. Rev Neurol (Paris) 148: 317–327

16. Brown P, Cathala F, Castaigne P, Gajdusek DC (1986) Creutzfeldt-Jakob disease: clinical analysis of a consecutive series of 230 neuropathologically verified cases. Ann Neurol 20: 597–602

17. Brown P, Cathala F, Sadowsky D, Gajdusek DC (1979) Creutzfeldt-Jakob disease in France: II. Clinical characteristics of 124 consecutive verified cases during the decade 1968–1977. Ann Neurol 6: 430–437

18. Brown P, Gajdusek DC, Gibbs Jr CJ, Asher DM (1985) Potential epidemic of Creutzfeldt-Jakob disease from human growth hormone therapy. N Engl J Med 313: 728–731

19. Brown P, Goldfarb LG, Kovanen J, Haltia M, Cathala F, Sulima M, Gibbs Jr CJ, Gajdusek DC (1992) Phenotypic characteristics of familial Creutzfeldt-Jakob disease associated with the codon 178^{Asn} *PRPNP* mutation. Ann Neurol 31: 282–285

20. Brown P, Goldfarb LG, McCombie WR, Nieto A, Squillacote D, Sheremata W, Little BW, Godec MS, Gibbs Jr CJ, Gajdusek DC (1992) Atypical Creutzfeldt-Jakob disease in an American family with an insert mutation in the PRNP amyloid precursor gene. Neurology 42: 422–427

21. Brown P, Jannotta F, Gibbs Jr CJ, Baron H, Guiroy DC, Gajdusek DC (1990) Coexistence of Creutzfeldt-Jakob disease and Alzheimer's disease in the same patient. Neurology 40: 226–228
22. Brown P, Rohwer RG, Gajdusek DC (1984) Sodium hydroxide decontamination of Creutzfeldt-Jakob disease virus. N Engl J Med 310: 727
23. Brown P, Wolff A, Gajdusek DC (1990) A simple and effective method for inactivating virus infectivity in formalin-fixed tissue samples from patients with Creutzfeldt-Jakob disease. Neurology 40: 887–890
24. Bruce ME, Dickinson AG (1987) Biological evidence that scrapie agent has an independent genome. J Gen Virol 68: 79–89
25. Bruce ME, McConnell I, Fraser H, Dickinson AG (1991) The disease characteristics of different strains of scrapie in sinc congenic mouse lines. Complications for the nature of the agent and host control of pathogenesis. J Gen Virol 72: 595–604
26. Buchanan CR, Preece MA, Milner RDG (1991) Mortality, neoplasia, and Creutzfeldt-Jakob disease in patients treated with human pituitary growth hormone in the United Kingdom. Br Med J 302: 824–828
27. Bugiani O, Tagliavini F, Giaconne G, Boeri R (1989) Creutzfeldt-Jakob disease: astrocytosis and spongiform changes of the white matter. In: Court LA, Dormont D, Brown P, Kingsburg DT (eds) Unconventional virus disease of the central nervous system. Département de Protection Sanitaire, Fonteney-aux-Roses Cedex, pp 172–183
28. Büeler H, Fischer M, Lang Y, Bluethmann H, Lipp H-P, DeArmond SJ, Prusiner SB, Aguet M, Weissmann C (1992) Normal development and behaviour of mice lacking the neuronal cell-surface PrP protein. Nature 356: 577–582
29. Cathala F, Baron H (1987) Clinical aspects of Creutzfeldt-Jakob disease. In: Prusiner SB, McKinley MP (eds) Prions: novel infectious pathogens causing scrapie and Creutzfeldt-Jakob disease. Academic Press, New York, pp 467–509
30. Chou SM, Payne WN, Gibbs Jr CJ, Gajdusek DC (1980) Transmission and scanning electron microscopy of spongiform change in Creutzfeldt-Jakob disease. Brain 103: 885–904
31. Clinton J, Lantos PL, Rossor M, Mullan M, Roberts GW (1990) Immunocytochemical confirmation of prion protein. Lancet 336: 515
32. Cochius JI, Mack K, Burns RJ, Alderman CP, Blumbergs PC (1990) Creutzfeldt-Jakob disease in a recipient of human pituitary-derived gonadotrophin. Aust NZ J Med 20: 592–593
33. Collinge J, Harding AE, Owen F, Poulter M, Lofthouse R, Boughey AM, Shah T, Crow TJ (1989) Diagnosis of Gerstmann-Sträussler syndrome in familial dementia with prion protein gene analysis. Lancet 2: 15–17
34. Collinge J, Owen F, Poulter M, Leach M, Crow TJ, Rossor MN, Hardy J, Mullan MJ, Janota I, Lantos PL (1990) Prion dementia without characteristic pathology. Lancet 336: 7–9
35. Collinge J, Palmer MS, Dryden AJ (1991) Genetic predisposition to iatrogenic Creutzfeldt-Jakob disease. Lancet 337: 1441–1442
36. Connolly JH, Allen IV, Dermott E (1988) Transmissible agent in the amyotrophic form of Creutzfeldt-Jakob disease. J Neurol Neurosurg Psychiatry 51: 1459–1460
37. Creutzfeldt HG (1920) Über eine eigenartige herdförmige Erkrankung des Zentralnervensystems. Z Ges Neurol Psychiatr 57: 1–18
38. Croxson M, Brown P, Synek B, Harrington MG, Frith R, Clover G, Wilson J, Gajdusek DC (1988) A new case of Creutzfeldt-Jakob disease associated with human growth hormone therapy in New Zealand. Neurology 38: 1128–1130

39. Cuillé J, Chelle PL (1936) Pathologie animale – La maladie dite tremblante du mouton est-elle inoculable? C R Acad Sci (III) 203: 1552–1554

40. Daniel PM (1972) Creutzfeldt-Jakob disease. J Clin Pathol Suppl (R Coll Pathol) 6: 97–101

41. Davanipour Z, Goodman L, Alter M, Sobel E, Asher D, Gajdusek DC (1984) Possible modes of transmission of Creutzfeldt-Jakob disease. N Engl J Med 311: 1582–1583

42. DeArmond SJ, Mobley WC, DeMott DL, Barry RA, Beckstead JH, Prusiner SB (1987) Changes in the localization of brain prion proteins during scrapie infection. Neurology 37: 1271–1280

43. Devillemeur TB, Gourmelen M, Beauvais P, Rodriguez D, Vaudour G, Deslys JP, Dormont D, Richard P, Richardet J-M (1992) Maladie de Creutzfeldt-Jakob chez quatre enfants traités par hormone de croissance. Rev Neurol (Paris) 148: 328–334

44. Diedrich JF, Minnigan H, Carp RI, Whitaker JN, Race R, Frey II W, Haase AT (1991) Neuropathological changes in scrapie and Alzheimer's disease are associated with increased expression of apolipoprotein E and cathepsin D in astrocytes. J Virol 65: 4759–4768

45. Doh-ura K, Tateishi J, Sasaki H, Kitamoto T, Sakaki Y (1989) Pro → Leu change at position 102 of prion protein is the most common but not the sole mutation related to Gerstmann-Sträussler syndrome. Biochem Biophys Res Commun 163: 974–979

46. Duffy P, Wolf J, Collins G, DeVoe AG, Streeten B, Cowen D (1974) Possible person-to-person transmission of Creutzfeldt-Jakob disease. N Engl J Med 290: 692–693

47. Duguid JR, Dinauer MC (1990) Library subtraction of *in vitro* cDNA libraries to identify differentially expressed genes in scrapie infection. Nucleic Acids Res 18: 2789–2792

48. Duguid JR, Rohwer RG, Seed B (1988) Isolation of cDNAs of scrapie-modulated RNAs by subtractive hybridization of a cDNA library. Proc Natl Acad Sci USA 85: 5738–5742

49. Ferber RA, Wiesenfeld SL, Roos RP, Bobowick AR, Gibbs Jr CJ, Gajdusek DC (1974) Familial Creutzfeldt-Jakob disease: transmission of the familial disease to primates. In: Subirana A, Burrows JM (eds) Proceedings of the Xth International Congress of Neurology. Excerpta Medica, ICS No. 319, Amsterdam, pp 358–380

50. Flament-Durand J, Couck AM (1979) Spongiform alterations in brain biopsies of presenile dementia. Acta Neuropathol (Berl) 46: 159–162

51. Fradkin JE, Schonberger LB, Mills JL, Gunn WJ, Piper JM, Wysowski DK, Thomson R, Durako S, Brown P (1991) Creutzfeldt-Jakob disease in pituitary growth hormone recipients in the United States. JAMA 265: 880–884

52. Gajdusek DC, Gibbs CJ, Alpers M (1966) Experimental transmission of a kuru-like syndrome to chimpanzees. Nature 209: 794–796

53. Gajdusek DC, Zigas V (1957) Degenerative disease of the central nervous system in New Guinea. The endemic occurrence of "kuru" in the native population. N Engl J Med 257: 974–978

54. Gerstmann J (1928) Über ein noch nicht beschriebenes Reflexphänomen bei einer Erkrankung des zerebellären Systems. Wien Med Wochenschr 78: 906–908

55. Gerstmann J, Sträussler E, Scheinker I (1936) Über eine eigenartige hereditär-familiäre Erkrankung des Zentralnervensystems. Zugleich ein Beitrag zur Frage des vorzeitigen lokalen Alterns. Z Neurol 154: 736–762

56. Ghetti B, Tagliavini F, Masters CL, Beyreuther K, Giaccone G, Verga L, Farlow MR, Conneally PM, Dlouhy SR, Azzarelli B, Bugiani O (1989) Gerstmann-Sträussler-Scheinker disease. II. Neurofibrillary tangles and plaques with PrP-amyloid coexist in an affected family. Neurology 39: 1453–1461

57. Gibbs CJ Jr, Amyx HL, Bacote A, Masters CL, Gajdusek DC (1980) Oral transmission of kuru, Creutzfeldt-Jakob disease, and scrapie to nonhuman primates. J Infect Dis 142: 205–208

58. Gibbs CJ Jr, Gajdusek DC, Asher DM, Alpers MP, Beck E, Daniel PM, Matthews WP (1968) Creutzfeldt-Jakob disease (spongiform encephalopathy): transmission to the chimpanzee. Science 161: 388–389

59. Gibbs CJ Jr, Joy A, Heffner R, Franko M, Miyazaki M, Asher DM, Parisi JE, Brown PW, Gajdusek DC (1985) Clinical and pathological features and laboratory confirmation of Creutzfeldt-Jakob disease in a recipient of pituitary-derived human growth hormone. N Engl J Med 313: 734–738

60. Goldfarb LG, Brown P, Goldgaber D, Asher DM, Strass N, Graupera G, Piccardo P, Brown WT, Rubinstein R, Boellaard JW, Gajdusek DC (1989) Patients with Creutzfeldt-Jakob disease and kuru lack the mutation in the PRIP gene found in Gerstmann-Sträussler syndrome, but they show a different double allele mutation in the same gene. Am J Hum Genet 45 [Suppl]: A189 (Abstract)

61. Goldfarb LG, Brown P, Haltia M, Cathala F, McCombie WR, Kovanen J, Cervenáková L, Goldin L, Nieto A, Godec MS, Asher DM, Gajdusek DC (1992) Creutzfeldt-Jakob disease cosegregates with the condon 178^{Asn} PRNP mutation in families of European origin. Ann Neurol 31: 274–281

62. Goldfarb LG, Brown P, McCombie WR, Goldgaber D, Swergold GD, Wills PR, Cervenakova L, Baron H, Gibbs Jr CJ, Gajdusek DC (1991) Transmissible familial Creutzfeldt-Jakob disease associated with five, seven, and eight extra octapeptide coding repeats in the PRNP gene. Proc Natl Acad Sci USA 88: 10926–10930

63. Goldfarb LG, Brown P, Vrbovská A, Baron H, McCombie WR, Cathala F, Gibbs Jr CJ, Gajdusek DC (1992) An insert mutation in the chromosome 20 amyloid precursor gene in a Gerstmann-Sträussler-Scheinker family. J Neurol Sci 111: 189–194

64. Goldfarb LG, Korczyn AD, Brown P, Chapman J, Gajdusek DM (1990) Mutation in codon 200 of scrapie amyloid precursor gene linked to Creutzfeldt-Jakob disease in Sephardic Jews of Libyan and non-Libyan origin. Lancet 336: 637–638

65. Goldfarb LG, Mitrová E, Brown P, Toh BH, Gajdusek DC (1990) Mutation in codon 200 of scrapie amyloid protein gene in two clusters of Crcutzfcldt-Jakob disease in Slovakia. Lancet 336: 514–515

66. Goldgaber D, Goldfarb LG, Brown P, Asher DM, Brown WT, Lin S, Teener JW, Feinstone SM, Rubenstein R, Kascsak RJ, Boellaard JW, Gajdusek DC (1989) Mutations in familial Creutzfeldt-Jakob disease and Gerstmann-Sträussler-Scheinker's syndrome. Exp Neurol 106: 204–206

67. Gordon WS (1946) Advances in veterinary research. Looping-ill, tickborne fever and scrapie. Vet Res 58: 516–525

68. Gray EG (1986) Spongiform encephalopathy: a neurocytologist's viewpoint with a note on Alzheimer's disease. Neuropathol. Appl Neurobiol 12: 149–172

69. Hadlow WJ (1959) Scrapie and kuru. Lancet 2: 289–290

70. Haltia M, Kovanen J, Goldfarb LG, Brown P, Gajdusek DC (1991) Familial Creutzfeldt-Jakob disease in Finland: epidemiological, clinical, pathological and molecular genetic studies. Eur J Epidemiol 7: 494–500

71. Harrington MG, Merril CR, Asher DM, Gajdusek DC (1986) Abnormal proteins in the cerebrospinal fluid of patients with Creutzfeldt-Jakob disease. N Engl J Med 315: 279–283

72. Hecker R, Taraboulos A, Scott M, Pan K-M, Yang S-L, Torchia M, Jendroska K, DeArmond SJ, Prusiner SB (1992) Replication of distinct scrapie prion isolates is region specific in brains of transgenic mice and hamsters. Genes Dev 6: 1213–1228

73. Hirano A, Ghatak NR, Johnson AB, Partnow MJ, Gomori AJ (1972) Argento-philic plaques in Creutzfeldt-Jakob disease. Arch Neurol 26: 530–542

74. Hsiao K, Baker HF, Crow TJ, Poulter M, Owen F, Terwilliger JD, Westaway D, Ott J, Prusiner SB (1989) Linkage of a prion protein missense variant to Gerstmann-Sträussler syndrome. Nature 338: 342–345

75. Hsiao K, Dlouhy SR, Farlow MR, Cass C, Da Costa M, Conneally PM, Hodes ME, Ghetti B, Prusiner SB (1992) Mutant prion proteins in Gerstmann-Sträussler-Scheinker disease with neurofibrillary tangles. Nat Genet 1: 68–71

76. Hsiao K, Prusiner SB (1990) Inherited human prion diseases. Neurology 40: 1820–1827

77. Hsiao KK, Cass C, Schellenberg GD, Bird T, Devine-Gage E, Wisniewski H, Prusiner SB (1991) A prion protein variant in a family with the telencephalic form of Gerstmann-Sträussler-Scheinker syndrome. Neurology 41: 681–684

78. Hsiao KK, Cass C, Schellenberg G, Dwine-Gage E, Bird T, Prusiner SB (1990) Correlation of specific prion protein mutations with different forms of prion diseases. Neurology 40 [Suppl 1]: 388 (Abstract)

79. Hsiao KK, Scott M, Foster D, Groth DF, DeArmond SJ, Prusiner SB (1990) Spontaneous neurodegeneration in transgenic mice with mutant prion protein. Science 250: 1587–1590

80. Jakob A (1921) Über eigenartige Erkrankungen des Zentralnervensystems mit bemerkenswertem anatomischem Befunde (spastische Pseudosklerose-Encephalo-myelopathie mit disseminierten Degenerationsherden). Dtsch Z Nervenheilkd 70: 132–146

81. Jakob A (1923) Spastische Pseudosklerose. In: Jakob A (ed) Die extrapyra-midalen Erkrankungen. Springer, Berlin Heidelberg, pp 215–245

82. Kim JH, Manuelidis EE (1989) Neuronal alterations in experimental Creutzfeldt-Jakob disease: a Golgi study. J Neurol Sci 89: 93–101

83. Kimberlin RH (1990) Bovine spongiform encephalopathy. Taking stock of the issues. Nature 345: 763–764

84. Kimberlin RH, Walker CA, Fraser H (1989) The genomic identity of different strains of mouse scrapie is expressed in hamsters and preserved on reisolation in mice. J Gen Virol 70: 2017–2026

85. Kirschbaum WR (1968) Jakob-Creutzfeldt disease. Elsevier, New York

86. Kitamoto T, Muramoto T, Mohri S, Doh-ura K, Tateishi J (1991) Abnormal isoform of prion protein accumulates in follicular dendritic cells in mice with Creutzfeldt-Jakob disease. J Virol 65: 6292–6295

87. Kitamoto T, Ogomori K, Tateishi J, Prusiner SB (1987) Formic acid pretreatment enhances immunostaining of cerebral and systemic amyloids. Lab Invest 57: 230–236

88. Kitamoto T, Shin R-W, Doh-ura K, Tomokane N, Miyazono M, Muramoto T, Tateishi J (1992) Abnormal isoform of prion proteins accumulates in the synaptic structures of the central nervous system in patients with Creutzfeldt-Jakob disease. Am J Pathol 140: 1285–1294

89. Kitamoto T, Tateishi J (1988) Immunohistochemical confirmation of Creutzfeldt-Jakob disease with a long clinical course with amyloid plaque core antibodies. Am J Pathol 131: 435–443

90. Kitamoto T, Tateishi J, Tashima T, Takeshita I, Barry RA, DeArmond SJ, Prusiner SB (1986) Amyloid plaques in Creutzfeldt-Jakob disease stain with prion protein antibodies. Ann Neurol 20: 204–208

91. Klatzo I, Gajdusek DC, Zigas V (1959) Pathology of the kuru. Lab Invest 8: 799–847

92. Koch TK, Berg BO, De Armond SJ, Gravina RF (1985) Creutzfeldt-Jakob disease in a young adult with idiopathic hypopituitarism. Possible relation to the administration of cadaveric human growth hormone. N Engl J Med 313: 731–733

93. Kondo K, Kuroiwa Y (1982) A case control study of Creutzfeldt-Jakob disease: association with physical injuries. Ann Neurol 11: 377–381

94. Kretzschmar HA, Honold G, Seitelberger F, Feucht M, Wessely P, Mehraein P, Budka H (1991) Prion protein mutation in family first reported by Gerstmann, Sträussler, and Scheinker. Lancet 337: 1160

95. Kretzschmar HA, Kitamoto T, Doerr-Schott J, Mehraein P, Tateishi J (1991) Diffuse deposition of immunohistochemically labeled prion protein in the granular layer of the cerebellum in a patient with Creutzfeldt-Jakob disease. Acta Neuropathol (Berl) 82: 536–540

96. Kretzschmar HA, Kufer P, Riethmüller G, DeArmond SJ, Prusiner SB, Schiffer D (1992) Prion protein mutation at codon 102 in an Italian family with Gerstmann-Sträussler-Scheinker syndrome. Neurology 42: 809–810

97. Kretzschmar HA, Prusiner SB, Stowring LE, DeArmond SJ (1986) Scrapie prion proteins are synthesized in neurons. Am J Pathol 122: 1–5

98. Kretzschmar HA, Stowring LE, Westaway D, Stubblebine WH, Prusiner SB, DeArmond SJ (1986) Molecular cloning of a human prion protein cDNA. DNA 5: 315–324

99. Krüger H, Meesmann C, Rohrbach E, Müller J, Mertens HG (1990) Panencephalopathic type of Creutzfeldt-Jakob disease with primary extensive involvement of white matter. Eur Neurol 30: 115–119

100. Kuroiwa Y, Celesia GG (1980) Clinical significance of periodic EEG patterns. Arch Neurol 37: 15–20

101. Lantos PL (1992) From slow virus to prion: a review of transmissible spongiform encephalopathies. Histopathology 20: 1–11

102. Laplanche JL, Chatelain J, Launay JM, Gazengel C, Vidaud M (1990) Deletion in prion protein gene in a Moroccan family. Nucleic Acids Res 18: 6745–6746

103. Laszlo L, Lowe J, Self T, Kenward N, Landon M, McBride T, Farquhar C, McConnell I, Brown J, Hope J, Mayer RJ (1992) Lysosomes as key organelles in the pathogenesis of prion encephalopathies. J Pathol 166: 333–341

104. Liberski PP, Budka H, Sluga E, Barcikowska M, Kwiecinski H (1991) Tubulovesicular structures in human and experimental Creutzfeldt-Jakob disease. Eur J Epidemiol 7: 551–555

105. Liberski PP, Budka H, Sluga E, Barcikowska M, Kwiecinski H (1992) Tubulovesicular structures in Creutzfeldt-Jakob disease. Acta Neuropathol (Berl) 84: 238–243

106. Liberski PP, Kwiecinski H, Barcikowska M, Mirecka B, Kulczycki J, Kida E, Brown P, Gajdusek DC (1991) PrP amyloid plaques in Creutzfeldt-Jakob disease of short duration: immunohistochemical studies of 5 cases from Poland. Eur J Epidemiol 7: 505–510

107. Liberski PP, Yanagihara R, Gibbs Jr CJ, Gajdusek DC (1990) Appearance of tubulovesicular structures in experimental Creutzfeldt-Jakob disease and scrapie precedes the onset of clinical disease. Acta Neuropathol (Berl) 79: 349–354

108. Lo Russo F, Neri G, Figa Talamanca L (1980) Creutzfeldt-Jakob disease and sheep brain. A report from central and southern Italy. Ital J Neurol Sci 1: 171–174

109. Mackenzie A (1983) Immunohistochemical demonstration of glial fibrillary acidic protein in scrapie. J Comp Pathol 93: 251–259

110. Manaka H, Kato T, Kurita K, Katagiri T, Shikama Y, Kujirai K, Kawanami T, Suzuki Y, Nihei K, Sasaki H, Yamada S, Hirota K, Kusaka H, Imai T (1992) Marked increase in cerebrospinal fluid ubiquitin in Creutzfeldt-Jakob disease. Neurosci Lett 139: 47–49

111. Mancardi GL, Mandybur TI, Liwnicz BH (1982) Spongiform-like changes in Alzheimer's disease. An ultrastructural study. Acta Neuropathol (Berl) 56: 146–150

112. Manson J, West JD, Thomson V, McBride P, Kaufman MH, Hope J (1992) The prion protein gene: a role in mouse embryogenesis? Development 115: 117–122

113. Markus HS, Duchen LW, Parkin EM, Kurtz AB, Jacobs HS, Costa DC, Harrison MJ (1992) Creutzfeldt-Jakob disease in recipients of human growth hormone in the United Kingdom: a clinical and radiographic study. Q J Med 82: 43–51

114. Marsh RF, Sipe JC, Morse SS, Hanson RP (1976) Transmissible mink encephalopathy. Reduced spongiform degeneration in aged mink of the Chediak-Higashi genotype. Lab Invest 34: 381–386

115. Marzewski DJ, Towfighi J, Harrington MG, Merril CR, Brown P (1988) Creutzfeldt-Jakob disease following pituitary-derived human growth hormone therapy: a new American case. Neurology 38: 1131–1133

116. Masters CL, Beyreuther K (1988) Neuropathology of unconventional virus infections: molecular pathology of spongiform change and amyloid plaque deposition. In: Novel infectious agents and the central nervous system. Ciba Foundation Symposium 135. Wiley, Chichester, pp 24–36

117. Masters CL, Gajdusek DC (1982) Rec Adv Neuropathol 2: 139–163

118. Masters CL, Gajdusek DC, Gibbs Jr CJ (1981) The familial occurrence of Creutzfeldt-Jakob disease and Alzheimer's disease. Brain 104: 535–558

119. Masters CL, Gajdusek DC, Gibbs Jr CJ (1981) Creutzfeldt-Jakob disease virus isolations from the Gerstmann-Sträussler syndrome. With an analysis of the various forms of amyloid plaque deposition in the virus-induced spongiform encephalopathies. Brain 104: 559–588

120. Masters CL, Harris JO, Gajdusek DC, Gibbs Jr CJ, Bernoulli C, Asher DM (1979) Creutzfeldt-Jakob disease: patterns of worldwide occurrence and the significance of familial and sporadic clustering. Ann Neurol 5: 177–188

121. Masters CL, Richardson Jr EP (1978) Subacute spongiform encephalopathy (Creutzfeldt-Jakob disease). The nature and progression of spongiform change. Brain 101: 333–344

122. Masullo C, Pocchiari M, Macchi G, Alema G, Piazza G, Panzera MA (1989) Transmission of Creutzfeldt-Jakob disease by dural cadaveric graft. J Neurosurg 71: 954–955

123. Matthews WB (1975) The clinical aspects of slow virus infections of the human brain. In: Illis LS (ed) Viral diseases of the central nervous system. Baillière Tindall, London, pp 145–160

124. McKinley MP, Meyer RK, Kenaga L, Rahbar F, Cotter R, Serban A, Prusiner SB (1991) Scrapie prion rod formation *in vitro* requires both detergent extraction and limited proteolysis. J Virol 65: 1340–1351

125. McKinley MP, Taraboulos A, Kenaga L, Serban D, Stieber A, DeArmond SJ, Prusiner SB, Gonatas N (1991) Ultrastructural localization of scrapie prion proteins in cytoplasmic vesicles of infected cultured cells. Lab Invest 65: 622–630

126. Medori R, Montagna P, Tritschler HJ, LeBlanc A, Cortelli P, Tinuper P, Lugaresi E, Gambetti P (1992) Fatal familial insomnia: a second kindred with mutation of prion protein gene at codon 178. Neurology 42: 669–670

127. Medori R, Tritschler H-J, LeBlanc A, Villare F, Manetto V, Chen HY, Xue R, Leal S, Montagna P, Cortelli P, Tinuper P, Avoni P, Moghi M, Baruzzi A, Hauw JJ, Ott J, Lugaresi E, Autilio-Gambetti L, Gambetti P (1992) Fatal familial insomnia, a prion disease with a mutation at codon 178 of the prion protein gene. N Engl J Med 326: 444–449

128. Meyer N, Rosenbaum V, Schmidt B, Gilles K, Mirenda C, Groth D, Prusiner SB, Riesner D (1991) Search for a putative scrapie genome in purified prion fractions reveals a paucity of nucleic acids. J Gen Virol 72: 37–50

129. Miyashita K, Inuzuka T, Kondo H, Saito Y, Fujita N, Matsubara N, Tanaka R, Hinokuma K, Ikuta F, Miyatake T (1991) Creutzfeldt-Jakob disease in a patient with a cadaveric dural graft. Neurology 41: 940–941

130. Miyazono M, Kitamoto T, Iwaki T, Tateishi J (1992) Colocalization of prion protein and β protein in the same amyloid plaques in patients with Gerstmann-Sträussler syndrome. Acta Neuropathol (Berl) 83: 333–339

131. Mizutani T (1981) Neuropathology of Creutzfeldt-Jakob disease in Japan. With special reference to the panencephalopathic type. Acta Pathol Jpn 31: 903–922

132. Mizutani T, Morimatsu Y, Shiraki H (1984) Clinical pictures of Creutzfeldt-Jakob disease based on 97 autopsy cases in Japan – with special reference to clinico-pathological correlation of cerebellar symptoms. Rinsho Shinkei-gaku (Clin Neurol) 24: 23–32

133. Nakazato Y, Hirato J, Ishida Y, Hoshi S, Hasegawa M, Fukuda T (1990) Swollen cortical neurons in Creutzfeldt-Jakob disease contain a phosphorylated neurofilament epitope. J Neuropathol Exp Neurol 49: 197–205

134. Nevin S, McMenemey WH, Behrman D, Jones DP (1960) Subacute spongiform encephalopathy – a subacute form of encephalopathy attributable to vascular dysfunction (spongiform cerebral atrophy). Brain 83: 519–564

135. New MI, Brown P, Temeck JW, Owens C, Hedley-Whyte ET, Richardson EP (1988) Preclinical Creutzfeldt-Jakob disease discovered at autopsy in a human growth hormone recipient. Neurology 38: 1133–1134

136. Nisbet TJ, MacDonaldson I, Bishara SN (1989) Creutzfeldt-Jakob disease in a second patient who received a cadaveric dura mater graft. JAMA 261: 1118

137. Nochlin D, Sumi SM, Bird TD, Snow AD, Leventhal CM, Beyreuther K, Masters CL (1989) Familial dementia with PrP-positive amyloid plaques: a variant of Gerstmann-Sträussler syndrome. Neurology 39: 910–918

138. Oesch B, Westaway D, Wälchli M, McKinley MP, Kent SBH, Aebersold R, Barry RA, Tempst P, Teplow DB, Hood LE, Prusiner SB, Weissmann C (1985) A cellular gene encodes scrapie PrP 27–30 protein. Cell 40: 735–746

139. Owen F, Poulter M, Collinge J, Leach M, Shah T, Lofthouse R, Chen Y, Crow TJ, Harding AE, Hardy J, Rossor MN (1991) Insertions in the prion protein gene in atypical dementias. Exp Neurol 112: 240–242

140. Owen F, Poulter M, Lofthouse R, Collinge J, Crow TJ, Risby D, Baker HF, Ridley RM, Hsiao K, Prusiner SB (1989) Insertion in prion protein gene in familial Creutzfeldt-Jakob disease. Lancet 1: 51–52

141. Palmer MS, Dryden AJ, Hughes JT, Collinge J (1991) Homozygous prion protein genotype predisposes to spoardic Creutzfeldt-Jakob disease. Nature 352: 340–342

142. Piccardo P, Safar J, Ceroni M, Gajdusek DC, Gibbs Jr CJ (1990) Immunohisto-chemical localization of prion protein in spongiform encephalopathies and normal brain tissue. Neurology 40: 518–522

143. Pocchiari M, Masullo C, Salvatore M, Genuardi M, Galgani S (1992) Creutzfeldt-Jakob disease after non-commercial dura mater graft. Lancet 340: 614–615

144. Powell-Jackson J, Weller RO, Kennedy P, Preece MA, Whitcombe EM, Newson-Davis J (1985) Creutzfeldt-Jakob disease after administration of human growth hormone. Lancet 2: 244–246

145. Prusiner SB (1982) Novel proteinaceous infectious particles cause scrapie. Science 216: 136–144

146. Prusiner SB (1991) Molecular biology of prion diseases. Science 252: 1515–1522

147. Prusiner SB, Hsiao KK, Bredesen DE, Kingsbury DT (1989) Human slow infections caused by prions. In: Gilden DH, Lipton HL (eds) Clinical and molecular aspects of neurotropic virus infection. Kluwer, Dortrecht, pp 423–467

148. Prusiner SB, Scott M, Foster D, Pan K-M, Groth D, Mirenda C, Torchia M, Yang S-L, Serban D, Carlson GA, Hoppe PC, Westaway D, DeArmond SJ (1990) Transgenetic studies implicate interactions between homologous PrP isoforms in scrapie prion replication. Cell 63: 673–686

149. Puckett C, Concannon P, Casey C, Hood L (1991) Genomic structure of the human prion protein gene. Am J Hum Genet 49: 320–329

150. Renault F, Richard P (1991) Early electroretinogram alterations in Creutzfeldt-Jakob disease after growth hormone treatment. Lancet 338: 191

151. Roos R, Gajdusek DC, Gibbs CJ (1973) The clinical characteristics of transmissible Creutzfeldt-Jakob disease. Brain 96: 1–20

152. Salazar AM, Masters CL, Gajdusek DC, Gibbs Jr CJ (1983) Syndromes of amyotrophic lateral sclerosis and dementia: relation to transmissible Creutzfeldt-Jakob disease. Ann Neurol 14: 17–26

153. Seitelberger F (1962) Eigenartige familiär-hereditäre Krankheit des Zentralnervensystems in einer niederösterreichischen Sippe. Wien Klin Wochenschr 74: 687–691

154. Seitelberger F (1981) Spinocerebellar ataxia with dementia and plaque-like deposits. In: Vinken PJ, Bruyn GW (eds) Handbook of clinical neurology, vol 42. North-Holland, Amsterdam, pp 182–183

155. Seitelberger F (1981) Sträussler's disease. Acta Neuropathol (Berl) [Suppl] 7: 341–343

156. Serban D, Taraboulos A, DeArmond SJ, Prusiner SB (1990) Rapid detection of Creutzfeldt-Jakob disease and scrapie prion proteins. Neurology 40: 110–117

157. Smith TW, Anwer U, DeGirolami U, Drachman DA (1987) Vacuolar change in Alzheimer's disease. Arch Neurol 44: 1225–1228

158. Sparkes RS, Simon M, Cohn VH, Fournier REK, Lem J, Klisak I, Heinzmann C, Blatt C, Lucero M, Mohandas T, DeArmond SJ, Westaway D, Prusiner SB, Weiner LP (1986) Assignment of the human and mouse prion protein genes to homologous chromosomes. Proc Natl Acad Sci USA 83: 7358–7362

159. Spielmeyer W (1922) Histopathologie des Nervensystems. Springer, Berlin Heidelberg

160. Spielmeyer W (1922) Die histopathologische Forschung in der Psychiatrie. Wien Klin Wochenschr 1: 1817–1819

161. Stahl N, Borchelt DR, Prusiner SB (1990) Differential release of celllar and scrapie prion proteins from ceullular membranes by phosphatidylinositol-specific phospholipase C. Biochemistry 29: 5405–5412

162. Steiner I, Polachek I, Melamed E (1984) Dementia and myoclonus in a case of cryptococcal encephalitis. Arch Neurol 41: 216–217

163. Tagliavini F, Prelli F, Ghiso J, Bugiani O, Serban D, Prusiner SB, Farlow MR, Ghetti B, Frangione B (1991) Amyloid protein of Gerstmann-Sträussler-Scheinker disease (Indiana kindred) is an 11 kd fragment of prion protein with an N-terminal glycine at codon 58. EMBO J 10: 513–519

164. Taraboulos A, Jendroska K, Serban D, Yang S-L, DeArmond SJ, Prusiner SB (1992) Regional mapping of prion proteins in brain. Proc Natl Acad Sci USA 89: 7620–7624

165. Taraboulos A, Serban D, Prusiner SB (1990) Scrapie prion proteins accumulate in the cytoplasm of persistently infected cultured cells. J Cell Biol 110: 2117–2132

166. Tateishi J, Doh-ura K, Kitamoto T, Tranchant C, Warter MM, Boellaard JW (1992) Prion protein gene analysis and transmission studies of Creutzfeldt-Jakob disease. Ellis Horwood (Abstract)

167. Tateishi J, Tashima T, Kitamoto T (1991) Practical methods for chemical inactivation of Creutzfeldt-Jakob disease pathogen. Microbiol Immunol 35: 163–166

168. Thadani V, Penar PL, Partington J, Kalb R, Janssen R, Schonberger LB, Rabkin CS, Prichard JW (1988) Creutzfeldt-Jakob disease probably acquired from a cadaveric dura mater graft. Case report. J Neurosurg 69: 766–769

169. Tintner R, Brown P, Hedley-Whyte ET, Rappaport EB, Piccardo CP, Gajdusek DC (1986) Neuropathologic verification of Creutzfeldt-Jakob disease in the exhumed American recipient of human pituitary growth hormone: epidemiologic and pathogenetic implications. Neurology 36: 932–936

170. Tranchant C, Doh-ura K, Steinmetz G, Chevalier Y, Kitamoto T, Tateishi J, Warter JM (1991) Mutation of codon 117 of the prion gene in a family of Gerstmann-Sträussler-Scheinker disease. Rev Neurol (Paris) 147: 274–278

171. Traub RD, Pedley TA (1981) Virus-induced electrotonic coupling: hypothesis on the mechanism of periodic EEG discharges in Creutzfeldt-Jakob disease. Ann Neurol 10: 405–410

172. Tsuji S, Kuroiwa Y (1983) Creutzfeldt-Jakob disease in Japan. Neurology 33: 1503–1506

173. Watson CP (1979) Clinical similarity of Alzheimer and Creutzfeldt-Jakob disease. Ann Neurol 6: 368–369

174. Weissmann C (1991) A "unified theory" of prion propagation. Nature 352: 679–683

175. Wietgrefe S, Zupancic M, Haase A, Chesebro B, Race R, Frey II W, Rustan T, Friedman RL (1985) Cloning of a gene whose expression is increased in scrapie and in senile plaques in human brain. Science 230: 1177–1179

176. Will RG, Matthews WB (1982) Evidence for case-to-case transmission of Creutzfeldt-Jakob disease. J Neurol Neurosurg Psychiatry 45: 235–238

177. Will RG, Matthews WB (1984) A retrospective study of Creutzfeldt-Jakob disease in England and Wales 1970–79. I: Clinical features. J Neurol Neurosurg Psychiatry 47: 134–140

178. Willison HJ, Gale AN, McLaughlin JE (1991) Creutzfeldt-Jakob disease following cadaveric dura mater graft. J Neurol Neurosurg Psychiatry 54: 940

Author's address: Dr. H.A. Kretzschmar, Abteilung Neuropathologie, Universität Göttingen, Robert-Koch-Strasse 40, D-37075 Göttingen, Federal Republic of Germany.

Arch Virol (1993) [Suppl] 7: 295–301

Distribution of cytopathogenic and noncytopathogenic bovine virus diarrhea virus in tissues from a calf with experimentally induced mucosal disease using antigenic and genetic markers

Brief Report

I. Greiser-Wilke[1], E. Liebler[2], L. Haas[1], B. Liess[1], J. Pohlenz[2],
and V. Moennig[1]

Institute of [1] Virology and [2] Pathology, Hannover Veterinary School, Hannover,
Federal Republic of Germany

Summary. A comparative analysis of the distribution of cytopathogenic (cp) and noncytopathogenic (ncp) bovine virus diarrhea disease (BVD) virus in tissues from a calf with experimentally induced mucosal disease was performed using immunohistology and polymerase chain reaction after reverse transcription (RT-PCR) of viral RNA. For immuno-histology, an antigenic marker on the superinfecting cp BVD virus defined by a monoclonal antibody (mab) was used, and overall presence of antigen was assessed with a pestivirus specific mab. The primers selected for RT-PCR detected the genomic insertion in the p125 region of the superinfecting cp BVD virus. Both methods gave consistent results.

*

Bovine viral diarrhea (BVD) virus is the causative agent of fatal mucosal disease (MD) which is characterized by severe lesions of the gastrointestinal barrier and lymphoid tissues. From naturally occurring cases of MD, both the cytopathogenic (cp) and the noncytopathogenic (ncp) biotypes of BVD virus can be isolated, whereas persistently infected animals harbour only the ncp biotype virus [8]. This fact and the successful induction of experimental MD by superinfection of persistently infected cattle using cp BVDV suggest that the cp biotype plays a crucial role in the pathogenesis of MD [1, 2, 12].

BVD virus is a member of the pestivirus genus in the family *Flaviviridae* [17]. On the molecular level the cp biotype is characterized by the expression of the nonstructural (ns) protein p80 that is antigenically related to the p125 of ncp pestiviruses [4, 14, 15]. Molecular analysis of the corresponding genes has revealed that either cellular RNA insertions

[5, 9, 10] or gene duplications and rearrangements with or without cellular insertions may lead to the expression of p80 [10, 11, 16].

For induction of experimental MD in persistently infected animals antigenic homology of both biotypes involved is a prerequisite [1, 2]. It was shown that monoclonal antibodies directed against the major viral glycoprotein gp53 allowed the identification of "matching" cp BVD virus for inducing MD in persistently infected animals [12]. In addition, this approach facilitated the definition of single marker epitopes on the cp BVD virus allowing the immunohistologic localization of the virus in infected tissues [7]. However, the possibility cannot be excluded that recombinations between the cp and the ncp BVD viruses occur during the course of superinfection, leading to the generation of "new" populations of cp BVD viruses lacking the antigenic marker.

In order to exclude this possibility, polymerase chain reaction (PCR) after reverse transcription of viral RNA (RT-PCR) was used to analyze the distribution of both biotypes in organ tissues of a calf suffering from MD [5]. In the present communication we demonstrate that the presence of cp BVD virus RNA correlates with the immunohistological demonstration of cp BVD viral antigen.

A panel of 15 monoclonal antibodies (mabs) was used for selection of cp BVD virus strains for induction of MD in a persistently infected calf [12]. Strains MD1 and Indiana showed homology with the persisting ncp BVD virus. In addition, Indiana reacted with mab BVD/CT3, and MD1 with mab BVD/CT2. Both viruses were shown to carry insertions in the genomic region coding for the ns protein p125 [5]. The persistently viremic calf was inoculated intranasally with a mixture of Indiana and MD1 grown in fetal bovine kidney (FBCK) cells, containing 10^7 tissue culture infectious doses 50 in 6 ml. Blood was collected daily beginning at day 2 post infection and the lymphocyte fraction was isolated [6]. The calf developed erosions of the oral mucosa and diarrhea 13 days post inoculation and was euthanized on day 14. At necropsy, tissues were collected from cerebrum, parotis, palatine tonsils, spleen, ileal Peyer's patch, lymphoid tissue at the ileocecal entrance and in the proximal colon and mid colon for RT-PCR analysis and immunohistochemistry.

Distribution of BVD virus antigen was analyzed with mabs using the indirect immunoperoxidase method on frozen sections [7]. Reactivity with BVD/CT3, the marker for Indiana virus, was found in tonsillar and intestinal epithelial cells, and in mononuclear cells in the lamina propria. Viral antigen was diffusely present in epithelium of ileum, ileocecal entrance and proximal colon, and multifocally in epithelium of tonsils and mid colon (Fig. 1A). In the mucosa-associated lymphoid tissue in the tonsils, ileum, ileocecal entrance and proximal colon, viral antigen was found in cells with dendritic or macrophage-like morphology pre-

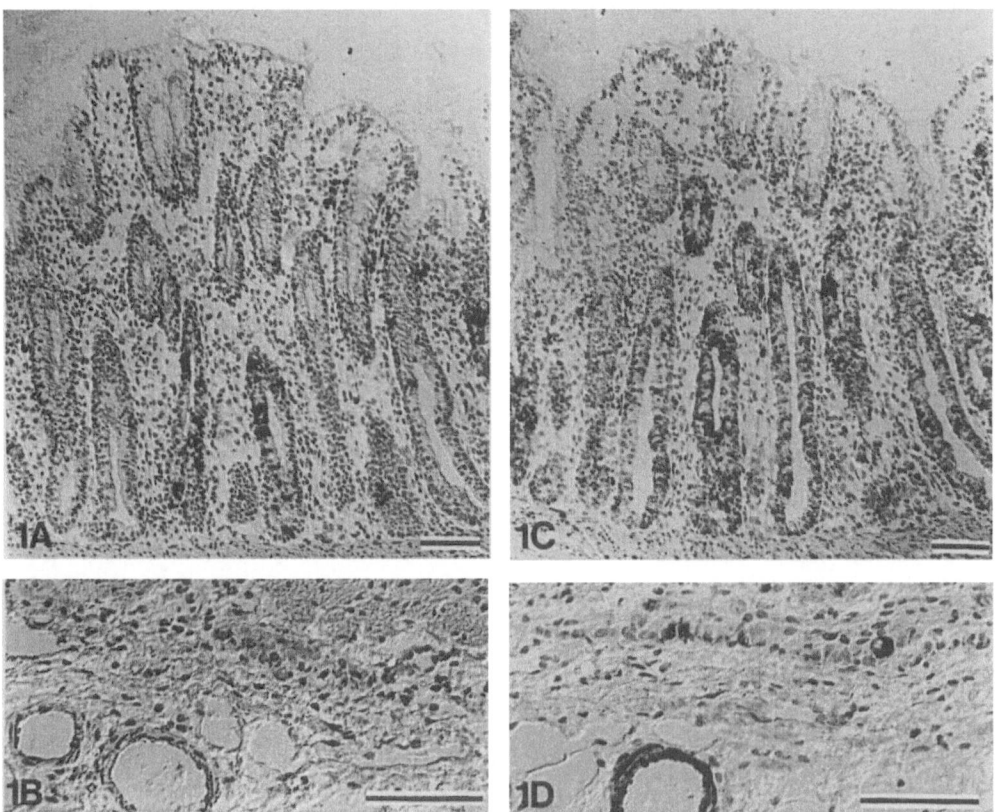

Fig. 1. Consecutive sections of mid colon stained with BVD/CT3 (**A** and **B**) and BVD/C16 (**C** and **D**). **A** Viral antigen was detected in groups of epithelial cells and in mononuclear cells in the lamina propria. **B** Intramural neurons and vascular walls were not stained with BVD/CT3. **C** In addition, weak staining with BVD/C16 was observed in most epithelial cells. **D** Viral antigen was also detected in vascular walls and in intramural neurons in the submucosa. Bar = 100 µm

dominantly in the depleted lymphoid follicles and less in subepithelial and interfollicular areas. Mononuclear cells of perivascular infiltrates contained viral antigen. In the spleen, viral antigen was present only in a few dendritic cells in lymphoid follicles (Fig. 2A). With mab BVD/CT3, viral antigen was not detected in cerebrum and parotis. No specific reactivity with BVD virus antigen was found with BVD/CT2, indicating that MD1 virus had not replicated. The overall distribution of BVD virus antigen was assessed using the pestivirus-specific mab BVD/C16 [13]. Viral antigen was detected in numerous glandular epithelial cells of the parotis and neurons of the cerebrum. In addition to the staining observed with mab BVD/CT3, viral antigen was detected with mab BVD/C16 in vascular walls and intramural neurons. In mid colon and tonsil, a weak staining was observed in most epithelial cells, besides the intense staining of groups of epithelial cells (Fig. 1B). In the spleen, viral antigen could

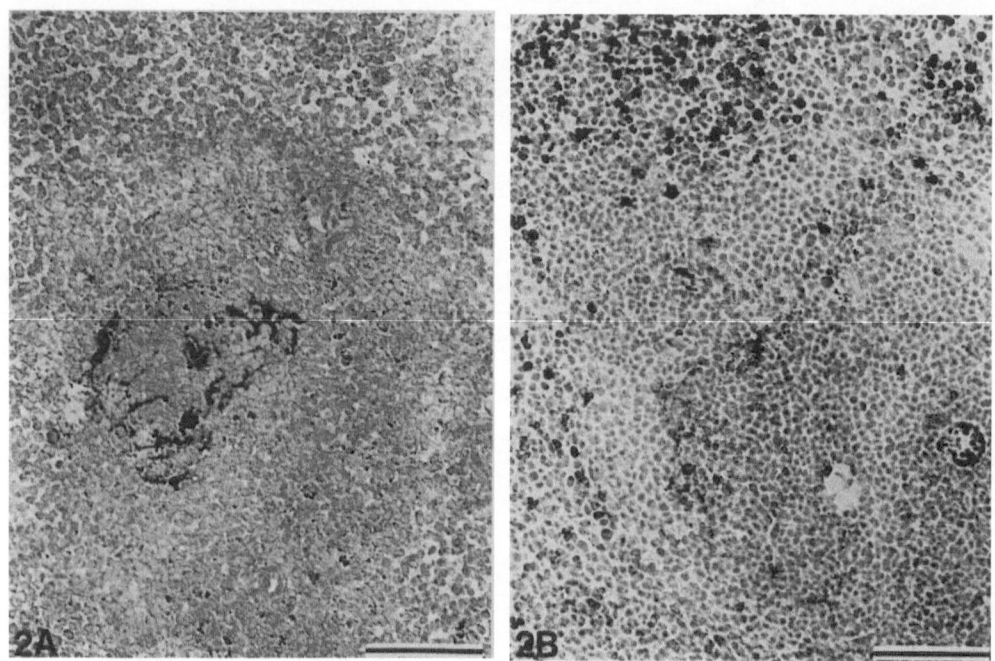

Fig. 2. Consecutive sections of spleen stained with BVD/CT3 (**A**) and BVD/C16 (**B**). **A** Viral antigen was present in a few dendritic cells in lymphoid follicles. **B** In addition, numerous positive lymphocytes were present in the marginal zone. Bar = 100 µm

be demonstrated with mab BVD/C16 in numerous lymphocytes in the marginal zone and a few in the periarteriolar lymphocyte sheaths and in the lymphoid follicles (Fig. 2B).

In order to confirm the immunohistological results, RT-PCR was employed taking advantage of the cellular insertions located in the p125 genes of strains Indiana and MD1, respectively. Sequences flanking the genomic insertion of cp strain NADL [3] and conserved in all pestiviral strains and isolates tested so far were selected as primers. RNA isolation from leucocytes, organs and from BVD virus-infected primary or secondary (FBCK) cells and RT-PCR were performed as described previously [5]. FBCK and PK(15) cells infected with 2 m.o.i. of the BVD or hog cholera (HC) virus strains, respectively, were used. Cells were lysed after 48 h of incubation. The upper primer was a 23-mer between bases 4,937 and 4,960 on the positive strand, and the lower primer an 18-mer from bases 5,591 to 5,609 on the negative strand of NADL cDNA. The calculated sizes for the amplified products were 672 bp for NADL, 630 bp for Osloss/2,498 and 402 bp for the ncp hog cholera strains Alfort and Brescia (Fig. 3A). The specificity of these fragments was verified by restriction enzyme analysis. The restriction enzyme HaeIII generated

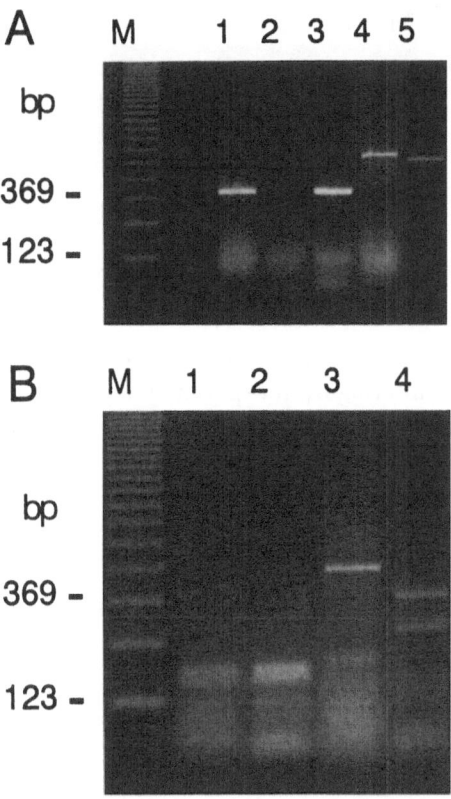

Fig. 3. Electrophoretic analysis of RT-PCR products from RNAs isolated from cells infected with BVD and hog cholera viruses, (**A**) undigested and (**B**) digested with restriction enzymes. Restriction enzyme analysis was done as suggested by the suppliers (BRL). **A** TAE-1.7% agarose gel. (*M*) 123 bp marker (BRL); (*1*) and (*3*) hog cholera strains Brescia and Alfort, respectively; (*2*) uninfected FBCK cells (negative control); (*4*) NADL; (*5*) Osloss/2,482. **B** 2% TAE-agarose gel. (*M*) 123 bp marker (BRL); (*1*) Brescia × HaeIII; (*2*) Alfort × HaeIII; (*3*) NADL × BamH1; (*4*) Osloss/2,482 × PvuII

three fragments from the amplicon of Brescia and Alfort with 181 bp, 162 bp and 59 bp, respectively. BamH1 digestion generated two fragments with 193 and 479 bp from NADL, and PvuII two fragments (265 bp and 365 bp) from Osloss/2,482 (Fig. 3B).

When RNAs isolated from the calf with experimentally induced MD were analyzed by RT-PCR, amplicons corresponding to the ncp and the cp BVD viruses were observed in an organ specific pattern [5]. Whereas in tissues from cerebrum and parotis only the 402 bp fragment amplified from the ncp BVD virus was detected, both amplicons were detectable in palatine tonsils, spleen, ileal Peyer's patch, lymphoid tissue at the ileocecal entrance and in the proximal colon, and in mid colon. This showed that both the cp and the ncp BVD viruses replicated in these or-

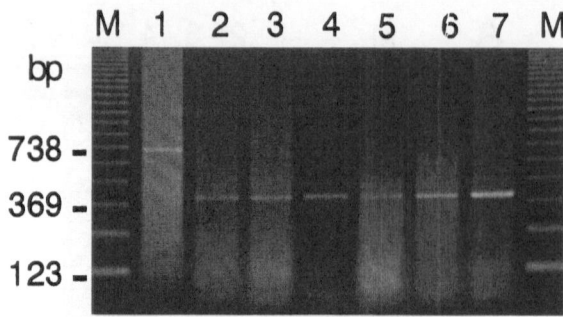

Fig. 4. Electrophoretic analysis (1.7% TAE-agarose gel) of RT-PCR products of lymphocytes from a calf suffering from experimentally induced MD. RNA isolation and RT-PCR were performed as described in the text. (*M*) 123 bp marker (BRL); (*1*) cpBVD virus strain Indiana used for superinfection, grown in FBCK cells; lymphocytes (*2*) 2 days, (*3*) 5 days, (*4*) 6 days, (*5*) 9 days, (*6*) 11 days and (*7*) 14 days post infection

gans. Specificity of the RT-PCR products was demonstrated by Southern blot analysis. In addition, using restriction enzyme analysis with BamH1, which generated two fragments from the amplified sequence from Indiana but did not cleave MD1, it could be shown that only strain Indiana had replicated in the organs tested [5].

Analysis of RNAs extracted from the lymphocyte fraction from blood obtained daily after infection yielded only the 402 bp band corresponding to the ncp BVD virus population (Fig. 4). These results indicate that the superinfecting cp BVD virus either did not replicate efficiently in the white blood cell fraction, or that this phase was extremely short and therefore not detectable in the time intervals analysed.

Using two different methods the distribution of persistent ncp BVD virus and the organ tropism of superinfecting cp BVD virus were analyzed in a calf suffering from experimentally induced MD. Immuno-histology with mabs revealed the localization of viral proteins in organ tissues, indicating that virus had replicated in the respective cells. Using an epitopic marker defined by a mab attempts are currently made to define the target cells of both biotypes. RNA analysis using RT-PCR did not allow localization of biotype-specific genomic material in individual cells. However, a semiquantitative analysis of the distribution of both biotypes revealed that the cp BVD virus preferentially resided in those organs involved in pathological lesions.

Providing that cp BVD viruses with suitable genetic markers are selected for the induction of MD, the combination of both methods offers a powerful tool for further studies of the pathogenesis of MD.

Acknowledgements

The authors wish to thank Drs. K. Dittmar and H.-R. Frey for their help, Ester Barthel for excellent technical assistance, and the "Niedersächsische Tierseuchenkasse" for financial support.

References

1. Bolin SR, McClurkin AW, Cutlip RC, Coria MF (1985) Severe clinical disease induced in cattle persistently infected with noncytopathic bovine viral diarrhea virus by superinfection with cytopathic bovine viral diarrhea virus. Am J Vet Res 46: 573–576
2. Brownlie J, Clarke MC, Howard CJ (1984) Experimental production of fatal mucosal disease in cattle. Vet Rec 114: 535–536
3. Collett MS, Larson R, Gold C, Strick D, Anderson DK, Purchio AF (1988) Molecular cloning and nucleotide sequence of the pestivirus bovine viral diarrhea virus. Virology 165: 191–199
4. Donis RO, Dubovi EJ (1987) Differences in virus-induced polypeptides in cells infected by cytopathic and noncytopathic biotypes of bovine virus diarrhea-mucosal disease virus. Virology 158: 168–173
5. Greiser-Wilke I, Haas L, Dittmar KE, Liess B, Moennig V (1993) RNA insertions and gene duplications in the nonstructural protein p125 region of pestivirus strains and isolates in vitro and in vivo. Virology 193: 977–980
6. Hooft van Iddekinge BJL, van Wamel JLB, van Gennip HGP, Moormann RJM (1992) Application of the polymerase chain reaction to the detection of bovine viral diarrhea virus infections in cattle. Vet Microbiol 30: 31–34
7. Liebler EM, Waschbüsch J, Pohlenz JF, Moennig V, Liess B (1991) Distribution of antigen of noncytopathogenic and cytopathogenic bovine virus diarrhea virus biotypes in the intestinal tract of calves following experimental production of mucosal disease. In: Liess B, Moennig V, Pohlenz J, Trautwein G (eds) Ruminant pestivirus infections. Springer, Wien, New York, pp 109–124 (Arch Virol [Suppl] 3)
8. McClurkin AW, Bolin SR, Coria MF (1985) Isolation of cytopathic and non-cytopathic bovine viral diarrhea virus from the spleen of cattle acutely and chronically affected with bovine viral diarrhea. J Am Vet Med Assoc 186: 568–569
9. Meyers G, Rümenapf T, Thiel H-J (1989) Ubiquitin in a togavirus. Nature 341: 491
10. Meyers G, Tautz N, Dubovi EJ, Thiel H-J (1991) Viral cytopathogenicity correlated with integration of ubiquitin-coding sequences. Virology 180: 602–616
11. Meyers G, Tautz N, Stark R, Brownlie J, Dubovi EJ, Collett MS, Thiel H-J (1992) Rearrangement of viral sequences in cytopathogenic pestiviruses. Virology 191: 368–386
12. Moennig V, Frey H-R, Liebler E, Pohlenz J, Liess B (1990) Reproduction of mucosal disease with cytopathogenic bovine viral diarrhoea virus selected in vitro. Vet Rec 127: 200–203
13. Peters W, Greiser-Wilke I, Moennig V, Liess B (1986) Preliminary serological characterization of bovine viral diarrhoea virus strains using monoclonal antibodies. Vet Microbiol 12: 195–200

14. Pocock DH, Howard CJ, Clarke MC, Brownlie J (1987) Variation in the intracellular polypeptide profiles from different isolates of bovine virus diarrhoea virus. Arch Virol 94: 43–53

15. Purchio AF, Larson R, Collett MS (1984) Characterization of bovine viral diarrhea virus proteins. J Virol 50: 666–669

16. Qi F, Ridpath JF, Lewis T, Bolin SR, Berry ES (1992) Analysis of the bovine viral diarrhea virus genome for possible cellular insertions. Virology 189: 285–292

17. Wengler G (1991) Family Flaviviridae. In: Francki RIB, Fauquet CM, Knudson DL, Brown F (eds) Classification and nomenclature of viruses. Springer, Wien New York, pp 223–233 (Arch Virol [Suppl] 2)

Authors' address: Dr. V. Moennig, Institute of Virology, Hannover Veterinary School, Bischofsholer Damm 15, D-30173 Hannover, Federal Republic of Germany.

Arch Virol (1993) [Suppl] 7: 303–308

Establishment of cell lines from bovine brain

Brief Report

A. Uysal and **O.-R. Kaaden**

Institute for Medical Microbiology, Infectious and Epidemic Diseases, Veterinary
Faculty, Ludwig-Maximilians-University Munich, Munich,
Federal Republic of Germany

Summary. As shown by immunocytochemistry, 16 cell lines of neuronal,
oligodendroglial or neuronoglial origin have been established from
bovine fetal brain by immortalization with SV40 virus and cloning in soft
agar. The cell lines were characterized according to their cell surface
markers using mono- and polyclonal antibodies.

*

Cell cultures are of great importance in all fields of virus research.
Cell lines are particularly interesting because they can be propagated
indefinitely and made be available for everyone. They offer relatively
stable properties and therefore are good candidates for in vitro models.

One of the most challenging fields in recent microbiology are the so-
called "unconventional agents" like those causing scrapie in sheep or
Creutzfeldt-Jakob disease (CJD) in man. The structure and replication
mechanisms of "unconventional agents" are mostly unknown and one of
the hypotheses is that they may consist of infectious protein components
solely ("prions"). There is circumstantial evidence that the aetiological
agent of the recently appeared bovine spongiform encephalopathy (BSE)
is closely related to or even identical with the scrapie agent. Although
numerous papers have been published (see Table 1), cell culture methods
remain unsuitable for diagnosis of unconventional agents: with most cell
lines only transient, only low yield of infectivity or no agent replication
could be achieved. The exceptions which proved to be very useful for the
propagation of unconventional agents are mouse neuroblastoma and rat
pheochromocytoma cell lines [3, 15, 19–21]. Except for the infection
of PC12 rat cells using mouse brain homogenates, the infection of tissue
culture cells has exhibited evidence of species-specificity. It is likely that
efficient adaption of unconventional agents to tissue culture has the
minimum requirement that the agent and the tissue culture target cell
have a homology as close as possible.

Table 1. Cell cultures for the examination of unconventional agent properties

Culture	Remarks	Authors
Explant cultures	brain of scrapie-infected mouse	Field & Windsor 1965 [10], Gustafson & Kanitz 1965 [11], Haig & Pattison 1967 [12], Caspary & Bell 1971 [4], Buening & Gustafson 1971 [2]
SMB cell line	little information regarding its characterization; agent replication over 150 passages, low agent yield.	Clarke & Haig 1970 [7]
L	Mouse fibroblasts (!)	Clarke & Millson 1976 [8]
Explant culture, transformed with SV40	brain of scrapie-infected mouse, loss of infectivity after 12 passages	Asher et al. 1979 [1]
PC12	rat pheochromocytoma cells, terminal differentiation after NGF treatment, thereafter no passages	Rubenstein et al. 1984 [21]
Glia cell monolayer	from rat Gasserian ganglion	Roikhel et al. 1987 [20]
L23	unspecified	Cherednichenko et al. 1985 [6]
NIE-115, N2a	Mouse-Neuroblastoma cells, continuous recloning necessary	Markovits et al. 1981 [15], Race et al. 1987 [19], Butler et al. 1988 [3]
Explant culture, infected with Scrapie agent	brain of scrapie-infected hamster	Taraboulos et al. 1990 [22]

Therefore, studies were initiated to establish cell cultures suitable to propagate unconventional agents which have not to be mouse- or hamster-adapted. Because of the importance of transmissible spongiform encephalopathy in feed stock cattle which was clearly brought to light by the BSE epidemy in England, it was decided to work with bovine cells for the first instance.

As the brain is the only affected organ detected in BSE, the aim of this study was to establish permanent cultures of cloned brain cells. The target cell and location of multiplication of unconventional agents in the brain has yet to be determined. Therefore, it was aimed to get cell lines of neuronal, glial as well as neuronoglial origin of the bovine brain.

Bovine fetuses of different ages where processed under sterile conditions. The brain tissue was prepared, cut into small pieces and dissociated using standard protocols (0.25% trypsine or 200 IE collagenase/ml in buffered saline/glucose solution).

The plating concentration of primary brain cells was between 1×10^5 and 5×10^5 viable cells per ml. MEM with 5% fetal calf serum was used for cell growth. As for the preferential treatment of neuronal cells two selection methods where tested: (1) selective medium described by Mizuguchi et al. [16]; (2) addition of nerve growth factor (50 and 100 ng/ml of medium). Primary cell cultures where incubated at 37°C with 5% CO_2 and a relative humidity of 100%.

The immortalization of primary bovine brain cells was attained by infection with SV40. Cell culture supernate from BHK21 cells was obtained 24 h after SV40 infection, filtered, diluted 1:10 and added for 12 hours to the primary brain cell cultures. After the appearance of visible foci, single colonies were picked and the cells were passaged and subjected to cloning in soft agar. Emerging cell clones were picked and propagated further.

The characterization of cell culture cells was done by immunocytochemistry using specific antibodies for neuronal and glial markers. Anti-neurofilament 200 (anti-NF 200), anti-neuron specific enolase (anti-NSE) and anti-neuronal cell adhesion molecule (anti-N CAM) have been used to identify neuronal markers. The astrocyte-specific, anti-glial fibrillary acidic protein (anti-GFAP) and the oligodendrocyte-specific anti-galactocerebroside (anti-GALC) have been used for identification of glia cell markers [14, 17]. Cells were grown on multitest slides. The fixation of cells was done in acetone (10 min at room temperature). All washing steps were done in TBS (0.05 M Tris/HCl, pH 7.6, 0.15 M NaCl). The primary antibodies were incubated 1 h at 37°C. The secondary antibodies (conjugated to peroxidase or alkaline phosphatase) were incubated 1 h at room temperature. DAB (3,3-diaminobenzidine tetrahydrochloride) and NBT/BCIP (nitro blue tetrazolium and 5-bromo-4-chloro-3-indolyl phosphate) were used as substrates. The control preparations were made of BHK21 (baby hamster kidney) cells.

As expected cultures with a high portion of immunocytochemically identifiable neurons could only be propagated from young fetuses with a maximal crown-rump-length of about 3 inches (age: 2–3 months). The dissociation of brain cells with trypsine was the better method compared to the use of collagenase. A higher plating density of cells was associated with a higher possible passage number of the primary cultures, as already shown by Pixley and Pun [18]. The temporary application of selective medium as well as the addition of NGF (100 ng/ml) led to an increase of neurofilament-reactive cells in the primary cultures as described by

Table 2. Marker expression on immortalized and cell type restricted cell cultures from bovine brain

Cell culture type groups	Marker expression (immunocytochemistry)	Interpretation of immunochemistry results
Group I 11 cultures	all cells: NF 200 (in most cultures of this group: NSE and N CAM, too)	cells of neuronal origin; cultures of different developmental levels
Group II 1 culture	all cells: GALC	cells of oligodendroglial origin
Group III 4 cultures	all cells: GALC, NF 200 (in most cultures of this group: GFAP, NSE and N CAM, too)	neuronoglial cell cultures with different percentage of glial and neuronal cells; neuronal cells of different developmental levels
Group IV 4 cultures	0 to 50% of cells: NF 200	cell cultures of (partly) unknown origin

NF 200 Neurofilament 200, *NSE* neuron specific enolase, *N CAM* neuronal cell adhesion molecule, *GFAP* glial fibrillary acidic protein, *GALC* galactocerebroside

Hartikka and Hefti [13]. This effect was shown only in brain cell cultures from young fetuses. Depending on the age of the fetus primary brain, the cells could be passaged 2–5 times. They showed cells and cell formations of different morphology. The younger the fetus the more uniform was the morphology of the primary cultures.

SV40 infection of parallel cultures was done in the first passage. Immortalized cultures showed a more uniform morphology with epithelial character. A portion of cells could be stained regularly with anti-NF 200. Cultures with a high proportion of such cells (more than 20%) were chosen for soft agar cloning.

Altogether 20 cell type-restricted cultures from two fetuses could be established and immunocyto-chemically analysed. Four of these cultures could not be completely characterized by application of the above mentioned antibodies. The other 16 clones obviously originated from neuronal and/or glial cells (see Table 2).

Contrary to our expectations most of the cell cultures reacted with one or more of the applied neuronal markers and only one culture reacted exclusively with the oligodendrocyte marker anti-GALC. In contrary, the glial part of the four neuronoglial cultures mostly consists of cells which are positive with the astrocyte marker anti-GFAP. An even increased percentage of cells with neuronal characteristics was formerly described in primary cultures from the embryonal rat striatum by Cattaneo and McKay [5].

The growth of the cell lines of suspected neuronal origin is likely to be dependent on the addition of conditioned medium (5%) from non-cell type restricted cultures. It is assumed that a dependence on neuro-trophic effects is exerted by glial cells as described by Engele and Churchill Bohn [9].

In conclusion: we established two different categories of bovine brain cell cultures from two fetuses: (1) immortalized mixed brain cell cultures (at present being in the 20–30 passage), and (3) immortalized and cell type restricted cultures of neuronal, oligodendroglial and neuronoglial origin (at present in the 17–25 passage). As there are different cell types as well as mixed cell cultures from the same fetus available, the described cell lines will be studied as possible target cells for the BSE agent. Herewith, the first step towards the establishment of a bovine brain cell culture model is done and it has to be determined to what extend it will be useful for the propagation and detection of unconventional agents such as scrapie and BSE.

References

1. Asher DM, Yanagiha RT, Rogers NG, Gibbs CJ, Gajdusek DC (1979) Studies of the viruses of spongiform encephalopathies. In: Prusiner SB, Hadlow WJ (eds) Slow transmissible diseases of the nervous system. Academic Press, New York, pp 235–242
2. Buening GM, Gustafson GP (1971) Growth characteristics of scrapie agent infected mouse brain cell cultures. Am J Vet Res 32: 953–963
3. Butler DA, Scott MRD, Bockman JM, Borchelt DR, Taraboulos A, Hsiao KK, Kingsbury DT, Prusiner SB (1988) Scrapie-infected murine neuroblastoma cells produce protease-resistent prion proteins. J Virol 62: 1558–1564
4. Caspary EA, Bell TM (1971) Growth potential of scrapie mouse brain in vitro. Nature 229: 269–271
5. Cattaneo E, McKay R (1990) Proliferation and differentiation of neuronal stem cells regulated by nerve growth factor. Nature 347: 762–765
6. Cherednichenko YN, Mikhailova GR, Rajcani J, Zhdanov VM (1985) In vitro studies with the scrapie agent. Acta Virol 29: 285–293
7. Clarke MC, Haig DA (1970) Evidence for the multiplication of scrapie agent in cell culture. Nature 225: 100–101
8. Clarke MC, Millson GC (1976) Infection of a cell line of mouse L fibroblasts with scrapie agent. Nature 261: 144–145
9. Engele J, Churchill Bohn M (1991) The neurotrophic effects of fibroblast growth factors on dopaminergic neurons in vitro are mediated by mesencephalic glia. J Neurosci 11: 3070–3078
10. Field EJ, Windsor GD (1965) Culture characteristics of scrapie mouse brain. Res Vet Sci 6: 130–132
11. Gustafson DP, Kanitz CL (1965) Evidence of the presence of scrapie in cell culture of brain. In: Gajdusek DC, Gibbs CJ, Alpers M (eds) Slow, latent and temperate

virus infection. National Institute of Neurological Diseases and Blindness, Monograph No. 2. Bethesda, pp 221–236

12. Haig DA, Pattison IH (1967) *In vitro* growth of pieces of brain from scrapie-affected mice. J Pathol Bact 93: 724–727
13. Hartikka J, Hefti F (1988) Development of septal cholinergic neurons in culture: plating density and glial cells modulate effects of NGF on survival, fiber growth, and expression of transmitter-specific enzymes. J Neurosci 8: 2967–2985
14. Hewicker M, Trautwein G (1991) Identifizierung von Zellen des Zentralnervensystems mittels zellspezifischer Marker. Tierärztl Prax 19: 241–246
15. Markovits P, Dormont D, Delpeck B, Court L, Lararjet R (1981) Essais de propagation *in vitro* de l'agent scrapie dans des cellules nerveuses de souris. Comptes rendus hebdomadaires des seances de l'Academie des sciences 293: 413–417
16. Mizuguchi M, Yamada M, Kim SU, Rhee SG (1991) Phospholipase C isozymes in neurons and glial cells in culture: an immunocytochemical and immunochemical study. Brain Res 548: 35–40
17. Osborn M, Weber K (1983) Tumor diagnosis by intermediate filament typing: A novel tool for surgical pathology. Lab Invest 48: 372–394
18. Pixley SK, Pun RK (1990) Cultured rat olfactory neurons are excitable and respond to odors. Dev Brain Res 53: 125–130
19. Race RE, Fadness LH, Chesebro B (1987) Characterization of scrapie infection in mouse neuroblastoma cells. J Gen Virol 68: 1391–1399
20. Roikhel VM, Fokina GI, Lisak VM, Kondakova LI, Korolev MB, Pogodina VV (1987) Persistence of the scrapie agent in glial cells from rat Gasserian ganglion. Acta Virol Praha 31: 36–42
21. Rubenstein R, Carp R, Callahan SM (1984) *In vitro* replication of scrapie agent in a neuronal model: infection of PC 12 cells. J Gen Virol 65: 2191–2198
22. Taraboulos A, Serban D, Prusiner SB (1990) Scrapie prion proteins accumulate in the cytoplasm of persistently infected cultured cells. J Cell Biol 110: 2117–2132

Authors' address: Dr. O.-R. Kaaden, Institut für Medizinische Mikrobiologie, Infektions- und Seuchenmedizin, Tierärztliche Fakultät der Ludwig-Maximilians-Universität München, Veterinärstrasse 13, D-80539 Munich, Federal Republic of Germany.

P. P. Liberski
The Enigma of Slow Viruses
Facts and Artefacts

1993. 56 figures. XVI, 277 pages.
ISBN 3-211-82427-8
Soft cover DM 250,–, öS 1750,–*
(Archives of Virology / Supplementum 6)

O. W. Barnett (ed.)
Potyvirus Taxonomy

1992. 57 figures. IX, 450 pages.
ISBN 3-211-82353-0
Soft cover DM 290,–, öS 2030,–*
(Archives of Virology / Supplementum 5)

C. De Bac, W. H. Gerlich, G. Taliani
(eds.)
Chronically Evolving Viral Hepatitis

1992. 72 figures. XIV, 348 pages.
ISBN 3-211-82350-6
Soft cover DM 260,–, öS 1820,–*
(Archives of Virology / Supplementum 4)

B. Liess, V. Moennig, J. Pohlenz,
G. Trautwein (eds.)
Ruminant Pestivirus Infections

Virology, Pathogenesis, and Perspectives
of Prophylaxis

1991. 78 figures. VIII, 271 pages.
ISBN 3-211-82279-8
Soft cover DM 220,–, öS 1540,–*
(Archives of Virology / Supplementum 3)

R. I. B. Francki, C. M. Fauquet,
D. L. Knudson, F. Brown (eds.)
Classification and Nomenclature of Viruses

Fifth Report of the International Committee
on Taxonomy of Viruses
Virology Division of the International Union
of Microbiological Societies

1991. IV, 450 pages.
ISBN 3-211-82286-0
Soft cover DM 110,–, öS 770,–*
(Archives of Virology / Supplementum 2)

C. H. Calisher (ed.)
Hemorrhagic Fever with Renal Syndrome, Tick- and Mosquito-Borne Viruses

1991. 75 figures. VII, 347 pages.
ISBN 3-211-82217-8
Soft cover DM 258,–, öS 1800,–*
(Archives of Virology / Supplementum 1)

* *10 % price reduction for subscribers to the journal
"Archives of Virology"*

Springer-Verlag Wien New York
Sachsenplatz 4–6, P.O.Box 89, A-1201 Wien · 175 Fifth Avenue, New York, NY 10010, USA
Heidelberger Platz 3, D-14197 Berlin · 37-3, Hongo 3-chome, Bunkyo-ku, Tokyo 113, Japan